Advances in Polyelectrolytes and Polyelectrolyte Complexes

Advances in Polyelectrolytes and Polyelectrolyte Complexes

Guest Editors

Paulina Mocchiutti
Anupama Sharma
Norma E. Marcovich

MDPI

Basel • Beijing • Wuhan • Barcelona • Belgrade • Novi Sad • Cluj • Manchester

Guest Editors
Paulina Mocchiutti
Instituto de Tecnología Celulósica
Universidad Nacional del Litoral
Santa Fe
Argentina

Anupama Sharma
SSB University Institute of Chemical Engineering and Technology
Panjab University
Chandigarh
India

Norma E. Marcovich
Departamento de Ingeniería Química y en Alimentos
Universidad Nacional de Mar del Plata
Mar del Plata
Argentina

Editorial Office
MDPI AG
Grosspeteranlage 5
4052 Basel, Switzerland

This is a reprint of the Special Issue, published open access by the journal *Polymers* (ISSN 2073-4360), freely accessible at: https://www.mdpi.com/journal/polymers/special_issues/03Q9BVX9OU.

For citation purposes, cite each article independently as indicated on the article page online and as indicated below:

Lastname, A.A.; Lastname, B.B. Article Title. *Journal Name* **Year**, *Volume Number*, Page Range.

ISBN 978-3-7258-4787-7 (Hbk)
ISBN 978-3-7258-4788-4 (PDF)
https://doi.org/10.3390/books978-3-7258-4788-4

Contents

About the Editors

Paulina Mocchiutti

Paulina Mocchiutti holds a Ph.D. in Chemistry and is a Professor at the Faculty of Chemical Engineering of the National University of the Littoral and a Researcher at the National Scientific and Technical Research Council (CONICET), Santa Fe, Argentina. Her research projects focus on chemistry and colloidal chemistry applied to the pulp and paper industry, as well as the development of innovative materials mainly based on lignocellulosic biomass. These materials include active packaging for food contact and hydrogels for controlled drug release. She has authored or co-authored more than 36 international scientific papers and participated in more than 70 conferences. She has acted as peer reviewer for several international journals, and alongside Norma Marcovich, served as Guest Editor of the Special Issue "Advances in Polyelectrolytes and Polyelectrolyte Complexes".

Anupama Sharma

Anupama Sharma is Chair and Prof. SSB University Institute of Chemical Engineering and Technology and Director, Energy Research Centre at Panjab University, India. She completed her PhD in Chemical Engineering in 2005 and specializes in Polymer nanocomposites, biomass valorization, and agricultural residue derived biopolymers-based eco-nanomaterials for wastewater remediation. She has received numerous accolades, including a Visiting/Mobility Fellowship (PROWIS-II) under the Indo-French Programme, a Young Scientist Award from the Asian Polymer Association, and a Career Award for Young Teachers by the All India Council for Technical Education, Government of India. She has co-authored 109 publications with an H-index of 28 and i10- i10-index of 59, along with 15 book chapters and two books. She has successfully completed 14 National and International projects, and one is in the pipeline. She has successfully organized 9 International conferences and 3 Faculty Development programmes.

Norma E. Marcovich

Norma E. Marcovich is a full Professor at the Department of Chemical and Food Engineering, Faculty of Engineering, National University of Mar del Plata, Argentina, as well as a Principal Researcher for the National Council for Scientific and Technical Research of Argentina. She earned her PhD in Materials Science in 1996 with the highest honors, and from that moment on began receiving international grants and recognition due to the quality of her scientific output. She has been included in the prestigious group of the top 2% most influential scientists in the world according to the "Stanford World's Top 2% Scientists" international ranking prepared by Professor John P. A. Ioannidis, appearing in the two most recent versions (2020 and 2024). Her scientific interests include the development and study of functional materials based on biopolymers and natural polymers, as well as the design of nanomaterial structures from green polymers and bio-based nanoparticles. She has authored or co-authored more than 90 scientific papers and more than 20 international book chapters. She also serves as Senior Editor, Guest Editor, and Peer Reviewer for several international journals and was one of the Guest Editors of the Special Issue "Advances in Polyelectrolytes and Polyelectrolyte Complexes".

Preface

This reprint intends to compile recent developments achieved in the field of materials based on polyelectrolyte complexes (PECs), published recently in the "Advances in Polyelectrolytes and Polyelectrolyte Complexes" Special Issue of the MDPI journal *Polymers*. This very exciting and active research area is actively strengthening due to the unusual properties that these materials present, which enable their application in fields as diverse as biomedicine, materials science, and environmental engineering. PECs, formed through electrostatic interactions between oppositely charged polyelectrolytes (polyanions and polycations), display unique properties such as tunable size, shape, and charge density, making them ideal for diverse applications and offering a versatile platform for creating biocompatible materials, drug- and gene-delivering active systems, adhesives, emulsifiers, and more. The formation of PEC supramolecular structures, which will ultimately define the materials' properties and therefore their most promising applications, is influenced by numerous factors, including the charge ratio between the polyions, polyelectrolyte concentration, pH, and the salt concentration of the medium, as well as preparation parameters such as mixing mode, mixing order, and ratio. Therefore, even using the same polyelectrolyte pair to obtain a PEC, the resulting complex materials can exhibit completely different, and therefore tunable, properties and behaviors. Thanks to the development of more advanced methods for analyzing nanoscale systems, the relationships between these parameters are now beginning to be elucidated. The current moment presents an ideal time to begin to deeply understand the characteristics of these surprising materials and apply them efficiently. We hope this reprint will also serve as inspiration for new researchers to direct their studies toward the development and advancement of this vibrant and fascinating field.

Paulina Mocchiutti, Anupama Sharma, and Norma E. Marcovich
Guest Editors

polymers

MDPI

Article

Effect of Degree of Substitution and Polymer Ratio on the Structure of Chitosan: Carboxymethyl Starch (Bio)Polyelectrolyte Complexes

David Castro, Valentina Brovina, Mikhail Litvinov and Aleksandr Podshivalov *

Center for Chemical Engineering, ITMO University, Kronverkskiy Prospekt, 49, 197101 Saint-Petersburg, Russia; dcastrovargas@itmo.ru (D.C.); vbrovinaa@gmail.com (V.B.); mikhail.litvinov.1996@mail.ru (M.L.)
* Correspondence: podshivalov@itmo.ru

Abstract: In this work, three carboxymethyl starches (CMS) were obtained by the two-step reaction process of carboxymethylation with different degrees of substitution (0.16, 0.33, and 0.36). From these samples, (bio)polyelectrolyte complexes ((bio)PECs) were obtained with chitosan (Chit) by the mixing of individual solutions of polymers (0.25 wt.%) at different volume ratios. The effect of the biopolymer and ionized groups of z ratios, pH, and degree of substitution of CMS in the formation of PEC were evaluated by turbidimetry and dynamic light scattering. The results showed that increasing the amount of CMS samples (ratio of z) led to an increase in the efficiency of the formation of (bio)PEC using CMS with a high DS value. Using the turbidimetry method for the chitosan and CMS mixtures, it was observed that the formation of (bio)PEC is divided into four transition zones delimited by pH transition points, and the stoichiometric complexation (z = 1) is achieved at a pH that displayed morphological changes "pH_{morph}", which is a single point for Chit:CMS 1, and for Chit:CMS 2 and Chit:CMS 3, this is a range of 4.9–6.4 and 4.3–6.4, respectively. Analysis of the structural properties of the structures of (bio)PECs by dynamic light scattering was characterized by monomodal distribution, and the main observed effect was associated with an increase in the value of D_{avg} with an increase in the ratio of Chit:CMS.

Keywords: carboxymethyl starch; chitosan; polyelectrolyte complex; degree of substitution

1. Introduction

The development and study of novel supramolecular structures, such as polyelectrolyte complexes (PECs), have notably increased in recent years. The obtention of these complexes is based on electrostatic interactions between individual solutions of polyelectrolytes (polycations and polyanions) forming a dense phase separated from the solvent of z [1,2]. There are some types of PECs such as water-soluble PECs, which are homogenous systems with smaller PEC particles, formed by large difference molar mass ionic group; turbid colloidal stable PECS, which are in the transition state to phase separation; and complex coacervates, which are two-phase systems [2,3]. The main factors that influence the formation of PECs are the ratio of charges of polycations and polyanions; media parameters such as the concentration of polyelectrolytes, pH, and salt concentration; and the preparation parameters like the mode of mixing, mixing order, and ratio [4]. About the applications, PECs are used in many fields because of their chemical properties and macromolecular structure. Adhesives; drug and gene delivery; emulsifiers; functionalization as therapeutic, targeting, or imaging agents; and rheology modifiers are some of the most common applications of PECs [5,6]. In addition, some self-assembly structures such as adhesives [7,8], films [9,10], gels [11,12], and microcapsules are produced from PECs [13,14].

Some of the polymers used for preparing materials based on PECs are chitosan, a widely available natural, non-toxic, biodegradable, and abundant polysaccharide obtained from chitin deacetylation (polycation) [15], and cassava carboxymethyl starch (CMS). Cassava starch is an abundant biopolymer prepared from the root of the cassava plant (Manihot esculenta) cultivated in tropical regions. It is the third largest source of carbohydrates among vegetables and one of the most important crops worldwide [16,17]. CMS is an ether derivate of the starch product of the reaction of carboxymethylation between activated starch, by sodium hydroxide, with monochloroacetic acid (MCA). The CMS presents negatively charged functional groups (CH_2COO^-), high swelling capacity, solubility at ambient temperature, low internal viscosity, and less retrogradation [18–20].

Chitosan is actively used in agriculture, aquaculture, and agronomy as an antimicrobial agent and as a soil fertilizer [21,22]. In addition, chitosan is also utilized in the production of food and nutritional supplements. It is also appreciated in medicine and cosmetology, the textile and paper industries, and in packaging, biotechnology, and many other fields [23,24]. In another hand, carboxymethyl starch is widely used as a thickener in food products, as a binder and emulsifying agent in industrial applications, and also in textile industries as a printing thickener. Nowadays, the most recent advances and applications of the CMS are in the pharmaceutical applications as a drug carrier [19,25].

Interest in PECs began in the 1930s with the discovery that some natural polyelectrolytes could interact in aqueous media to form colloidal complexes known as complex coacervates in work carried out by Bungenberg de Jong H. and Kruyt H. [26]. In the 1960s, insoluble PECs were identified using synthetic polymers, and the swelling and plasticizing properties of aqueous electrolytes were described [27]. One of the first works related to PECs was carried out by Michaels A. and Miekka R. [28] in 1961 studying the formation and properties of the stoichiometric complexes of synthetic PEs—poly(4—vinylbenzyltrimethylammonium chloride) associated with sodium polystyrene sulfonate. PECs based on chitosan and CMS are of great importance in the development of new-generation environmentally friendly materials with low toxicity, high biocompatibility, and biodegradability [29,30]. Their formation occurs mainly due to electrostatic interactions between the positively charged amino groups of chitosan and negatively charged carboxymethyl groups of CMS. The advantages of the complexes are due to their improved stability, mechanical strength, and film formation properties compared to individual polymers. Also, such complexes can form gels or microcapsules, which open up opportunities for their application in biomedicine, food preservation, chemical detectors, membrane technology, fuel cells, and pharmaceuticals as effective carriers for drug delivery [30].

Some works have prepared and studied the use of PECs in drug delivery applications. For example, Quadrado F. and Fajardo A. [31] synthesized microparticles based on rice CMS with a degree of substitution of 0.5 and chitosan at room temperature, with a ratio of 3:4 for chitosan and CMS, respectively. These microparticles presented better chemical and thermal properties than the microparticles prepared by conventional ionotropic crosslinking. In addition, microparticles based on PEC showed more efficiency for drug encapsulation, leading to their use as vehicles for drug delivery systems. In another work, Assad E. et al. [32] prepared and characterized PECs of CMS and chitosan at a 1:1 ratio with a final pH of 5, and used them for drug delivery. The results of fluid diffusion and swelling in simulated gastric fluid or simulated intestinal fluid solutions were lower with PEC compounds in comparison with CMS:chitosan powder mixture. This allows the use of PECs as a drug carrier since it can prolong drug release time. In other research [33], polyelectrolyte complexation and ionotropic gelation methods were compared for obtaining hydrogels based on CMS and Chitosan. The hydrogels were prepared by mixing polyelectrolytes with a ratio of 16:1 for chitosan:CMS, respectively. Then, the hydrogel was kept for 24 h at room temperature until the equilibrium. Wang Y. et al. [34] demonstrated that the tablets made of PEC chitosan:CMS had lower swelling in acid media, and slower water uptake in comparation with the tablets prepared from independent polymers. In the study [35], devoted to the creation of PEC nanoparticles based on chitosan and carboxymethyl starch, their ability to package aminosalicylic acid (5-ASA) and release this substance in vitro was studied. The obtained chitosan:CMS nanoparticles containing 5-ASA had a high loading rate and 5-ASA could be released under controlled conditions, further proving the great potential of chitosan:CMS nanoparticles in targeted drug delivery, especially for the treatment of colorectal diseases. Also, in the work [36], polyelectrolyte complexes of chitosan with carboxymethyl starch were investigated and used as monolithic matrices with acetaminophen as excipients for directly pressed tablets. The mixture also included 5% bromocresol green as pH indicator to study drug behavior in gastric acid. It was shown that after the interaction of the tablets with simulated gastric fluid, the integrity of the tablets was maintained, thus achieving higher drug loading, providing useful properties for drug delivery. This study confirmed that chitosan-based polyelectrolyte complexes can be effectively utilized in biopharmaceuticals, therapeutic enzymes, and for probiotic delivery to the colon.

As the literature shows, most of the studies are focused on the study of the application of PECs based on chitosan:CMS in drug delivery [31–36]. Perhaps there are not enough studies focused on the processes of the formation of PECs based on chitosan and CMS, and the influence of the molecular weight of CMS, DS, and biopolymer ratio has been poorly studied. Studies of the PEC formation process are important to know the optimal conditions such as pH, ratio, and polymer concentration to obtain PECs. For this reason, the main aim of the present research was to obtain a polyelectrolyte complex based on chitosan and carboxymethyl starch by the titration method depending on the type of carboxymethyl starch (CMS) and its ratio and study the process of formation. For this purpose, carboxymethyl starches were synthesized by the two-step reaction process of carboxymethylation with a constant amount of NaOH and different concentrations of monochloroacetic acid to obtain products with different degrees of substitution. The obtention of the PECs starch was carried on by mixing a diluted solution of CMS and chitosan and adjusting the pH until the obtention of the maximum formation of the (bio)PEC dispersions. The obtained (bio)PEC dispersions were characterized by dynamic light scattering and turbidimetry methods. The influence of the ratio of biopolymers, molecular weight, and degree of substitution were studied.

2. Materials and Methods

2.1. Reagents and Chemicals

Cassava starch (CS) (Thai Food King, Samutprakarn, Thailand) was used for the obtention of carboxymethyl starch (CMS), and chitosan (Chit) with low molecular weight (Bioprogress, Biokombinata, Russia) was used for the obtention of polyelectrolyte complex. Monochloroacetic acid (MCA) (NevaReactiv, Saint-Petersburg, Russia) was used for the carboxymethylation of starch. Distilled water was used as a solvent. Solutions of 0.1 M NaOH and HCl 0.1 M were used to set the pH.

2.2. Attenuated Total Reflectance Fourier-Transform Infrared Spectroscopy Analysis

The chemical composition of the used chitosan, cassava starch, CMS products, and dried (bio)PEC dispersions of Chit:CMS were analyzed by Fourier-transform infrared spectroscopy (FTIR) using an FTIR spectrometer Tensor 37 (Bunker, Karlsruhe, Germany) equipped with the attenuated total reflection (ATR) unit MIRacle (Pike, Madison, WI, USA) with a crystal of ZnSe. The ATR-FTIR spectra of the samples were obtained with a wavenumber range of 4000–500 cm^{-1} with an increment of 2 cm^{-1} as averages over 32 measurements.

The degree of acetylation (DA) of the used chitosan was calculated from the ATR-FTIR spectrum of chitosan powder using Equation (1) [37]

$$DA = \frac{\frac{A_{1320}}{A_{1420}} - 0.3822}{0.03133}, \quad (1)$$

where A_{1320} and A_{1420} are the normalized absorbance at wavenumbers 1320 and 1420 cm^{-1}, respectively. The degree of deacetylation (DD) was calculated by the difference in DA (Equation (2))

$$DD = 100\% - DA. \tag{2}$$

2.3. Determination of Acid Dissociation Constant of Chitosan

The acid dissociation constant pK_a was determined by potentiometric titration using a pH meter S213-kit (Mettler Toledo, Giesen, Germany) in the range from 2 to 12. For this, a solution of 0.1 g of chitosan was prepared in 50 mL of 0.1 N HCl solution at 40 °C and mixed for 1 h. After this, the solution was titrated by separating 0.1 M NaOH solution until pH = 12. The pH and pK_a were calculated using the modified Henderson–Hasselbalch Equations (3) and (4) [38]

$$pH = pK_{1/2} - p\, ln\left(\frac{1-\alpha}{\alpha}\right), \tag{3}$$

$$pK_a = pK_{1/2} - (p-1)\, ln\left(\frac{1-\alpha}{\alpha}\right), \tag{4}$$

where p is an empirical parameter related to the free energy of change during titration and $\alpha = 0.5$ is the degree of dissociation (based on the volume of solution used in the titration). In addition, $pK0$ is the protonation constant at $\alpha \rightarrow 0$ [38].

2.4. Carboxymethylation of Cassava Starch

For the obtention of CMS, 4 g of cassava starch were dissolved in isopropanol water solution (75 v/v%) in a 250 mL three-necked flask and mixed by mechanical stirring under room temperature for 30 min. Then, 50 mL of NaOH (30 wt.%) solution was added at 50 °C for 60 min. Different amounts of 2, 2.5, and 3 g of monochloroacetic acid (MCA) were dissolved in 20 mL of isopropanol at room temperature for 30 min in order to synthesize three types of CMS. Then, 20 mL of the solution of MCA was gradually introduced into the reaction mixture at a feed rate of 40 mL/h using a syringe pump Sino MDT, SN-506 (Shenzhen, China) for 30 min. After that, the reaction progressed to 3 h at 50 °C. Finally, 200 mL of isopropanol was added to stop the reaction, and the product was sedimented for 24 h at room temperature. For the purification, the supernatants were separated and collated followed by removal and neutralization with HCl solution. After that, dialysis was carried out for 48 h with distilled water using an M-Cel dialysis bag with a pore diameter of 14 kDa (Rosmedbio, Saint-Petersburg, Russia) to remove salt from the products. Finally, the products were dried and named CMS 1, CMS 2, and CMS 3 in accordance with the amount of MCA used for their production. A detailed scheme for synthesizing the CMS products was drawn using the BioRender® platform and presented in Figure 1.

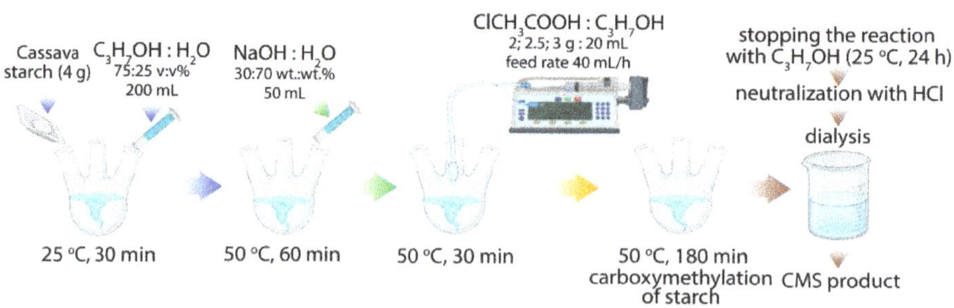

Figure 1. Scheme for synthesizing carboxymethyl starch products.

2.4.1. Establishing the Degree of Substitution of Carboxymethyl Starch

Direct titration was used for the determination of the degree of substitution (DS) of the obtained CMS products. For this purpose, 1 g of CMS for each type of product was dispersed in 50 mL of NaCl solution (2 wt.%), and then titrated with a solution of NaOH (1 M) until pH = 12 [39]. The *DS* values were calculated by Equation (5)

$$DS = \frac{162 \times \eta_{COOH}}{m_{ds} - 58 \times \eta_{COOH}}, \tag{5}$$

where 162 is the molar mass of anhydroglucose unit (AGU), g/mol; η_{COOH} amount of –COOH, mol, calculated from the equivalent volume of the known molarity solution of NaOH; 58 is the net increase in the mass of an AGU for each carboxymethyl group substituted, g/mol; and m_{ds} is the dry mass of sample, g.

2.4.2. Determination of Molecular Weight of Polymers

To establish the molecular weight of polymers, the method of capillary viscometry of dilute polymer solutions was used followed by the calculation of the viscosity-average molecular weight M_η using the Mark–Kuhn–Houwink Equation (6) [40]

$$[\eta] = K \times M_\eta^\alpha, \tag{6}$$

where $[\eta]$ is the intrinsic viscosity; K and α are constants that characterize the length and shape of the macromolecule, respectively. To determine the values of the intrinsic viscosity, dilute chitosan solutions in 0.33 M acetic acid/0.3 M NaCl mixture [41] and CMS solutions in 0.01 M NaCl with a concentration of 0.1–0.5 g/dL with an increment of 0.1 g/dL were prepared. Then, the flow time of the solutions and solvent through a glass thermostatic capillary with a diameter of 0.92 mm at a constant temperature of 25 °C was determined. Next, the values of specific and reduced viscosity were calculated. The values of $[\eta]$ for chitosan and all the CMS products were determined using the Huggins approach, Equation (7), described in detail in the work [40]

$$\eta_r = [\eta] + K_h \times [\eta]^2 \times C, \tag{7}$$

where η_r is the reduced viscosity and K_h is the Huggins constant. Finally, the M_η values were calculated according to Equation (6). For the used chitosan, the values of K and α constants were calculated by Equations (8) and (9) [41]

$$Exp\ \alpha = \frac{0.6202 \cdot x + 0.699 \cdot x}{0.4806 + x}, \tag{8}$$

$$logK \cdot 10^{-5} = -5.7676 \cdot Exp\ a + 5.9232, \tag{9}$$

where $x = DA/pH \cdot \mu$ and μ is the ionic strength of the solvent. For the CMS products, the values of K and α constants were selected based on the data from work [31] and were $K = 2.28$ mL/g and $\alpha = 0.277$.

Also, using the values of $[\eta]$ and M_η, the apparent radius of gyration $(R_g{}^2)^{1/2}$ of chitosan and CMS macromolecules was calculated using Equation (10) [42]

$$\left(R_g^2\right)^{1/2} = \sqrt[3]{\frac{3}{10\pi} \cdot \frac{[\eta] \cdot M_\eta}{N_a}}, \tag{10}$$

where $N_a = 6.22 \times 10^{23}$ is the Avogadro number.

2.5. Obtention of the Chitosan:Carboxymethyl Starch (Bio)Polyelectrolyte Complex Dispersions

First, individual solutions of chitosan and CMS with a concentration of 0.25 wt.% were prepared by mixing in distilled water at 25 °C and 40 °C, respectively. The temperature with an accuracy of ±0.2 °C was controlled with a thermostatic thermometer ETS–D6 (IKA–Werke, Staufen, Germany). After mixing, the pH of the CMS solution was measured and readjusted until pH = 7 with a solution of 0.1 M HCl. The (bio)polyelectrolyte complex ((bio)PEC) dispersions between chitosan and all the CMS products were obtained by the mixing of individual solutions of polymers at different volume ratios of Chit:CMS: 1:1, 1:2, 1:3, 1:4, 1:5, and 1:6.

2.6. Characterization of the Chitosan:Carboxymethyl Starch (Bio)Polyelectrolyte Complex Dispersions

For the study of the pH influence in the formation of the Chit:CMS (bio)PEC dispersions, the change in the absorbance of all the individual solutions was analyzed. For this, each dispersion was titrated in the pH range from 2 to 12 by adding an aliquot of 0.1 M NaOH and 0.1 M HCl solutions, respectively. After each aliquot addition, the dispersion was mixed for 1 min, and the absorbance was measured with a spectrophotometer (Unico, Dayton, NJ, USA) at a constant wavelength of λ = 364 nm [43,44] and temperature of 25 °C. For the study of the particle size distribution of the initial chitosan, cassava starch, CMS products solutions, and Chit:CMS dispersions, the dynamic light scattering (DLS) method was used. In this connection, the refractive index n was determined with a refractometer IRF-454B2 M KOMZ (Kazan, Russia) for all the solutions. Using the values of n, the obtained solutions were diluted to 1:2 and measured using the particle size analyzer LB 550, Horiba (Edison New Jersey, NJ, USA) at 25 °C five times for each sample of solution. The final particle diameter distributions for the Chit:CMS (bio)PEC dispersions were obtained by averaging over the five measurements.

3. Results and Discussion

3.1. Physical and Chemical Characteristics and Structure of the Carboxymethyl Cassava Starch and Chitosan Solutions

Figure 2 shows the ATR-FTIR spectra of the initial chitosan and cassava starch powder samples and the obtained CMS powder samples.

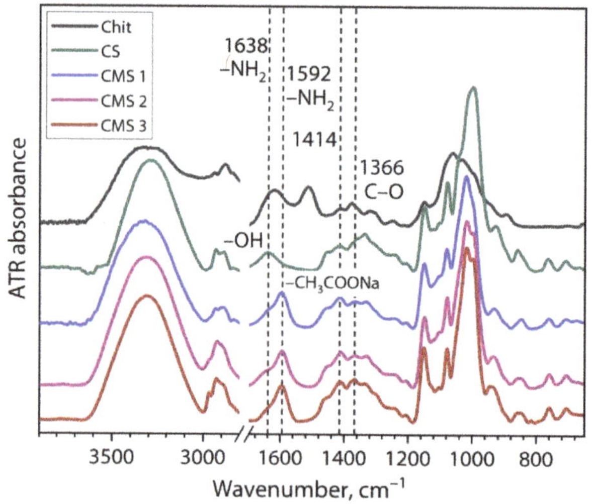

Figure 2. ATR-FTIR spectra of used chitosan, cassava starch, and obtained carboxymethyl starches.

The presented spectra of polysaccharides show a region at 3700–3000 cm^{-1} that represents the stretching of hydroxyl groups. Another region at 3000–2800 cm^{-1} corresponds to C–H stretching. For the Chit, the spectrum is observed as a broad band with a center at 3349 cm^{-1} associated with –NH and –OH groups' stretching vibration. The C–H stretching vibration of CH_2 and CH_3 is represented by the bands in the region of 2928–2856 cm^{-1}. The bands at 1622 cm^{-1} and 1516 cm^{-1} correspond to the C=O stretching vibrating and N–H deformation, receptively. Peaks associated with the amino group in chitosan around

1592 cm^{-1} and *N*-acetylated chitin at 1638 cm^{-1} can be observed [31,43]. For the cassava starch, the band at 1638 cm^{-1} corresponds to the scissoring of two –OH bonds of absorbed water molecules [7,11]. In the case of the CMS samples, the substitution with the –CH_3COONa salt radicals on the cassava starch is represented by the bands at 1592 cm^{-1} and 1414 cm^{-1} [12]. Also, the new bands between 1400 cm^{-1} and 1300 cm^{-1} correspond to the anti-symmetrical and symmetrical vibrations of the –COO^- structure due to two C–O bonds that would receive resonance effects as a result of the ionization [11].

Using Equations (1) and (2) based on the ATR-FTIR spectrum, the value of the degree of deacetylation *DD* of the used chitosan was established. Also, using Equation (5), the values of the degree of substitution *DS* of the obtained CMS products were calculated. The specified values were collected and presented in Table 1.

Table 1. Molecular characteristics of the used chitosan and CMS products.

Polymer	MCA, g	DS (or DD for Chit)	K_h	R^2	[η], mL/g	M_η, kDa	$(R_g{}^2)^{1/2}$, nm	R_h, nm
Chit	–	0.82	0.16	0.90	66.1	30	7	–
CMS 1	2.0	0.16	0.24	0.98	54.0	92	9	17
CMS 2	2.5	0.33	0.25	0.99	56.4	107	10	17
CMS 3	3.0	0.36	0.23	0.99	62.0	151	11	22

From Table 1, the value of the *DD* of the chitosan is greater than the *DS* values of the CMS products. This difference indicates that the chitosan, in proportion, has more available side amino groups for ionization than the CMS. Therefore, it can be seen that an increase in the amount of MCA in the carboxymethylation process leads to an increase in the *DS*. We believe that this increase is due to the fact that during CMS synthesis, there is a greater amount of carboxymethyl groups that are substituted to the starch chain, generating new active sites. In addition, the acid dissociation calculated constants of the chitosan from Equations (3) and (4) are $pK_a = 6.41$ and $pK_0 = 6.34$. These values are consistent with the literature data when pK_a varies in the range of 6.20–6.50 and represents, for pK_a, the pH at which half of the amino groups in chitosan are protonated, while the pK_0 is the pH at which the net charge density of the polymer is altered in complex formation [38,44].

Figure 3 shows the dependence of the reduced viscosity of the solutions of chitosan and CMS products on their concentration in salt buffer medium.

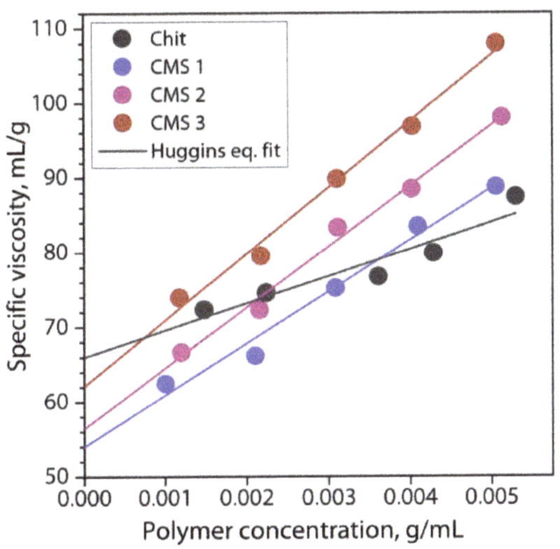

Figure 3. Reduced viscosity of chitosan and carboxymethyl starch solutions versus concentration.

Dependence demonstrates a linear increase in the reduced viscosity of the solution with increasing polymer concentration, which indicates a high affinity of the polymers and the solvent system. Also, the increasing nature of the dependence indicates a suppressed effect of the electrolytic swelling of polyelectrolyte macromolecules in the presence of a salt buffer and allows us to establish the unperturbed size of uncharged macromolecules and their molecular weight. The obtained dependences are well described by the Huggins equation with R^2 values greater than 0.9 (Table 1), which allowed us to establish the values of the Huggins constant K_h and intrinsic viscosity $[\eta]$, as well as calculate the viscosity-average molecular weight of the polymer M_η and mean-square radius of gyration of macromolecule $(R_g{}^2)^{1/2}$. The parameter values are presented in Table 1. It is clear from the table that the K_h value for the chitosan and CMS samples is less than 0.5, which is usually typical for good solvents [45]. It is also evident that the molecular weight of the CMS products is several times higher than the value for the low molecular weight water-soluble chitosan used. At the same time, with an increase in the amount of MCA used in the carboxymethylation of cassava starch, the formation of a CMS product with an increase in DS and M_η occurs. These results can be explained by the presence of negatively charged carboxymethyl groups in CMS, the fraction of which increases with an increase in the degree of substitution. In addition, the lowest value of $(R_g{}^2)^{1/2}$ for chitosan molecules in the buffer solution is also characterized due to the compact chain and the small molecular weight of chitosan in comparison with CMS.

The dynamic particle diameter distribution for the individual solutions of chitosan and CMS products in distilled water is shown in Figure 4. The refractive index n values of the solutions are presented in the boxes. The particle size distribution (frequency distribution %) vector is represented by q.

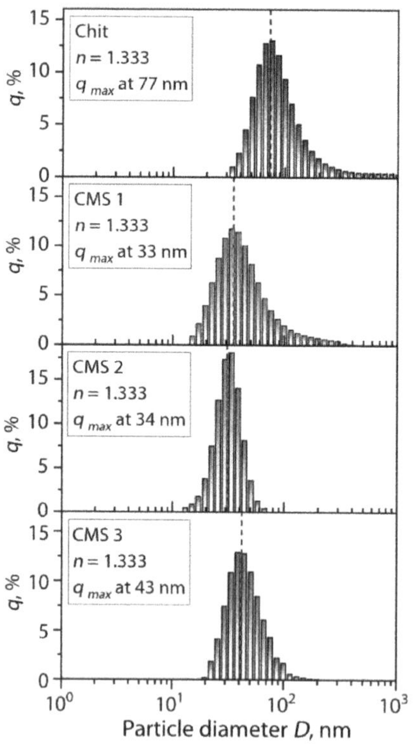

Figure 4. Particle diameter distributions of chitosan and carboxymethyl starch solutions in distilled water at 0.25 wt.%.

From Figure 4 it can be seen that the used chitosan and CMS product solutions are characterized by the monomodal distribution of particle diameters with different widths. It is noticeable that a small number of submicron-sized particles are also observed in the chitosan solution, which significantly exceeds the size of chitosan macromolecules as well as a large average particle diameter D_{avg} (average diameter of particles at q_{max}) compared to its low molecular weight (Table 1). This trend is explained by the aggregation of chitosan macromolecules in a water solution. The main causes of chitosan aggregation in aqueous solutions can be both the association of *N*-acetyl-*D*-glucosamine units due to hydrophobic interactions and the hydrophilic interaction of *D*-glucosamine units with the formation of hydrogen bonds [46]. Aggregation occurs when the hydrophobic association of macromolecules overcomes the electrostatic repulsion of charged groups. It is evident that the D_{avg} values and the width of the distributions for the CMS samples are significantly smaller. The reason for this is probably the increased solubility of CMS macromolecules in water due to the presence of side carboxymethyl groups [47–49]. This shows that CMS macromolecules in water are not prone to aggregation, like chitosan molecules, and can be separate units. Therefore, for CMS solutions, it becomes possible to determine the hydrodynamic radius of macromolecule $R_h = D_{avg}/2$. The calculated values of R_h are presented in Table 1. The values of R_h demonstrate proportionality with the values of *DS*, M_η, and $(R_g{}^2)^{1/2}$ and increase with an increase in the mass of MCA used in the carboxymethylation of cassava starch. Another work [50] also showed that increasing the degree of substitution of CMS from potato starch from 0.40 to 0.81 causes an increase in $[\eta]$, R_g, and R_h due to the presence of carboxymethyl groups.

3.2. *Effect of the Biopolymer Ratio Chit:CMS v/v% in the Formation of PEC*

The appearance of individual solutions of chitosan and CMS, as well as their mixtures at different Chit:CMS ratios, are shown in Figure 5.

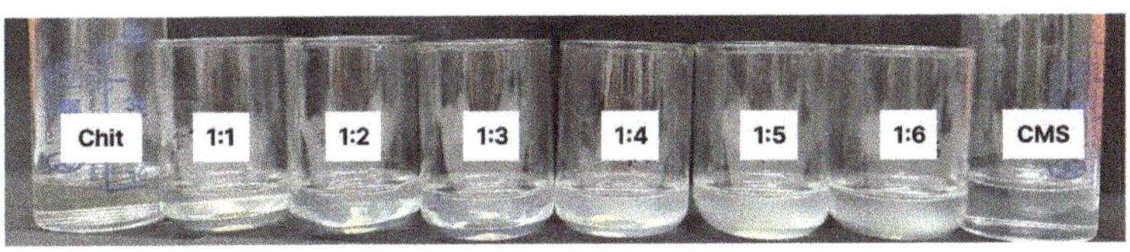

Figure 5. Visual state of chitosan, CMS, and their mixture solutions in distilled water at different Chit:CMS ratios.

The figure shows that individual polymer solutions are transparent, while their mixtures exhibit significant light scattering. It can be seen how an increase in the fraction of CMS in the mixtures leads to an increase in the turbidity of the solutions. This effect qualitatively reflects the process of polyelectrolyte interaction and may be associated with the formation and increase in the size of (bio)PEC structures between polyelectrolytes.

This interaction can be expressed using the ratio of ionized groups z, which depends on the content of ionized functional side groups in the polymer. At $z = 1$, the PEC is stoichiometric and a balance of charges in the system is observed (the most effective polyelectrolyte interaction). On the contrary, under conditions $z < 1$ and $z > 1$, non-stoichiometric interactions between polyelectrolytes are presented, and there is an excess of positive and negative charge, respectively. The ratio of z was determined using Equation (11) [51]

$$z = \frac{[COO^-]}{[NH_3^+]}, \quad (11)$$

where $[COO^-]$ and $[NH_3{}^+]$ are a molar concentration of charged units. The values of z were calculated taking into consideration the *DD* of chitosan and *DS* values of CMS. The results of the calculation of the z values for various compositions of (bio)PEC are presented in Table 2.

Table 2. The effect of the ratio of ionized groups on the composition of PEC with different types of CMS.

Chit:CMS	Ratio of Ionized Groups z		
	CMS 1	CMS 2	CMS 3
1:1	0.10	0.21	0.32
1:2	0.20	0.42	0.65
1:3	0.30	0.64	0.98
1:4	0.40	0.85	1.30
1:5	0.50	1.06	1.63
1:6	0.60	1.27	1.96

It can be seen from Table 2 that for all the mixtures of Chit:CMS 1 with the lowest *DS*, the formation of a (bio)PEC with an excess of positive charge $z < 1$ is observed (non-stoichiometric PEC). On the contrary, for mixtures with CMS 2 and CMS 3 with higher *DS* values, there is a transition from an excess of positive to an excess of negative charge with an increase in the proportion of CMS in the mixture observed ($z < 1$ and $z > 1$). A similar charge transition was also observed in work [52] for oligochitosan (M_η = 5 kDa) and CMS PEC based on zeta potential measurement results. For example, it has been shown that when the ratio of polymers changes from 1:4 to 3:7, the zeta potential increases from −11.2 to 4.8 mV, correspondingly. This is correlated with the results of calculating the ratio of ionized groups in PEC.

The formation of non-stoichiometric polyelectrolyte complexes has been reported by some works [53–59]. Typically, these PECs are prepared by mixing polyelectrolytes of significantly different molecular weights in non-stoichiometric proportions. The obtained particles consist of a long host macromolecule complexed with shorter molecules of a guest polyelectrolyte with an opposite charge [60]. Drogoz et al. [56] studied the formation of polyelectrolyte complexes of polysaccharides using chitosan and dextran sulfate. The obtained dispersions had a negative or positive charge depending on the excess polyelectrolyte, and it was shown that to maintain the stability of the dispersions, the z value should be less than 0.6. In addition, the non-stochiometric PEC of modified chitosan with polystyrenesulfonate was investigated by Izumrudo A. [57]. In this work, the critical ratio (z_{cr} = 0.3) was determined by titration; at this point, a slight turbidity was observed and it was also shown that the addition of one of the polyelectrolytes allows the formation and accumulation of insoluble PEC. These results confirm the formation of the insoluble complexes obtained in our work when $z \neq 1$.

Optical density was measured to study the effect of the CMS type and the ratio of ionized groups on the formation of (bio)PEC dispersions. The relationship between the initial pH value of the Chit:CMS mixtures after preparation pH_0 and their optical density versus Chit:CMS ratio are shown in Figure 6.

For a better understanding of the influence of the biopolymer ratio in solution, it is important to analyze the change in the initial pH_0. Figure 6a shows that the pH values of the initial chitosan and CMS solutions are approximately 4.3 and 7.17, correspondingly. Therefore, with the increase in the fraction of the CMS solution, the pH of the mixture increases. However, it is noteworthy that the character of the change in pH values depends on the *DS* value of CMS. For example, for mixtures with CMS 3 with a higher *DS* the pH values increase more intensively compared to mixtures with CMS 1. This happens because with an increase in *DS* of CMS, many negatively charged carboxymethyl ions become available to positively charged amino groups of chitosan which reduces the amount of H^+ ions producing an increase in pH [33,61,62]. Figure 6b shows that mixing individual solutions of chitosan and CMS, which have no optical density, results in a sharp increase in optical density, which is confirmed by the visual observations of the mixtures (Figure 5). Moreover, when using CMS with a high *DS* value, the optical density of the mixture solutions increases sharply to plateau values, for example, compared to the mixture with CMS 1. This observation confirms the increase in *DS* for CMS products with an increase in the proportion of MCA during their production, and also indicates a more efficient formation of (bio)PEC structures when using CMS 2 and 3 [2,63,64].

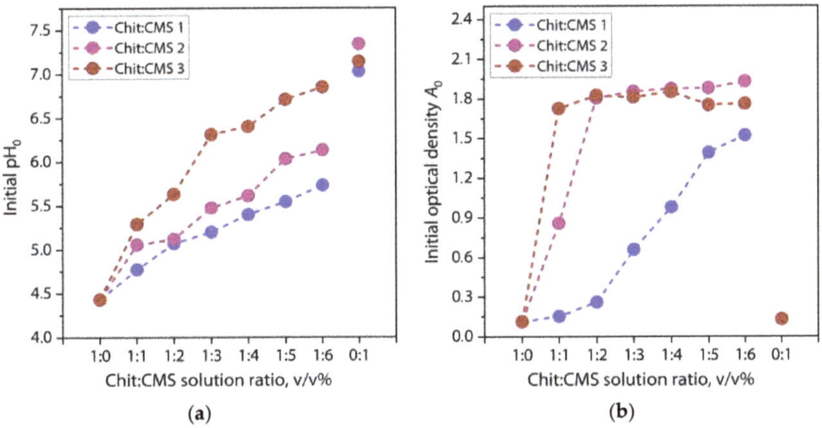

Figure 6. Dependences of (**a**) pH_0 and (**b**) optical density of solutions on the Chit:CMS ratio.

3.3. Characterization of (Bio)Polyelectrolyte Complex

Figure 7 shows the ATR-FTIR spectra of the chitosan, CMS 1, and dried (bio)PEC dispersion samples at different ratios of Chit:CMS 1.

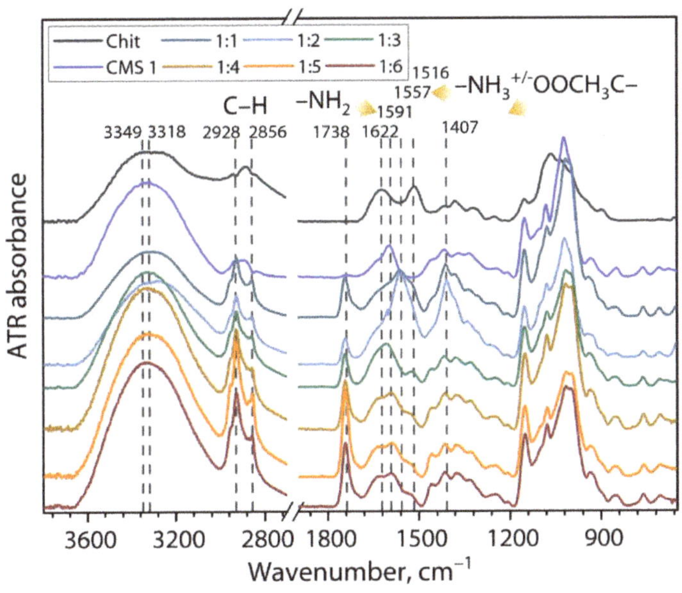

Figure 7. ATR-FTIR spectra of dried (bio)PEC dispersion samples at different ratios of Chit:CMS 1.

On the spectrum of Chit:CMS 1 (bio)PEC samples, new absorption bands and more intensive peaks are shown in comparison with the Chit and CMS 1 spectra. One significant change is the narrowing of the bands with a center around 3349–3318 cm^{-1}. This can be associated with the formation of hydrogen bonds between the Chit and CMS macromolecules, as well as with the electrostatic interaction between the amino groups of chitosan and the carboxymethyl groups of CMS in the formation of (bio)PECs [43,65,66]. In addition, the absorption peaks at 2928 and 2865 cm^{-1} became sharper. This could be attributed to the C–H stretching vibration becoming more intensive, likely due to the CH_2 groups of CMS [31,67,68]. Another change observed in the samples with a high fraction of CMS (1:4, 1:5, and 1:6) is the appearance of shoulder bands in the region of 1635–1591 cm^{-1} related to the stretching vibration of the C=O from CMS. In addition, the bands at 1557 cm^{-1} are likely due to $-NH_3^+$ groups in chitosan. Moreover, the interaction between chitosan and CMS may happen by hydrogen bonds and ionic interactions to form (bio)PEC which is shown in the bands at around 1738 and 1557 cm^{-1} [31,32].

3.4. Effect of the pH on the (Bio)Polyelectrolyte Complex

The turbidimetric titration of the mixtures of chitosan and CMS solutions was carried out to clarify the ranges of formation of the (bio)PEC structures and their stability. The influence of pH on the optical density of chitosan, CMS, and Chit:CMS mixtures with different CMS products is illustrated in Figure 8.

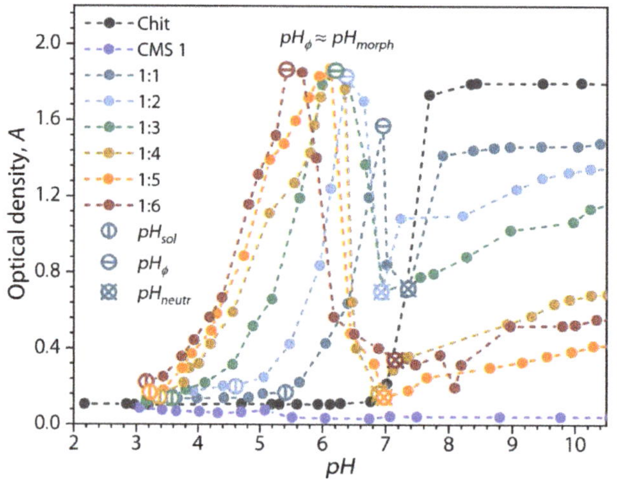

(**a**)

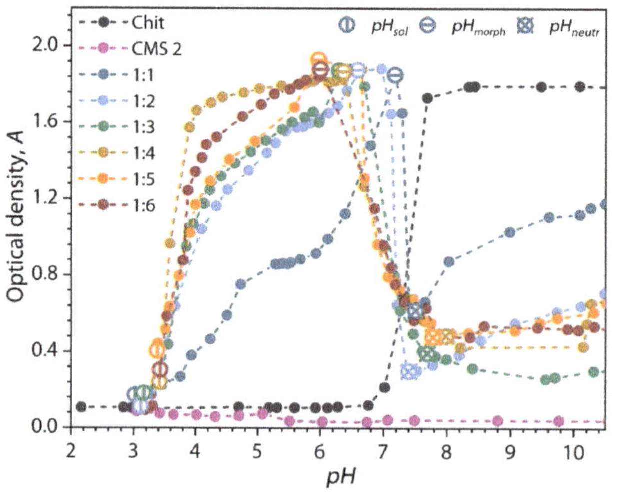

(**b**)

Figure 8. *Cont.*

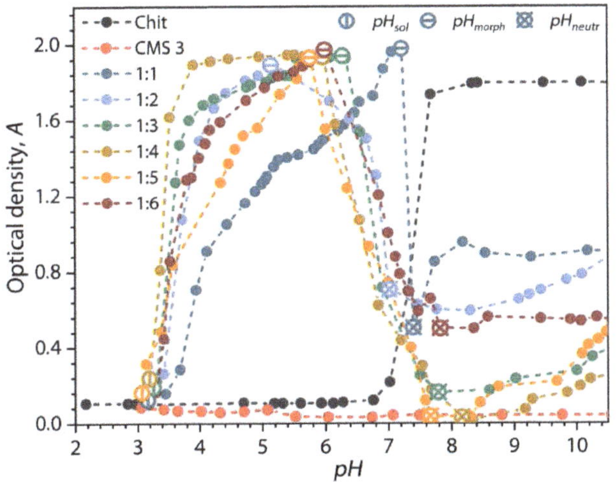

(c)

Figure 8. Effect of the pH on the optical density with (**a**) CMS 1, (**b**) CMS 2, and (**c**) CMS 3.

At first, for the chitosan solution, it can be observed that there are no changes in the optical density (*A*) up to a pH of seven (more than pK_0 = 6.34). Above this point, there is a sharp increase in the values of *A*, which indicates the formation of a microphase dispersion of chitosan and causes its degree of protonation to be reduced [69,70]. On the other hand, for the CMS solutions, the *A* has not significantly changed because of their chemical structure. It is known that at the degree of substitution (carboxymethylation) of starch $DS \geq 0.2$, CMS dissolves well in water [71]. Moreover, from the dependencies, it is clear that even under strongly acidic conditions, CMS remains in a single-phase solution and does not precipitate into a microphase or sediment. In the case of mixtures of chitosan and CMS solutions, we observe an increase in the *A* values uncharacteristic for individual polymer solutions. This, as previously shown, is due to the formation and accumulation of (bio)PEC particles.

For the Chit:CMS (bio)PEC dispersions, four zones delimited by pH transition points can be identified according to the method described in works [51,72]. The first zone is delimited by the $pH \leq pH_{sol}$. In this zone, the *A* is constant and the solutions are almost transparent which is associated with the formation of unstable and soluble PEC structures. The second zone is pH_{ϕ}–pH_{morph}, which corresponds to the formation of insoluble PEC dispersion, where a strong increase in *A* value is seen compared to the individual polymer solutions. It is noteworthy that the position of the pH_{ϕ} change depends on the *DS* of CMS. For example, for Chit:CMS 1 mixtures (Figure 8a), as the ratio increases from 1:1 to 1:6, the pH_{ϕ} values decrease from 5.6 to 3.4, respectively. However, for mixtures with CMS 2 and CMS 3 the pH_{ϕ} values are weakly dependent on the polymer ratio. We believe that this is due to an increase in the content of ionized carboxymethyl groups in CMS with an increase in *DS*, which leads to a decrease in pH values necessary for the effective formation of PEC particles. With a further increase in pH, we observe a maximum optical density, which corresponds to the value of pH_{morph}. At this point, the stoichiometric complexation is achieved ($n^{+} \approx n^{-}$) and the polyelectrolyte interaction process occurs most efficiently.

It is important to note that in cases where both oppositely charged polyelectrolytes have a relatively high degree of substitution with charged functional groups, pH_{morph} may not be a single point but a range of pH values (plateau of *A* values). In such cases, the dispersion of (bio)PEC particles may be effectively formed over a wide pH range. It can be seen from the figures that the pH_{morph} range changes depending on the type of CMS. So, for example, for mixtures of Chit:CMS 1, pH_{morph} is a single point, but for mixtures of Chit:CMS 2 and Chit:CMS 3, this is a range of 4.9–6.4 and 4.3–6.4, respectively. This phenomenon may be because at higher *DS* of CMS, more carboxymethyl groups are present for complexation with the amino groups of chitosan. As a result, this leads to an increase in the stoichiometry of the complex structures when the *z* value is close to one or higher. In this regard, for mixtures of Chit:CMS 1, the values of *z* increase slightly with increasing polymer ratio and the pH_{morph} point decreases, while the area under the curve *A* (pH) increases strongly, which is associated with an increase in the concentration of ionized carboxymethyl groups in the solution despite the low value of *DS* of this CMS. At the same time, in mixtures with CMS 2 and 3 with higher values of *DS*, with an increase in the proportion of CMS in the solution, the value of *z* increases more intensively, which leads to an increase in the range of pH_{morph} and a large area under the curve *A*(pH) even at low Chit:CMS ratios.

In the next zone, when pH_{morph}–pH_{neutr} the (bio)PEC structures start to collapse and degrade. As can be seen from Figure 8, the pH values are close to pK_0 for chitosan, which links the effect of decreasing *A* for dispersions with the process of the deprotonation of chitosan amino groups under alkaline conditions, leading to the destruction of (bio)PEC structures. In this range, the dispersion contains soluble (bio)PEC and CMS particles, and insoluble chitosan particles [73–75]. For the last range at pH > pK_0, the dispersion contains insoluble chitosan microphase, as mentioned above, and dissolved CMS macromolecules.

3.5. Structure of (Bio)PEC Dispersions Based on Chitosan and Carboxymethyl Starch

The study of the effect of pH on Chit:CMS mixtures gives a qualitative idea of the formation of (bio)PEC dispersions and shows the pH range necessary to achieve stable dispersions. However, by means of this analysis, it is not possible to analyze the structure and particle size of the (bio)PEC particles. For this reason, the dynamic light scattering method was used in order to analyze the particle diameter distribution of the (bio)PEC dispersions at pH_{morph} (Figure 9). The *n* values of the solutions are presented in the boxes.

From Figure 9, it can be seen that all of the ratios are characterized by monomodal distribution and this distribution reminds the result for the individual biopolymer solutions (Figure 4). In addition, it is observed that the average particle diameter D_{avg} of the (bio)PEC dispersions is bigger than the sum of it for the independent macromolecules of chitosan and CMS. This effect is probably due to the intermolecular interaction of the polymers following its complexation. The main observed effect is associated with an increase in the value of D_{avg} with an increase in the ratio of Chit:CMS in dispersions for all types of the obtained CMS. At the same time, it is evident that the relative width of all the distributions (within one order of diameter) is similar, which means the isotropy of the formed particles and the absence of their coagulation. The observed effects indicate that an increase in D_{avg} is mainly associated with an increase in *DS* and the proportion of CMS in the composition of the mixtures.

Some works [76–80] report the same form of distribution for PECs, attributing this to the neutralization of the charges of the polyelectrolytes, which decreases the repulsive forces and leads to an increase in particle size. For example, in the work [76], microgels from different substituted CMS showed a monomodal distribution of the particle diameter that became larger with the increase in *DS* because of the larger amounts of carboxymethyl groups. Li et al. [77] obtained complex nanogels from chitosan hydrochloride and carboxymethyl starch. It was noted that the increase in CMS content might induce an increase in particle size because of an increase in interaction between polymers.

For a better understanding of the influence of the Chit:CMS ratio and the type of CMS on the particle diameter, Figure 10 shows the effect of the ratio at pH_{morph} on the average diameter q_{max} for the dispersion of (bio)PEC.

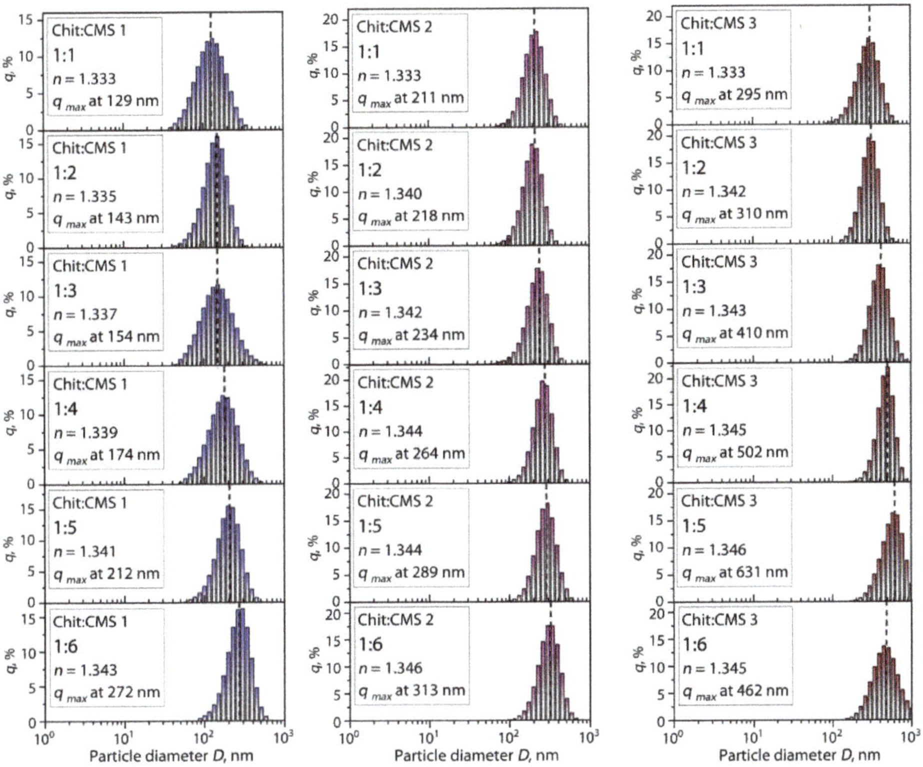

(a) (b) (c)

Figure 9. Particle diameter distributions for Chit:CMS (bio)PEC dispersions at pH_{morph} with different ratios of chitosan and (**a**) CMS 1, (**b**) CMS 2, and (**c**) CMS 3.

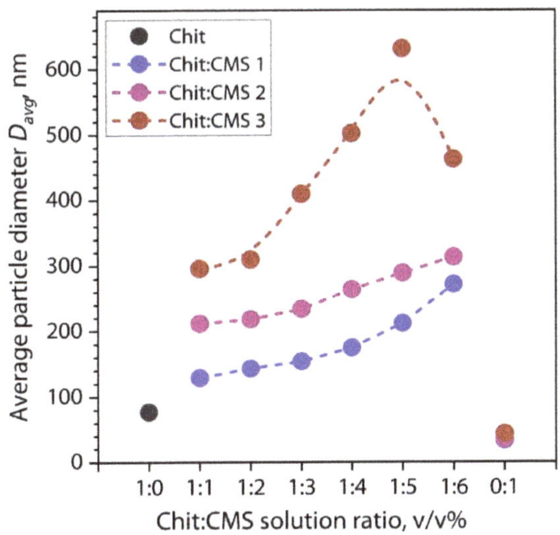

Figure 10. Effect of the Chit:CMS ratio and type of CMS at pH_{morph} on the average diameter at qmax for the dispersions of (bio)PEC.

From Figure 10, it can be seen that the D_{avg} value increases with the fraction of CMS for mixtures of Chit:CMS with all types of CMS. In addition, with an increase in the *DS* and M_η values for CMS, a general increase in D_{avg} is observed regardless of the proportion of CMS in the mixture. With an increase in the proportion of CMS in the Chit:CMS 1 and Chit:CMS 2 mixtures from 1:1 to 1:6, the D_{avg} value increases by 143 nm and 102 nm, respectively. Interestingly, in the Chit:CMS 3 mixture, this effect is more intense, and an increase in the proportion of CMS 3 in the mixture from 1:1 to 1:5 leads to an increase in D_{avg} by 335 nm. This is probably because this mixture is in the studied range of polymer ratios $z \geq 1$ compared to the use of CMS 1 and 2 (Table 2). However, with an increase in the Chit:CMS 3 ratio from 1:5 to 1:6, there is a decrease in the D_{avg} value, which is explained by a strong excess of negative charge of carboxyl groups of CMS at $z > 1$ (Table 2). This leads to the fact that the formed negatively charged particles of primary PEC aggregate worse and do not form larger particles as in the case of $z \approx 1$. A similar dependence of the decrease in the diameter of the PEC particles was also observed in a number of works [50,78,81–83].

Comparing our results with other studies carried out, for example, those by Saikia C. et al. [84] who obtained carboxymethyl starch-chitosan-coated iron oxide magnetic nanoparticles with a particle size distribution of 219–247 nm, it can be seen that our particles (CMS 2) can also be used in encapsulation applications for drug delivery. Furthermore, Wu D. and Delair T. [85] obtained electrolyte complexes based on chitosan/hyaluronan and studied the effect of factors such as pH and polymer ratio. It can be observed that, as in our work, an increase in the polymer ratio leads to an increase in the average particle size, but with a DA 16% chitosan, an excess in this ratio decreases the particle size due to the charge shift in the electrolytes. Moreover, in the work [86], the stability of the PECs based on chitosan:alginate with respect to pH changes was also evaluated. For these dispersions, the most significant changes were noted in the pH range 6.4 to 7, while at pH > *pKa* weak electrostatic interactions occur, causing aggregate formation. This is in agreement with our optical density results which show a reduction in its value at pH higher than *pKa*.

4. Conclusions

In the present work, biopolyelectrolyte complex (bio)PECs based on chitosan (Chit) and carboxymethyl starch (CMS) were obtained by mixing independent biopolymer solutions. The effect of the ratio of biopolymer and of ionized groups *z*, pH, and degree of substitution of CMS in the formation of the (bio)PECs were studied by the turbidimetry and dynamic light scattering (DLS) methods. For this purpose, three carboxymethyl starches were synthesized. The ATR-FTIR spectra of these products showed new bands at 1592 cm^{-1} and 1414 cm^{-1}, representing the substitution with –CH_3COONa salt radicals on the cassava starch. In addition, the values of *DS* were 0.16, 0.33, and 0.36 for CMS 1, CMS 2, and CMS 3, respectively.

For the preparation of (bio)PECs, independent solutions of Chit and CMS with 0.25 wt% were mixed in ratios from 1:1 to 1:6 v/v%, with an increment of 1 in the amount of CMS. The formation of the (bio)PEC was confirmed by the increase in optical density after mixing the individual solutions of Chit and CMS. Furthermore, the increase in the fraction of CMS in the mixtures leads to an increase in the size of (bio)PEC structures between polyelectrolytes. Therefore, the use of CMS 2 and CMS 3 with a higher *DS* increases the efficient formation of (bio)PEC. In addition, in the ATR-FTIR spectra appeared new bands at around 1738 and 1557 cm^{-1}, which confirmed the interaction between Chit and CMS.

The formation of (bio)PECs is divided into four zones delimited by pH transition points. The first zone corresponds to the formation of unstable and soluble PEC structures. The second zone corresponds to the formation of insoluble PEC dispersion, where *DS* of the CMS influences the pH_ϕ. The final pH point of this second zone is the pH_{morph}, the point where the stoichiometric complexation is achieved (z = 1) and the polyelectrolyte interaction process occurs most efficiently. This pH_{morph} depends on the type of CMS. For Chit:CMS 1, pH_{morph} is a single point, and for CMS 2 and CMS 3, this is a range of 4.9–6.4 and 4.3–6.4, respectively. The third zone corresponds to the collapse and degradation of (bio)PECs. And, the last zone is characterized by an insoluble chitosan dispersion with dissolved CMS. The results of the dynamic light scattering method have shown a monomodal distribution, similar to the initial biopolymers, for all ratios of polymers and for all types of CMS. Moreover, an increase in the ratio of Chit:CMS in dispersions for all types of the obtained CMS leads to an increase in the value of D_{avg}. But, in the dispersion with the higher *DS* (CMS 3) at a value of $z \geq 2$, this effect is annulated, and a slight decrease is appreciated due to the too-low density of ionic bonds between polyelectrolyte radicals.

Author Contributions: Conceptualization, D.C. and A.P.; methodology, D.C. and V.B.; validation, A.P. and M.L.; investigation, D.C. and V.B.; writing—original draft preparation, D.C.; writing—review and editing, A.P. and M.L.; supervision, A.P.; funding acquisition, A.P. All authors have read and agreed to the published version of the manuscript.

Funding: This work was supported by the Ministry of Science and Higher Education of the Russian Federation (agreement No. 075-15-2021-1349).

Institutional Review Board Statement: Not applicable.

Data Availability Statement: Data are contained within the article.

Conflicts of Interest: The authors declare no conflicts of interest.

References

1. Kulkarni, A.D.; Vanjari, Y.H.; Sancheti, K.H.; Patel, H.M.; Belgamwar, V.S.; Surana, S.J.; Pardeshi, C.V. Polyelectrolyte Complexes: Mechanisms, Critical Experimental Aspects, and Applications. *Artif. Cells Nanomed. Biotechnol.* **2016**, *44*, 1615–1625. [CrossRef] [PubMed]
2. Meka, V.S.; Sing, M.K.G.; Pichika, M.R.; Nali, S.R.; Kolapalli, V.R.M.; Kesharwani, P. A Comprehensive Review on Polyelectrolyte Complexes. *Drug Discov. Today* **2017**, *22*, 1697–1706. [CrossRef]
3. Patwekar, S.L.; Potulwar, A.P.; Pedewad, R.; Gaikwad, M.S.; Khan, S.A.; Suryawanshi, A.B. Review on Polyelectrolyte Complex as Novel Approach for Drug Delivery System. *Int. J. Pharm. Pharm. Res.* **2016**, *5*, 98–109.
4. Ghimire, Y.; Bhattarai, A. A Review on Polyelectrolytes (PES) and Polyelectrolyte Complexes (PECs). *IJERT* **2020**, *V9*, 877–899. [CrossRef]
5. Rumyantsev, A.M.; Jackson, N.E.; De Pablo, J.J. Polyelectrolyte Complex Coacervates: Recent Developments and New Frontiers. *Annu. Rev. Condens. Matter Phys.* **2021**, *12*, 155–176. [CrossRef]
6. Lankalapalli, S.; Kolapalli, V.R.M. Polyelectrolyte Complexes: A Review of Their Applicability in Drug Delivery Technology. *Indian. J. Pharm. Sci.* **2009**, *71*, 481. [CrossRef]
7. Lin, W.-S.; Catchmark, J.M. Development of Sustainable Adhesives Based on Polysaccharide–Polyelectrolyte Complexes for Natural Fiber-Based Materials. *ACS Appl. Polym. Mater.* **2024**, *6*, 7947–7955. [CrossRef]
8. Muller, M.; Vehlow, D.; Torger, B.; Urban, B.; Woltmann, B.; Hempel, U. Adhesive Drug Delivery Systems Based on Polyelectrolyte Complex Nanoparticles (PEC NP) for Bone Healing. *Curr. Pharm. Des.* **2018**, *24*, 1341–1348. [CrossRef] [PubMed]
9. Jamróz, E.; Janik, M.; Juszczak, L.; Kruk, T.; Kulawik, P.; Szuwarzyński, M.; Kawecka, A.; Khachatryan, K. Composite Biopolymer Films Based on a Polyelectrolyte Complex of Furcellaran and Chitosan. *Carbohydr. Polym.* **2021**, *274*, 118627. [CrossRef] [PubMed]
10. Delgado, J.D.; Schlenoff, J.B. Polyelectrolyte Complex Films from Blends Versus Copolymers. *Macromolecules* **2019**, *52*, 7812–7820. [CrossRef]
11. Huang, J.; Wu, D.; Liu, W.; Sun, J.; Zhao, X.; Zhang, Y.; Feng, Y. Polyelectrolyte Complex Gels Used for CO2 Absorption. *J. Polym. Sci.* **2024**, *62*, 3275–3286. [CrossRef]
12. Wanasingha, N.; Dorishetty, P.; Dutta, N.; Choudhury, N. Polyelectrolyte Gels: Fundamentals, Fabrication and Applications. *Gels* **2021**, *7*, 148. [CrossRef]
13. Musin, E.V.; Kim, A.L.; Dubrovskii, A.V.; Tikhonenko, S.A. New Sight at the Organization of Layers of Multilayer Polyelectrolyte Microcapsules. *Sci. Rep.* **2021**, *11*, 14040. [CrossRef] [PubMed]
14. Paşcalău, V.; Bogdan, C.; Pall, E.; Matroş, L.; Pandrea, S.-L.; Suciu, M.; Borodi, G.; Iuga, C.A.; Ştiufiuc, R.; Topală, T.; et al. Development of BSA Gel/Pectin/Chitosan Polyelectrolyte Complex Microcapsules for Berberine Delivery and Evaluation of Their Inhibitory Effect on Cutibacterium Acnes. *React. Funct. Polym.* **2020**, *147*, 104457. [CrossRef]
15. Sutharsan, J.; Boyer, C.A.; Zhao, J. Physicochemical Properties of Chitosan Edible Films Incorporated with Different Classes of Flavonoids. *Carbohydr. Polym. Technol. Appl.* **2022**, *4*, 100232. [CrossRef]
16. Abotbina, W.; Sapuan, S.M.; Ilyas, R.A.; Sultan, M.T.H.; Alkbir, M.F.M.; Sulaiman, S.; Harussani, M.M.; Bayraktar, E. Recent Developments in Cassava (*Manihot esculenta*) Based Biocomposites and Their Potential Industrial Applications: A Comprehensive Review. *Materials* **2022**, *15*, 6992. [CrossRef]
17. Parmar, A.; Sturm, B.; Hensel, O. Crops That Feed the World: Production and Improvement of Cassava for Food, Feed, and Industrial Uses. *Food Sec.* **2017**, *9*, 907–927. [CrossRef]

18. Verma, D.K.; Srivastav, P.P. Isolation, Modification, and Characterization of Rice Starch with Emphasis on Functional Properties and Industrial Application: A Review. *Crit. Rev. Food Sci. Nutr.* **2022**, *62*, 6577–6604. [CrossRef] [PubMed]
19. Pooresmaeil, M.; Namazi, H. Developments on Carboxymethyl Starch-Based Smart Systems as Promising Drug Carriers: A Review. *Carbohydr. Polym.* **2021**, *258*, 117654. [CrossRef]
20. Das, S.; Das, M.K.; Jamatia, T.; Bhattacharya, B.; Mazumder, R.; Yadav, P.K.; Ghose Bishwas, N.R.; Deka, T.; Roy, D.; Sinha, B.; et al. Advances of Cassava Starch-Based Composites in Novel and Conventional Drug Delivery Systems: A State-of-the-Art Review. *RSC Pharm.* **2024**, *1*, 182–203. [CrossRef]
21. Oh, J.-W.; Chun, S.C.; Chandrasekaran, M. Preparation and In Vitro Characterization of Chitosan Nanoparticles and Their Broad-Spectrum Antifungal Action Compared to Antibacterial Activities against Phytopathogens of Tomato. *Agronomy* **2019**, *9*, 21. [CrossRef]
22. Riseh, R.S.; Vazvani, M.G.; Kennedy, J.F. The Application of Chitosan as a Carrier for Fertilizer: A Review. *Int. J. Biol. Macromol.* **2023**, *252*, 126483. [CrossRef] [PubMed]
23. Kanmani, P.; Aravind, J.; Kamaraj, M.; Sureshbabu, P.; Karthikeyan, S. Environmental Applications of Chitosan and Cellulosic Biopolymers: A Comprehensive Outlook. *Bioresour. Technol.* **2017**, *242*, 295–303. [CrossRef] [PubMed]
24. Hu, W.; Niu, Y.; Zhu, H.; Dong, K.; Wang, D.; Liu, F. Remediation of Zinc-Contaminated Soils by Using the Two-Step Washing with Citric Acid and Water-Soluble Chitosan. *Chemosphere* **2021**, *282*, 131092. [CrossRef]
25. Zdanowicz, M.; Markowska-Szczupak, A.; Spychaj, T. Carboxymethyl Starch/Medium Chain Fatty Acid Compositions: Rheological Changes During Storage and Selected Film Properties. *Starch Stärke* **2020**, *72*, 1900240. [CrossRef]
26. Bungenberg de Jong, H.G.; Kruyt, H.R. Coacervation (Partial Miscibility in Colloid Systems). *Proc. K. Ned. Akad. Van. Wet.* **1932**, *32*, 849–856.
27. Michaels, A.S. Polyelectrolite Complex. *Ind. Eng. Chem.* **1965**, *57*, 32–40. [CrossRef]
28. Michaels, A.S.; Miekka, R.G. Polycation-Polyanion Complexes: Preparation and Properties of Poly (Vinylbenzyltrimethylammonium) Poly-(Styrenesulfonate). *J. Phys. Chem.* **1961**, *65*, 1765–1773. [CrossRef]
29. Facchi, D.P.; Da Cruz, J.A.; Bonafé, E.G.; Pereira, A.G.B.; Fajardo, A.R.; Venter, S.A.S.; Monteiro, J.P.; Muniz, E.C.; Martins, A.F. Polysaccharide-Based Materials Associated with or Coordinated to Gold Nanoparticles: Synthesis and Medical Application. *Curr. Med. Chem.* **2017**, *24*, 2701–2735. [CrossRef] [PubMed]
30. Lin, Y.; Wang, X.; Liu, Q.; Fang, Y. Preparation and Application of Chitosan-Based Polyelectrolyte Complex Materials: An Overview. *Pap. Biomater.* **2022**, *7*, 1. [CrossRef]
31. Quadrado, R.F.N.; Fajardo, A.R. Microparticles Based on Carboxymethyl Starch/Chitosan Polyelectrolyte Complex as Vehicles for Drug Delivery Systems. *Arab. J. Chem.* **2020**, *13*, 2183–2194. [CrossRef]
32. Assaad, E.; Wang, Y.J.; Zhu, X.X.; Mateescu, M.A. Polyelectrolyte Complex of Carboxymethyl Starch and Chitosan as Drug Carrier for Oral Administration. *Carbohydr. Polym.* **2011**, *84*, 1399–1407. [CrossRef]
33. Henao, E.; Delgado, E.; Contreras, H.; Quintana, G. Polyelectrolyte Complexation versus Ionotropic Gelation for Chitosan-Based Hydrogels with Carboxymethylcellulose, Carboxymethyl Starch, and Alginic Acid. *Int. J. Chem. Eng.* **2018**, *2018*, 3137167. [CrossRef]
34. Wang, Y.J.; Assaad, E.; Ispas-Szabo, P.; Mateescu, M.A.; Zhu, X.X. NMR Imaging of Chitosan and Carboxymethyl Starch Tablets: Swelling and Hydration of the Polyelectrolyte Complex. *Int. J. Pharm.* **2011**, *419*, 215–221. [CrossRef] [PubMed]
35. Saboktakin, M.R.; Tabatabaie, R.M.; Maharramov, A.; Ramazanov, M.A. Synthesis and in Vitro Evaluation of Carboxymethyl Starch–Chitosan Nanoparticles as Drug Delivery System to the Colon. *Int. J. Biol. Macromol.* **2011**, *48*, 381–385. [CrossRef] [PubMed]
36. Leonida, M.; Ispas-Szabo, P.; Mateescu, M.A. Self-Stabilized Chitosan and Its Complexes with Carboxymethyl Starch as Excipients in Drug Delivery. *Bioact. Mater.* **2018**, *3*, 334–340. [CrossRef]
37. Brugnerotto, J.; Lizardi, J.; Goycoolea, F.M.; Argüelles-Monal, W.; Desbrières, J.; Rinaudo, M. An Infrared Investigation in Relation with Chitin and Chitosan Characterization. *Polymer* **2001**, *42*, 3569–3580. [CrossRef]
38. Wang, Q.Z.; Chen, X.G.; Liu, N.; Wang, S.X.; Liu, C.S.; Meng, X.H.; Liu, C.G. Protonation Constants of Chitosan with Different Molecular Weight and Degree of Deacetylation. *Carbohydr. Polym.* **2006**, *65*, 194–201. [CrossRef]
39. Stojanović, Ž.; Jeremić, K.; Jovanović, S.; Lechner, M.D. A Comparison of Some Methods for the Determination of the Degree of Substitution of Carboxymethyl Starch. *Starch Stärke* **2005**, *57*, 79–83. [CrossRef]
40. Manna, P.K.; Choudhury, P.K. Molecular Parameters of Sodium Carboxymethyl Amylose in Dilute Solution. *J. Macromol. Sci. Part. A Chem.* **1974**, *8*, 909–917. [CrossRef]
41. Kasaai, M.R. Calculation of Mark–Houwink–Sakurada (MHS) Equation Viscometric Constants for Chitosan in Any Solvent–Temperature System Using Experimental Reported Viscometric Constants Data. *Carbohydr. Polym.* **2007**, *68*, 477–488. [CrossRef]
42. Flory, P.J. *Principles of Polymer Chemistry*; 19. Print; Cornell University Press: Ithaca, NY, USA, 2006; ISBN 978-0-8014-0134-3.
43. Lustriane, C.; Dwivany, F.M.; Suendo, V.; Reza, M. Effect of Chitosan and Chitosan-Nanoparticles on Post Harvest Quality of Banana Fruits. *J. Plant Biotechnol.* **2018**, *45*, 36–44. [CrossRef]
44. Mohammed, M.; Syeda, J.; Wasan, K.; Wasan, E. An Overview of Chitosan Nanoparticles and Its Application in Non-Parenteral Drug Delivery. *Pharmaceutics* **2017**, *9*, 53. [CrossRef]
45. Costa, C.N.; Teixeira, V.G.; Delpech, M.C.; Souza, J.V.S.; Costa, M.A.S. Viscometric Study of Chitosan Solutions in Acetic Acid/Sodium Acetate and Acetic Acid/Sodium Chloride. *Carbohydr. Polym.* **2015**, *133*, 245–250. [CrossRef]

46. Philippova, O.E.; Korchagina, E.V. Chitosan and Its Hydrophobic Derivatives: Preparation and Aggregation in Dilute Aqueous Solutions. *Polym. Sci. Ser. A* **2012**, *54*, 552–572. [CrossRef]
47. Adeyanju, O.; Olademehin, O.P.; Hussaini, Y.; Nwanta, U.C.; Adejoh, A.I.; Plavec, J. Synthesis and Characterization of Carboxymethyl Plectranthus Esculentus Starch. A Potential Disintegrant. *J. Pharm. Appl. Chem.* **2016**, *2*, 189–195. [CrossRef]
48. Wang, W.; Liu, W.; Wu, J.; Liu, M.; Wang, Y.; Liu, H.; Liu, J. Preparation and Characterization of Particle-Filled Microgels by Chemical Cross-Linking Based on Zein and Carboxymethyl Starch for Delivering the Quercetin. *Carbohydr. Polym.* **2024**, *323*, 121375. [CrossRef] [PubMed]
49. Vishakha, V.; Abdel-Mohsen, A.M.; Michalicka, J.; White, P.B.; Lepcio, P.; Tinoco Navarro, L.K.; Jančář, J. Carboxymethyl Starch as a Reducing and Capping Agent in the Hydrothermal Synthesis of Selenium Nanostructures for Use with Three-Dimensional-Printed Hydrogel Carriers. *R. Soc. Open Sci.* **2023**, *10*, 230829. [CrossRef] [PubMed]
50. Stojanović, Ž.; Jeremić, K.; Jovanović, S. Molecular Structure of Carboxymethyl Starch in Dilute Aqueous Sodium Chloride Solutions. *Starch Stärke* **2013**, *65*, 902–911. [CrossRef]
51. Litvinov, M.; Kashurin, A.; Aleksandr, P. Study of the Influence of the Composition and pH of the Solution on the Structure and Morphology of Particle Dispersions of a (Bio)Polyelectrolyte Complex between Chitosan and Gelatin. *Colloid. Polym. Sci.* **2024**. [CrossRef]
52. Chen, X.; Yan, Y.; Li, H.; Wang, X.; Tang, S.; Li, Q.; Wei, J.; Su, J. Evaluation of Absorbable Hemostatic Agents of Polyelectrolyte Complexes Using Carboxymethyl Starch and Chitosan Oligosaccharide Both *in Vitro* and *in Vivo*. *Biomater. Sci.* **2018**, *6*, 3332–3344. [CrossRef] [PubMed]
53. Buchhammer, H.-M.; Kramer, G.; Lunkwitz, K. Interaction of Colloidal Dispersions of Non-Stoichiometric Polyelectrolyte Complexes and Silica Particles. *Colloids Surf. A Physicochem. Eng. Asp.* **1995**, *95*, 299–304. [CrossRef]
54. Klimaviciute, R.; Bendoraitiene, J.; Lekniute, E.; Zemaitaitis, A. Non-Stoichiometric Complexes of Cationic Starch and 4-Sulfophthalic Acid and Their Flocculation Efficiency. *Colloids Surf. A Physicochem. Eng. Asp.* **2014**, *457*, 180–188. [CrossRef]
55. Gao, J.; Zhan, Q.; Tang, Z.; Huang, Y. The Critical Transition from Soluble Complexes to Colloidal Aggregates of Polyelectrolyte Complexes at Non-Stoichiometric Charge Ratios. *Macromol. Rapid Commun.* **2022**, *43*, 2100880. [CrossRef]
56. Drogoz, A.; David, L.; Rochas, C.; Domard, A.; Delair, T. Polyelectrolyte Complexes from Polysaccharides: Formation and Stoichiometry Monitoring. *Langmuir* **2007**, *23*, 10950–10958. [CrossRef] [PubMed]
57. Izumrudov, V.A.; Volkova, I.F.; Grigoryan, E.S.; Gorshkova, M.Y. Water-Soluble Nonstoichiometric Polyelectrolyte Complexes of Modified Chitosan. *Polym. Sci. Ser. A* **2011**, *53*, 281–288. [CrossRef]
58. Mihai, M.; Dragan, E.S. Chitosan Based Nonstoichiometric Polyelectrolyte Complexes as Specialized Flocculants. *Colloids Surf. A Physicochem. Eng. Asp.* **2009**, *346*, 39–46. [CrossRef]
59. Dragan, E.S.; Mihai, M.; Schwarz, S. Complex Nanoparticles Based on Chitosan and Ionic/Nonionic Strong Polyanions: Formation, Stability, and Application. *ACS Appl. Mater. Interfaces* **2009**, *1*, 1231–1240. [CrossRef] [PubMed]
60. Paneva, D.; Mincheva, R.; Yancheva, E.; Manolova, N.; Stoilova, O.; Dubois, P.; Rashkov, I. N-Carboxyethylchitosan-Based Polymer Materials. In *Chitosan: Manufacture, Propierties and Usage*; Biotechnology in Agriculture, Industry and Medicine; Nova Science Publishers: Hauppauge, NY, USA, 2011; pp. 261–320, ISBN 978-1-61728-831-9.
61. Souza, C.J.F.; Da Costa, A.R.; Souza, C.F.; Tosin, F.F.S.; Garcia-Rojas, E.E. Complex Coacervation between Lysozyme and Pectin: Effect of pH, Salt, and Biopolymer Ratio. *Int. J. Biol. Macromol.* **2018**, *107*, 1253–1260. [CrossRef]
62. Mao, S.; Bakowsky, U.; Jintapattanakit, A.; Kissel, T. Self-Assembled Polyelectrolyte Nanocomplexes between Chitosan Derivatives and Insulin. *J. Pharm. Sci.* **2006**, *95*, 1035–1048. [CrossRef]
63. Tabandeh, S.; Leon, L. Engineering Peptide-Based Polyelectrolyte Complexes with Increased Hydrophobicity. *Molecules* **2019**, *24*, 868. [CrossRef]
64. Kulikouskaya, V.I.; Lazouskaya, M.E.; Agabekov, V.E. Features of the Formation of Interpolyelectrolyte Complexes Based on Chitosan and Pectin. *Theor. Exp. Chem.* **2019**, *54*, 375–385. [CrossRef]
65. Sunarti, T.C.; Febrian, M.I.; Ruriani, E.; Yuliasih, I. Some Properties of Chemical Cross-Linking Biohydrogel from Starch and Chitosan. *Int. J. Biomater.* **2019**, *2019*, 1–6. [CrossRef]
66. Nicolay, V.; Nina, S.; Yuliya, K.; Galina, B. Formation of Polyelectrolyte Complexes from Chitosan and Alkaline Gelatin. *KnE Life Sci.* **2020**, *5*, 109–119. [CrossRef]
67. Ibrahim, H.K.; Sorour, R.M.H.; Salah Ad-din, I. Application of Mathematical Modelling to Alginate Chitosan Polyelectrolyte Complexes for the Prediction of System Behavior with Venlafaxine HCl as a Model Charged Drug. *Saudi Pharm. J.* **2022**, *30*, 1507–1520. [CrossRef]
68. Tamburaci, S.; Kimna, C.; Tihminlioglu, F. Bioactive Diatomite and POSS Silica Cage Reinforced Chitosan/Na-Carboxymethyl Cellulose Polyelectrolyte Scaffolds for Hard Tissue Regeneration. *Mater. Sci. Eng. C* **2019**, *100*, 196–208. [CrossRef] [PubMed]
69. Rinaudo, M.; Pavlov, G.; Desbrières, J. Influence of Acetic Acid Concentration on the Solubilization of Chitosan. *Polymer* **1999**, *40*, 7029–7032. [CrossRef]
70. Abuelella, K.; Abd-Allah, H.; Soliman, S.; Abdel-Mottaleb, M. Chitosan Based Polyelectrolyte Complex Nanoparticles: Preparation and Characterization. *Bull. Pharm. Sci. Assiut Univ.* **2022**, *45*, 53–62. [CrossRef]
71. Lefnaoui, S.; Moulai-Mostefa, N. Synthesis and Evaluation of the Structural and Physicochemical Properties of Carboxymethyl Pregelatinized Starch as a Pharmaceutical Excipient. *Saudi Pharm. J.* **2015**, *23*, 698–711. [CrossRef] [PubMed]

72. Kaibara, K.; Okazaki, T.; Bohidar, H.B.; Dubin, P.L. pH-Induced Coacervation in Complexes of Bovine Serum Albumin and Cationic Polyelectrolytes. *Biomacromolecules* **2000**, *1*, 100–107. [CrossRef]
73. Botelho Da Silva, S.; Krolicka, M.; Van Den Broek, L.A.M.; Frissen, A.E.; Boeriu, C.G. Water-Soluble Chitosan Derivatives and pH-Responsive Hydrogels by Selective C-6 Oxidation Mediated by TEMPO-Laccase Redox System. *Carbohydr. Polym.* **2018**, *186*, 299–309. [CrossRef] [PubMed]
74. Aranaz, I.; Alcántara, A.R.; Civera, M.C.; Arias, C.; Elorza, B.; Heras Caballero, A.; Acosta, N. Chitosan: An Overview of Its Properties and Applications. *Polymers* **2021**, *13*, 3256. [CrossRef] [PubMed]
75. Ardean, C.; Davidescu, C.M.; Nemeş, N.S.; Negrea, A.; Ciopec, M.; Duteanu, N.; Negrea, P.; Duda-Seiman, D.; Musta, V. Factors Influencing the Antibacterial Activity of Chitosan and Chitosan Modified by Functionalization. *Int. J. Mol. Sci.* **2021**, *22*, 7449. [CrossRef]
76. Zhang, B.; Tao, H.; Wei, B.; Jin, Z.; Xu, X.; Tian, Y. Characterization of Different Substituted Carboxymethyl Starch Microgels and Their Interactions with Lysozyme. *PLoS ONE* **2014**, *9*, e114634. [CrossRef]
77. Li, X.-M.; Wu, Z.-Z.; Zhang, B.; Pan, Y.; Meng, R.; Chen, H.-Q. Fabrication of Chitosan Hydrochloride and Carboxymethyl Starch Complex Nanogels as Potential Delivery Vehicles for Curcumin. *Food Chem.* **2019**, *293*, 197–203. [CrossRef]
78. Kokufuta, E.; Ogawa, K.; Doi, R.; Kikuchi, R.; Farinato, R.S. Geometrical Characteristics of Polyelectrolyte Nanogel Particles and Their Polyelectrolyte Complexes Studied by Dynamic and Static Light Scattering. *J. Phys. Chem. B* **2007**, *111*, 8634–8640. [CrossRef] [PubMed]
79. Newham, G.; Mathew, R.K.; Wurdak, H.; Evans, S.D.; Ong, Z.Y. Polyelectrolyte Complex Templated Synthesis of Monodisperse, Sub-100 Nm Porous Silica Nanoparticles for Cancer Targeted and Stimuli-Responsive Drug Delivery. *J. Colloid. Interface Sci.* **2021**, *584*, 669–683. [CrossRef]
80. Ferreira, D.C.M.; Ferreira, S.O.; De Alvarenga, E.S.; Soares, N.D.F.F.; Coimbra, J.S.D.R.; De Oliveira, E.B. Polyelectrolyte Complexes (PECs) Obtained from Chitosan and Carboxymethylcellulose: A Physicochemical and Microstructural Study. *Carbohydr. Polym. Technol. Appl.* **2022**, *3*, 100197. [CrossRef]
81. De Vasconcelos, C.L.; Bezerril, P.M.; Dos Santos, D.E.S.; Dantas, T.N.C.; Pereira, M.R.; Fonseca, J.L.C. Effect of Molecular Weight and Ionic Strength on the Formation of Polyelectrolyte Complexes Based on Poly(Methacrylic Acid) and Chitosan. *Biomacromolecules* **2006**, *7*, 1245–1252. [CrossRef] [PubMed]
82. Hu, Y.; Yang, T.; Hu, X. Novel Polysaccharides-Based Nanoparticle Carriers Prepared by Polyelectrolyte Complexation for Protein Drug Delivery. *Polym. Bull.* **2012**, *68*, 1183–1199. [CrossRef]
83. Liu, X.; Haddou, M.; Grillo, I.; Mana, Z.; Chapel, J.-P.; Schatz, C. Early Stage Kinetics of Polyelectrolyte Complex Coacervation Monitored through Stopped-Flow Light Scattering. *Soft Matter* **2016**, *12*, 9030–9038. [CrossRef]
84. Saikia, C.; Hussain, A.; Ramteke, A.; Sharma, H.K.; Maji, T.K. Carboxymethyl Starch-Chitosan-Coated Iron Oxide Magnetic Nanoparticles for Controlled Delivery of Isoniazid. *J. Microencapsul.* **2015**, *32*, 29–39. [CrossRef]
85. Wu, D.; Delair, T. Stabilization of Chitosan/Hyaluronan Colloidal Polyelectrolyte Complexes in Physiological Conditions. *Carbohydr. Polym.* **2015**, *119*, 149–158. [CrossRef]
86. Sæther, H.V.; Holme, H.K.; Maurstad, G.; Smidsrød, O.; Stokke, B.T. Polyelectrolyte Complex Formation Using Alginate and Chitosan. *Carbohydr. Polym.* **2008**, *74*, 813–821. [CrossRef]

polymers

MDPI

Article

Investigation of Silk Fibroin/Poly(Acrylic Acid) Interactions in Aqueous Solution

Jelena Škrbić [1,*], Ljiljana Spasojević [1], Altynay Sharipova [2], Saule Aidarova [3], Alpamys Babayev [3], Raziya Sarsembekova [3], Ljiljana Popović [1], Sandra Bučko [1], Jelena Milinković Budinčić [1], Jadranka Fraj [1], Lidija Petrović [1] and Jaroslav Katona [1]

1 Faculty of Technology Novi Sad, University of Novi Sad, Bulevar cara Lazara 1, 21000 Novi Sad, Serbia; lj.spasojevic@tf.uns.ac.rs (L.S.); ljiljana04@tf.uns.ac.rs (L.P.); jelenamilinkovic@uns.ac.rs (J.M.B.); jadranka@uns.ac.rs (J.F.); lidijap@uns.ac.rs (L.P.); jaroslav.katona@uns.ac.rs (J.K.)
2 Mining and Metallurgical Institute, Satbayev University, Satbayev str. 22a, 050013 Almaty, Kazakhstan; a_sharipova85@mail.ru
3 Petroleum Engineering Institute "One Belt, One Road", Kazakh–British Technical University, Tole bi str. 59, 050000 Almaty, Kazakhstan; tempus_aidarova@mail.ru (S.A.); a_babayev@mail.ru (A.B.); r.sarsembekova@kbtu.kz (R.S.)
* Correspondence: jelenaskrbic94@uns.ac.rs; Tel.: +381-21-485-3786

Abstract: Silk fibroin (SF) is a protein with many outstanding properties (superior biocompatibility, mechanical strength, etc.) and is often used in many advanced applications (epidermal sensors, tissue engineering, etc.). The properties of SF-based biomaterials may additionally be tuned by SF interactions with other (bio)polymers. Being a weak amphoteric polyelectrolyte, SF may form polyelectrolyte complexes (PECs) with other polyelectrolytes of opposite charge, such as poly(acrylic acid) (PAA). PAA is a widely used, biocompatible, synthetic polyanion. Here, we investigate PEC formation between SF and PAA of two different molecular weights (MWs), low and high, using various techniques (turbidimetry, zeta potential measurements, capillary viscometry, and tensiometry). The colloidal properties of SF isolated from *Bombyx mori* and of PAAs (MW, overlap concentration, the influence of pH on zeta potential, adsorption at air/water interface) were determined to identify conditions for the SF-PAA electrostatic interaction. It was shown that SF-PAA PEC formation takes place at different SF:PAA ratios, at pH 3, for both high and low MW PAA. SF-PAA PEC's properties (phase separation, charge, and surface activity) are influenced by the SF:PAA mass ratio and/or the MW of PAA. The findings on the interactions contribute to the future development of SP-PAA PEC-based films and bioadhesives with tailored properties.

Keywords: silk fibroin; poly(acrylic acid); polymer/polymer interaction; polyelectrolyte complex (PEC); complex coacervation

1. Introduction

Silk is a natural fibrous protein that represents one of the most preferred and valued biomaterials from the period of its discovery until today. Silk is a biomaterial produced by various types of insects, but the greatest commercial importance is silk produced by silkworms *Bombyx mori* and spiders such as *Nephila clavipes* [1,2]. Silk fibers obtained from *Bombyx mori* silkworm cocoons usually contain 72–81% SF and 19–28% silk sericin and small amounts of ash, waxes, dyes, and other compounds. In *B. mori* silk, two parallel SF fibers are coated with a glue-like protein sericin. While SF provides mechanical strength, sericin enables the compactness of silk fibers and the entire cocoons [2,3]. In order to use silk as a biomaterial, sericin must be removed from the silk fibers by a process called degumming. There are many ways to extract the sericin, but one of the most commonly employed and examined degumming processes involves boiling cut cocoons in a sodium carbonate solution [3–5].

SF is composed of three fractions: heavy chain (H), light chain (L), and small P25 glycoprotein. The heavy chain is hydrophobic and rigid (~390 kg/mol), while the light chain is flexible and hydrophilic with a lower MW (~26 kg/mol). These two chains are linked by disulfide bonds and form an H-L complex, which is connected to the small glycoprotein P25 (~25 kg/mol) via hydrophobic interactions [6–8]. Due to its natural hierarchical structure and high degree of crystallinity, SF cannot be directly dissolved or separated in water, but it can be dissolved in some concentrated aqueous or aqueous/organic salt solutions [9]. Dissolving SF using an aqueous solution of lithium bromide is one of the most commonly used dissolution methods.

SF has unique properties such as mechanical stability, biocompatibility, biodegradability, and good handling characteristics. It can be fabricated into various materials such as hydrogels, films, sponges, scaffolds, mats, and micro- and nanoparticles [10,11]. SF is considered a very attractive biopolymer material for use in advanced technologies such as epidermal sensors [10,12], edible films and barrier coatings in the food industry [13,14], tissue engineering [5,15,16], drug carriers with controlled release [6,17], wound healing [1,11,18], natural emulsion and foam stabilizers [19,20], etc. Due to its favorable properties and potential clinical application, SF has been approved as a biomaterial by the FDA (Food and Drug Administration) since 1993 [8].

Although natural SF materials possess many outstanding properties, they rarely meet all the highly specific requirements necessary for a particular application. Different methods can be used to design new SF-based materials with improved functional and structural properties. Blending and cross-linking SF with other macromolecules are often used for this purpose [15].

Polymer/polymer interactions may occur when two polymers are found jointly in the mixture. Generally, several types of interactions occur between macromolecules, such as intermolecular hydrogen bonding, intermolecular covalent bonding, and electrostatic and hydrophobic interactions [21]. Complexes formed through electrostatic interactions between two oppositely charged polymer chains are called polyelectrolyte complexes (PECs) [22]. Kabanov and Zezin (1984) first showed that PEC formation is a fast process which takes place in less than 5 ms [23]. The initially formed complex undergoes subsequent intracomplex restructuration (the formation of new bonds and/or change in the conformation of polymer chains) to produce the ordered secondary complex. The second step proceeds within the order of an hour. The final, third step, involves the aggregation of the intercomplexes, mainly through hydrophobic interactions, leading to the formation of complex aggregates [24]. PEC formation is driven by electrostatic interactions and by the entropy gain attained through the release of counterions during complex formation between two oppositely charged polyelectrolytes [25]. PEC formation and stability depends on many factors, such as ionic strength, temperature, MW and the concentration of polyelectrolytes (PEs), the mixing ratio of polymers, the type of ionic groups, and the solution pH. The charge and charge density of PEs vary depending on the solution pH [22,26].

SF interactions with both natural and synthetic polymers have been reported. Malay et al. investigated the interaction of oppositely charged SF and hyaluronic acid and the formation of their complexes at different conditions [27,28]. Ghaeli et al. monitored the interactions between collagen and SF during the formulation of their stable and one-phase mixture [29]. Bahardwaj et al. studied the formation of an SF/chitosan polyelectrolyte complex for tissue engineering application, while the complexation of SF and tannic acid through hydrogen bonding for the preparation of wood adhesives was investigated by Kim et al. [30,31]. Wang et al. investigated the influence of SF/poly(lactic acid) interactions on the preparation of their biocomposites [32]. Gao et al. reported SF complex formation with the in-house synthetized PAA of medium MW when mixing dilute SF and PAA solutions, however, without pH adjustment [33].

PAA is a synthetic polymer made by polymerizing acrylic acid monomers. It is biocompatible, biodegradable, water-soluble, and a superabsorbent polymer [34]. PAA is also a non-toxic polymer with bioadhesive properties due to the large number of carboxylic groups [35]. PAA is a mucoadhesive material and has good adhesion properties to the mucosa cells, as well as the ability to adhere to skin, bone, and teeth [36]. Based on its properties, PAA has a wide range of applications, including adhesives, packaging, coatings, pharmacy, cosmetics, drug delivery systems, medicine, dentistry, etc. [34].

In this paper, electrostatic interactions between SF isolated from *B. mori* cocoons and commercially available PAA of low and high MWs were investigated. The colloidal properties of SF and PAA were first characterized to identify conditions for SF-PAA PEC formation. Furthermore, SF-PAA complex formation obtained at different interaction conditions was investigated. Different techniques, such as gel electrophoresis, turbidimetry, zeta potential measurements, capillary viscometry, and tensiometry were used for this purpose.

2. Materials and Methods

2.1. Materials

Bombyx mori silkworm cocoons were obtained from a farm cooperative in Wuhan, China. Sodium carbonate (Na_2CO_3, anhydrous, 99.5%) was purchased from Loba Chemie Pvt. Ltd. (Mumbai, India). Lithium bromide (LiBr, anhydrous) was purchased from Sigma-Aldrich Co. (St. Louis, MO, USA), and dialysis tubing cellulose membrane (molecular weight cut off 14 kg/mol) was purchased from Sigma-Aldrich, Co. (St. Louis, MO, USA). Hydrochloric acid (HCl, 37%) was purchased from ZORKA Pharma-HEMIJA d.o.o. (Šabac, Serbia), and sodium hydroxide micropearls (NaOH) were purchased from Lach-Ner, s.r.o. (Neratovice, Czech Republic). PAAs of high and low MWs were purchased from Sigma-Aldrich, Co. (St. Louis, USA). MW = 1033,000 and MW = 1800 were reported by the supplier for the high MW PAA and the low MW PAA, respectively. All other chemicals were of analytical grade, except as specified, and were used without further purification. Distilled water was used as the solvent in all experiments.

2.2. Preparation of SF Solution

Bombyx mori cocoons were cut into small pieces and boiled for 30 min in a 0.5% (*w*/*v*) sodium carbonate solution, then rinsed in distilled water heated to 40 °C with constant stirring for 30 min. This procedure was repeated three times in order to degum SF. The degummed SF was dried overnight in ambient conditions. The dry and degummed SF fibers were dissolved in a 9.3 M aqueous lithium bromide solution at 60 °C for 90 min. The resulting solution was dialyzed against distilled water using a dialysis tubing membrane for 3 days to remove the salt. Dialysis was performed at room temperature with constant stirring. After dialysis, the solution was centrifuged for 20 min at 5000 rpm and filtered through an 8.0 μm pore size filter to remove any remaining impurities and undissolved fibers. The final concentration of the aqueous SF solution was determined gravimetrically by measuring the dry matter content and calculated using Equation (1):

$$\omega(\%) = (m/m_0) \times 100 \quad (1)$$

where m is the dry matter mass and m_0 SF solution mass, averaged over at least three measurements. All SF solutions were stored at 4 °C before further use.

2.3. Preparation of PAA Solution

PAA aqueous solutions of different concentrations, namely 0.1%, 0.5%, and 1.0% (*w*/*v*), were prepared by dissolving PAA powder in distilled water with constant stirring one day before use. The pH of the solutions was adjusted using 0.1 M or 1.0 M HCl.

High MW PAA solution at a concentration of 0.5 g/100 mL and low MW PAA solution at a concentration of 10 g/100 mL in 2 M NaOH were prepared using the same method as in water and were used for determining the PAA viscosity average molar mass.

2.4. Preparation of SF-PAA Mixtures

SA-PAA mixtures were prepared at room temperature by adding PAA solution to SF solution in different proportions. The pH of both solutions was adjusted to 3 before mixing.

PAAs of two different MWs (the high MW PAA and the low MW PAA) were used for SF-PAA complex preparation. SF/high MW PAA mixtures were prepared by adding the required volume of 0.1% (*w*/*v*) PAA solution to 0.1% (*w*/*v*) SF solution to obtain mixtures with SF:PAA mass ratios of 100:0, 98:2, 95:5, 92:8, 90:10, 85:15, 80:20, 70:30, 60:40, 50:50, 40:60, 30:70, 20:80, 15:85, 10:90, 8:92, 5:95, 2:98, and 0:100. SF/low MW PAA mixtures were prepared by adding the required volume of 1.0% (*w*/*v*) PAA solution to 1.0% (*w*/*v*) SF solution to obtain mixtures with SF:PAA mass ratios of 100:0, 99:1, 98:2, 97:3, 95:5, 92:8, 90:10, 85:15, 80:20, 70:30, 60:40, 50:50, 40:60, 30:70, 20:80, 10:90, 5:95, and 0:100. The obtained mixtures were stirred on a magnetic stirrer for 3 min and left overnight to allow for enough time for the formation of the equilibrium SF-PAA complex.

A 0.5 g/100 mL high MW PAA solution at pH 3 was mixed with a 0.5 g/100 mL SF solution at pH 3 to obtain SF-PAA mixtures with various SF:PAA mass ratios and a total polymer concentration of 0.5 g/100 mL for viscosity characterization. The mixtures were left for 24 h to equilibrate before further use.

2.5. Viscosity Measurements

Viscosity measurements were employed to determine the overlap concentration of PAAs, the viscosity average molar mass of PAA samples, and to investigate SF-PAA complex formation. Viscosity measurements were performed using a Cannon capillary viscometer 50 E599 (Cannon Instrument Company, State College, PA, USA) immersed in a thermostatic bath set to a measuring temperature. Before measurements, each sample was allowed to equilibrate for at least 15 min at the measuring temperature to ensure it reached the desired temperature. The flow times of polymer solutions of different concentration and solvents were measured. Each measurement was repeated at least five times, and the average values were reported. The measurement results were expressed as relative viscosity (η_{rel}), specific viscosity (η_{sp}), and reduced viscosity (η_{red}):

$$\eta_{rel} = t/t_0 \quad (2)$$

$$\eta_{sp} = t - t_0/t_0 = \eta_{rel} - 1 \quad (3)$$

$$\eta_{red} = \eta_{sp}/c \ (100 \text{ mL/g}) \quad (4)$$

where t is the flow time of the polymer solution, t_0 the flow time of the solvent, and c is the concentration of the polymer solution.

The intrinsic viscosity $[\eta]$ of the PAA samples was obtained graphically, by extrapolating $ln\eta_{rel}/c$ and η_{sp}/c to zero polymer concentration, according to Equation (5) [37]:

$$[\eta] \equiv lim_{c\to 0} \ (ln\eta_{rel}/c) \equiv lim_{c\to 0} \ \eta_{sp}/c \quad (5)$$

The overlap concentration (c*) of PAA was estimated as $c^* = 1/[\eta]$.

The viscosity average molar masses (Mvs) of PAA samples were calculated from the Mark–Houwink equation:

$$[\eta] = K \times Mv^a \quad (6)$$

where K and a are constants characteristic for a particular solvent/polymer system at a given temperature. $K = 42.2 \times 10^{-5}$ 100 mL/g and $a = 0.64$ at 25 ± 0.1 °C as well as $K = 53.9 \times 10^{-4}$ 100 mL/g and $a = 0.43$ at 30 ± 0.1 °C were used for the low MW PAA and the high MW PAA Mv determination in 2 M NaOH aqueous solution, respectively [38,39].

Viscosimetric measurements of SF-PAA mixtures with different SF:PAA mass ratios, SF solutions of different concentrations, and PAA solutions of different concentrations were performed at 20 ± 0.1 °C and pH 3 to investigate SF-PAA complex formation. Viscosimetric measurements of supernatants were carried out in SF-PAA mixtures where sediment formation occurred.

2.6. Turbidity Measurements

Turbidity measurements were performed by using the UV-VIS HALO DB-20S Spectrophotometer (Dynamica Scientific Ltd., Livingston, UK). Transmittance at 500 nm in a 1 cm length cell in reference to distilled water was measured at room temperature. The turbidity of samples was calculated using Equation (7) [40]:

$$\text{turbidity} = 100 - \text{transmittance } (\%) \quad (7)$$

The turbidity of 1.0% (*w*/*v*) SF aqueous solutions at different pH values and the turbidity of SF-PAA mixtures at different SF:PAA mass ratios were determined. The total polymer concentration in SF-PAA mixtures was 0.1% (*w*/*v*) and 1.0% (*w*/*v*) for the high MW and low MW PAA mixtures, respectively. The turbidity of the supernatant was determined in all samples where sediment formation occurred.

All measurements were carried out in triplicate, and average values were calculated.

2.7. Zeta Potential Measurements

The zeta potential (ζ) of SF, PAA, and SF-PAA mixtures was measured using dynamic light scattering with the Zetasizer Nano ZS instrument (Malvern Instruments, Malvern, UK) at 25 °C. A folded capillary cell (DTS 1060) was used for zeta potential measurements.

The zeta potential of 0.1% (*w*/*v*) SF and PAA solutions at different pH (2.5–7.0) was determined. The pH of the samples was adjusted using 1.0 M or 0.1 M HCl.

The zeta potential of SF-PAA mixtures with total polymer concentration of 0.1% (*w*/*v*) and 1.0% (*w*/*v*), corresponding to the high and low MW PAA, respectively, was measured. These mixtures, prepared with different SF:PAA mass ratios, were characterized 24 h after preparation. The zeta potential of the supernatant was measured in mixtures where sediment formation occurred.

All measurements were performed in triplicate, and average values were reported.

2.8. Potentiometric Titration

The pI of SF was determined using potentiometric titration, where the pH of a 0.1% (*w*/*v*) SF solution was recorded during titration with 0.01 M HCl titrant at room temperature. The pH measurements were performed using 827 pH lab pH-meter (Metrohm, Herisau, Switzerland). The pI of SF was derived from the obtained titration curve based on the titration equivalent point. The equivalent point was determined from the obtained dependence of $\Delta pH/\Delta V$ on HCl volume.

2.9. Sodium Dodecyl Sulfate–Polyacrylamide Gel Electrophoresis (SDS-PAGE)

The MW of SF was determined by sodium dodecyl sulfate–polyacrylamide gel electrophoresis, described by the method of Laemmli [41] with certain corrections. Polyacrylamide gel electrophoresis (PAGE) in the presence of the anionic detergent, sodium dodecyl sulfate (SDS), has proven to be a useful method for protein separation and for the determination of their molecular weights. The gel system consists of a 4% (*w*/*v*) acrylamide stacking gel and a 12% (*w*/*v*) acrylamide separation gel. Samples were prepared by adding 0.5 mL of 2% (*w*/*v*) SF solution in 0.5 mL of Tris/Gly buffer (pH 6.8) which contains 20 g/L SDS and 50 g/L 2-mercaptoethanol. After the preparation of gels and samples, the apparatus for electrophoresis (Multi Drive XL, Pharmacia, Stockholm, Sweden) was put into operation at 50 mA, at 25 °C, as long as the dyed tracker reached the bottom of each gel. When electrophoresis was finished, the gels were stained with 0.2% (*w*/*v*) Coomassie brilliant blue R-250 in 10% (*v*/*v*) acetic acid. The gels were distained for 48 h with 10% (*v*/*v*) acetic acid which contains 40% (*v*/*v*) methanol.

2.10. Tensiometry

The surface tension measurements of 0.05% (*w*/*v*) PAA solution, SF solutions (0.01–2% (*w*/*v*)), and SF-PAA mixtures (0.05% (*w*/*v*) PAA and 0.01–2.00% (*w*/*v*) SF) were carried out by a KSV Sigma 703D tensiometer at 25 °C. The Du Noüy ring method was employed. Prior to the measurements, the air/liquid interface was disturbed by vigorous mixing, the ring was then immersed into the liquid (below the interface), and then the interface was left to equilibrate for 30 min. The reported values of the surface tension are average values of at least three measurements.

3. Results and Discussions

3.1. Colloidal Properties of SF and PAA

The colloidal properties of polymers (e.g., MW, charge, and pI) influence their behavior in solutions, as well as their interaction with other polymers when found jointly in mixtures. The MW, the influence of pH on zeta potential, and the pI of SF and PAA were determined.

3.1.1. Determination of Molecular Weights

The MW of SF was determined by SDS-PAGE electrophoresis (Figure 1). It can be seen from Figure 1 that the molar mass of the isolated SF ranged from ≈30 kg/mol to ≈116 kg/mol, which is significantly lower than that found in native SF [6–8].

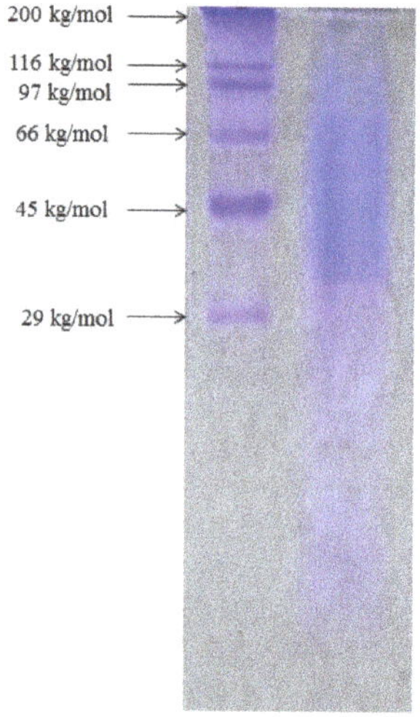

Figure 1. Sodium dodecyl sulfate–polyacrylamide gel electrophoresis (SDS-PAGE) for standard protein markers (lane 1) and for 2% (*w*/*v*) SF solution (line 2).

This suggests that the employed isolation procedure, which includes degumming in sodium carbonate solution, results in the cleavage of native SF chains and bonds between the SF heavy and light chains which eventually leads to a decrease in the MW of SF [42]. The native SF is therefore degraded during the isolation procedure, and the isolated SF represents a mixture of polypeptides of various MWs.

Capillary viscometry of both low MW PAA and high MW PAA was carried out to determine the intrinsic viscosity of PAAs and their Mvs. The intrinsic viscosity was determined using the double extrapolation method, where experimentally obtained η_{sp}/c and $ln\eta_{rel}/c$ values for PAA solutions of different concentrations were extrapolated to zero concentration, as illustrated in Figure 2A,B.

The intrinsic viscosity determined was 0.0485 100 mL/g (Figure 2A) for the low MW PAA and 1.965 100 mL/g (Figure 2B) for the high MW PAA. This gives c* = 20.62 g/100 mL and c* = 0.51 g/100 mL for the low and high MW PAA, respectively. PAA solutions with a concentration lower than c* are diluted, meaning that no interaction between individual PAA chains occurs [43].

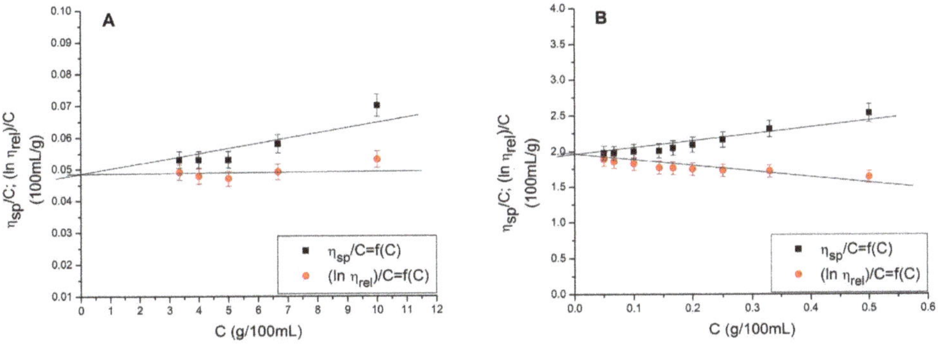

Figure 2. The double extrapolation method determination of intrinsic viscosity for (**A**) the low MW PAA and (**B**) the high MW PAA, both in 2 M NaOH solution.

The Mv of PAA samples was calculated from the experimentally obtained intrinsic viscosities using the Mark–Houwink equation (Equation (6)). It was determined that Mv = 1598 g/mol for the low MW PAA and Mv = 907,012 g/mol for the high MW PAA.

3.1.2. Determination of Isoelectric Point

The solution properties of weak polyelectrolytes such as SF and PAA are influenced by pH. Figure 3 shows photograph of SF solutions at different pH values (3–6). It can be seen from Figure 3 that the visual appearance of the solutions is influenced by the solution pH. The solutions with low and high pH values are clear and transparent, while the solutions with intermediate pH values appear turbid.

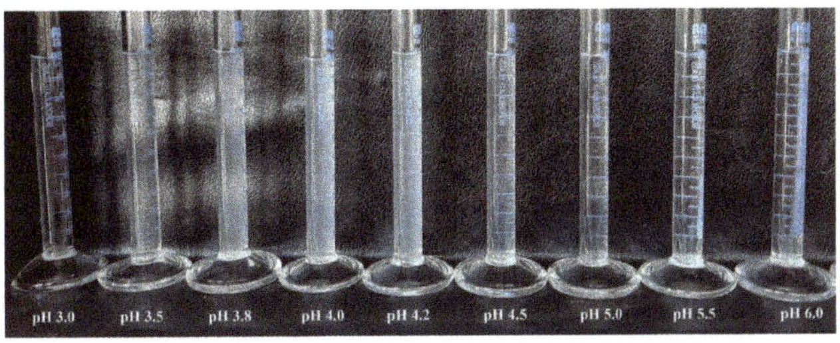

Figure 3. Photograph of 1.0% (w/v) SF solutions at different pHs.

The turbidity of SF solutions with different pHs was measured, and the influence of the solution pH on turbidity is shown in Figure 4.

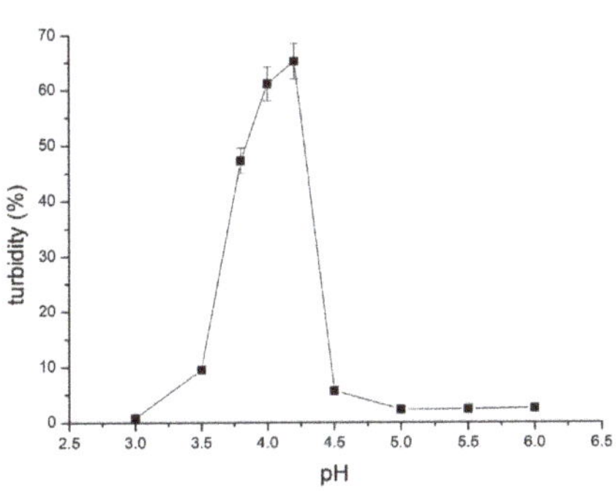

Figure 4. Influence of pH on turbidity of 1.0% (w/v) SF solutions.

From Figure 4, it can be seen that the turbidity of SF solutions increases with an increase in pH, reaching maximum turbidity at pH 4.2. A further increase in the solution pH brings about a sharp decrease in turbidity to reach a constant, low turbidity value for pH > 5. Unlike polymer solutions in good solvents, which do not scatter light, colloidal dispersions possess this ability, and as a result of light scattering, turbidity occurs. The increased turbidity of SF solutions indicates the phase separation of the SF solution, where the aggregation of SF fibrils takes place due to the low charge of SF molecules and the subsequently decreased solubility of SF. Namely, being a weak, amphoteric polyelectrolyte, the charge of SF molecules depends on the solution pH. At a low pH (lower than pI) SF is positively charged, facilitating good solubility and resulting in clear SF solutions. As the pH increases, the positive charge on SF molecules decreases while the negative charge increases, effectively reducing the net charge of SF. This reduction in charge decreases the solubility of SF and leads to an increase in the turbidity of the solutions. At pI, the amount of positive and negative charge in the SF molecule is balanced and SF becomes effectively uncharged and least soluble. This balance brings about phase separation, resulting in the maximum turbidity of the solution. Figure 3 shows the maximum in the turbidity of the SF solution and thereby pI = 4.2. As the pH increases above the pI, SF molecules become more negatively charged. This increased negative charge enhances the solubility of SF, thereby reducing the turbidity of the SF solution.

Such an influence of the solution pH on the charge of SF molecules was confirmed by zeta potential measurements (Figure 5). Freshly prepared SF solutions after dialysis typically have a pH ranging from 7 to 8, depending on the concentration of the SF solution. As shown in Figure 5, at this pH, SF is negatively charged and behaves as a polyanion. A decrease in the SF solution pH brings about a gradual increase in the SF zeta potential from −19.20 mV at pH 6.9 to 0 mV at pH 4.2. With a further decrease in pH, SF becomes positively charged with a zeta potential of 9.73 mV observed at pH 3.1. The zero value of the zeta potential was obtained at pI = 4.2. This is in agreement with the literature data, where reported values of SF's pI range from 3.6 to 5.2 [9,19,44].

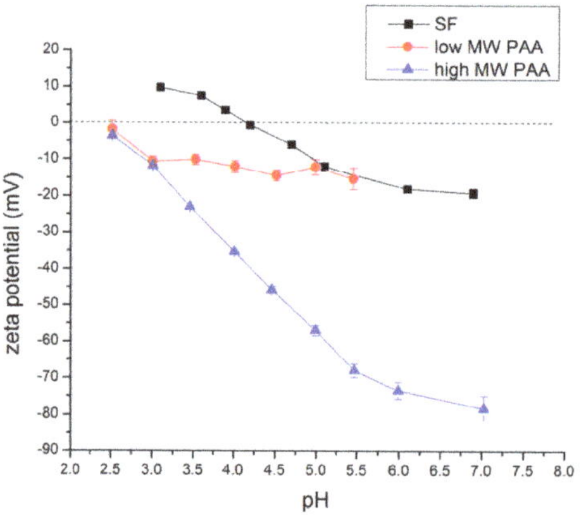

Figure 5. Influence of pH value on zeta potential of 0.1% (w/v) SF and PAA aqueous solution.

Figure 5 also shows the influence of the pH on the zeta potential of both the low MW and the high MW PAA. In both cases, the low MW and the high MW PAA solutions exhibited a negative zeta potential across the entire investigated pH range, typical for polyanions. At the lowest pH studied (pH 2.53), the zeta potential was close to zero due to the hindered dissociation of carboxylic groups. An increase in pH brings about the increased dissociation of carboxylic groups and thereby an increase in the negative charge of PAA. The increase in negative charge is significantly more pronounced for the high MW PAA compared to the low MW PAA, as shown in Figure 5.

Potentiometric titration with a suitable acid or base is another convenient method used for pI determination. The influence of HCl volume added to the SF solution on the pH of the SF solution is shown in Figure 6. The equivalent titration point was determined from the first derivative of the obtained dependence and was found to be 0.5 mL of HCl (inset to Figure 6). The obtained data for the equivalent titration point suggests that the pI of SF ranges from 5.50 to 6.25. The pI obtained by potentiometric titration is somewhat higher than the pI obtained by turbidity measurements (Figure 4) and by zeta potential measurements (Figure 5).

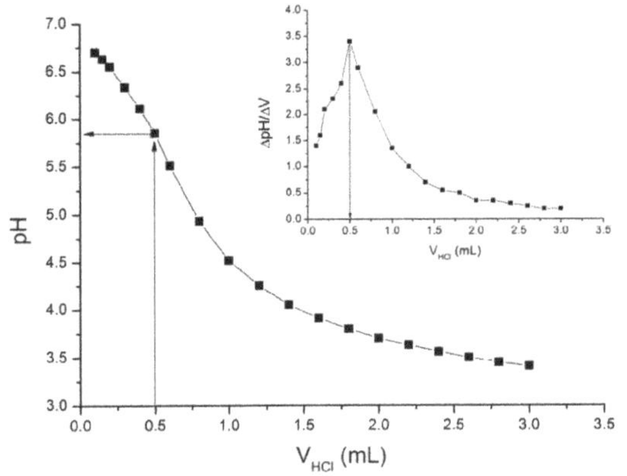

Figure 6. Titration curve of 0.1% (*w*/*v*) SF solution titrated with 0.01 M HCl, obtained at room temperature.

3.2. Investigation of SF-PAA Complexes Formation

The investigation of the zeta potential (Figure 5) showed that SF was in the protonated form and positively charged below pH 4.2. Conversely, above pH 2.5, both low and high MW PAAs were in the deprotonated forms, and therefore negatively charged. This suggests that the electrostatic interaction between SF and PAAs and PEC formation is expected to take place when both polymers are found in a solution at a pH which is higher than 2.5 and lower than 4.2. Therefore, mixtures of SF with both low MW PAA and high MW PAA were prepared at pH 3 and at different SF:PAA mass ratios.

3.2.1. Turbidity Measurements

Photographs of 1.0% (*w*/*v*) SF/low MW PAA mixtures at pH 3 and at different SF:PAA mass ratios, taken 3 min and 24 h after preparation, are shown in Figure 7.

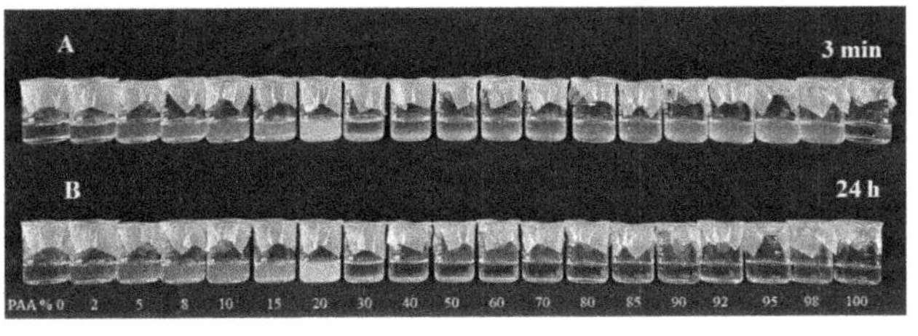

Figure 7. Photograph of 1.0% (*w*/*v*) SF/low MW PAA mixtures at different SF:PAA mass ratios (shown as PAA%), (**A**) 3 min after preparation and (**B**) 24 h after preparation.

It can be seen from Figure 7 that the 100% SF solution (0% PAA) and the 100% low MW PAA solution (0% SF) remained clear at both 3 min and 24 h after preparation. However, SF-PAA mixtures at all investigated SF:PAA mass ratios immediately became turbid after preparation, indicating the formation of SF/low MW PAA PECs. The complex formation is attributed to electrostatic interactions of positively charged SF and negatively charged PAA at pH 3 (Figure 5). PEC formation is also driven by the entropy gain attained through the release of counterions during complex formation between two oppositely charged polyelectrolytes [25]. Moreover, it was reported previously that hydrogen bonding between PAA's carboxylic groups and SF's amide groups can also contribute to SF-PAA complex formation [33].

Figure 8 shows that SF-PAA mixtures containing 2% to 20% of PAA remained turbid after 24 h. SF-PAA mixtures containing 30% to 80% PAA cleared overnight, with a significant decrease in turbidity observed 24 h after preparation (Figure 8). The decrease in turbidity was due to PEC precipitation and sediment formation, which were visually observed in the mixtures. The highest yield of precipitate occurred at about 30–40% PAA and could be easily resuspended by stirring with a glass rod. The precipitate formed at 50–70% PAA was sticky and could not be resuspended by stirring, while the precipitate at 80% was barely noticeable on the walls of the beaker. No sediment formation was observed for SF/low MW PAA mixtures containing >80% PAA.

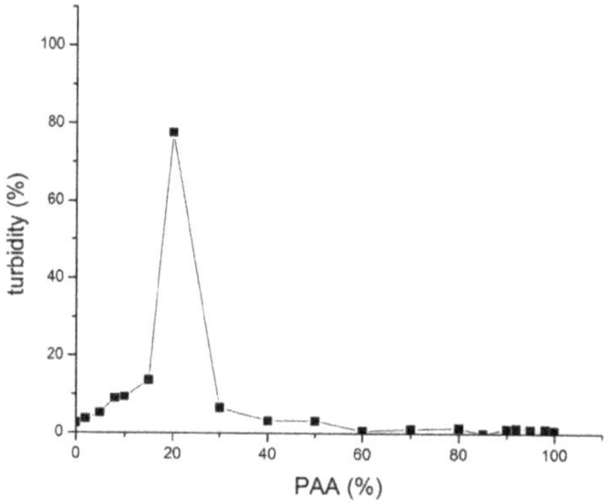

Figure 8. Turbidity of 1.0% (w/v) SF/low MW PAA mixtures at different SF:PAA mass ratios (shown as PAA%), at pH 3, 24 h after mixing.

SF/high MW PAA PEC formation was also investigated. Figure 9 shows 0.1% (w/v) SF/high MW PAA mixtures at pH 3 and at different SF:PAA mass ratios, 3 min and 24 h after preparation.

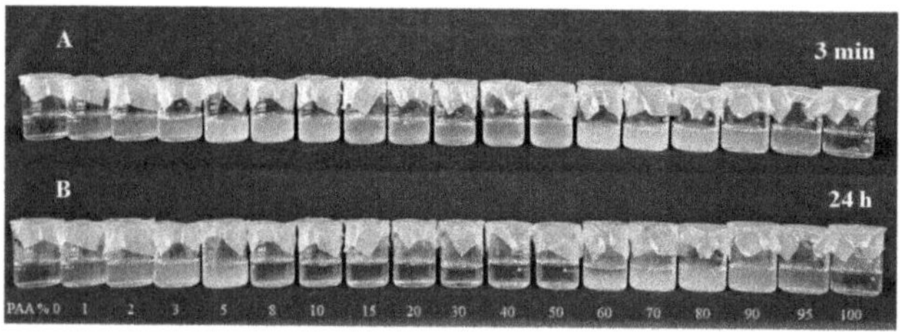

Figure 9. Photograph of 0.1% (w/v) SF/high MW PAA mixtures at different SF:PAA mass ratios (shown as PAA%), (**A**) 3 min after preparation and (**B**) 24 h after preparation.

It can be seen from Figure 9 that SF/high MW PAA mixtures for all SF:PAA mass ratios (except for 100% PAA and 100% SF) were turbid 3 min after preparation, suggesting SF/high MW PAA PEC formation. Mixtures with 8% to 50% of PAA cleared overnight due to PEC sediment formation. The precipitate yield was the highest at about 8–10% PAA and could be easily resuspended by stirring with a glass rod. The precipitate formed at 15–20%

PAA was sticky and could not be resuspended by stirring, while the precipitate at 30–50% was barely noticeable at the bottom and on the walls of the beaker.

The turbidity of 0.1% (*w*/*v*) SF/high MW PAA mixtures at different SF:PAA mass ratios (shown as PAA%), at pH 3, 24 h after mixing is shown in Figure 10.

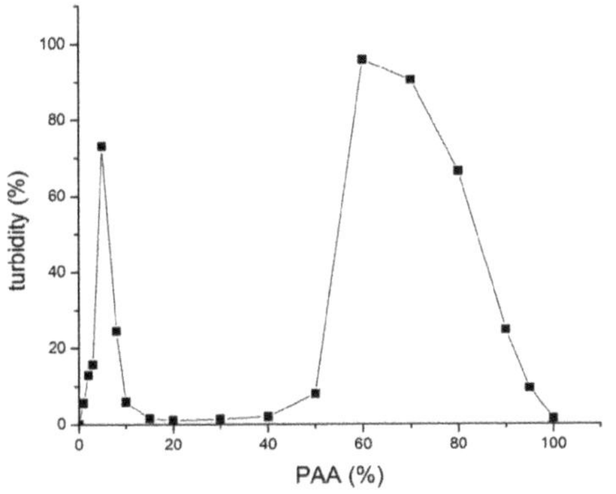

Figure 10. Turbidity of 0.1% (*w/v*) SF/high MW PAA mixtures at different SF:PAA mass ratios (shown as PAA%), at pH 3, 24 h after mixing.

Turbidity increases with an increase in the PAA% up to 5% of PAA. This increase is attributed to the formation of more PECs as more PAA is available for interacting with SF and/or a decrease in PEC solubility likely due to a decrease in the net positive charge of the PEC. The turbidity of SF-PAA mixtures with 10% to 50% of PAA drops to close to zero values due to the sedimentation of more or less electroneutral PEC formation. However, a further increase in the PAA% in the mixture gradually increases the net negative charge of the PEC, resulting in the redispersion of the complex and a subsequent sharp increase in the turbidity of the SF-PAA mixture with 60% of PAA. A further increase in the PAA% brings about a decrease in the turbidity of the mixtures (Figure 10). This decrease in turbidity may be attributed to an increase in the complex solubility due to an increase in the net negative charge of the complex and/or to a decrease in the amount of the complex formed, as less and less SF is present in the system when the PAA% approaches 100%.

3.2.2. Zeta Potential Measurements

The influence of the PAA content in SF-PAA mixtures on the zeta potential of the SF PEC with both the low MW and the high MW PAA is shown in Figures 11A and 11B, respectively.

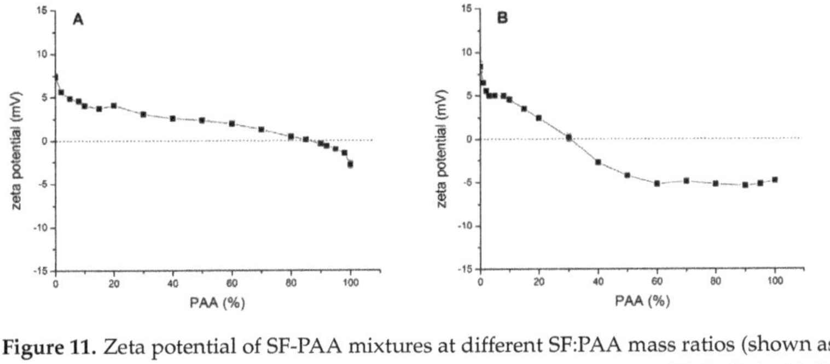

Figure 11. Zeta potential of SF-PAA mixtures at different SF:PAA mass ratios (shown as PAA%), at pH 3, 24 h after mixing: (**A**) 1.0% (*w*/*v*) SF/low MW PAA mixtures; (**B**) 0.1% (*w*/*v*) SF/high MW PAA mixtures.

A net positive zeta potential of SF/low MW PAA complexes was observed in all SF-PAA mixtures where the PAA% was less than 85 (Figure 11A). The net positive zeta potential gradually decreases with an increase in the PAA% until it reaches negative values at a PAA% higher than 90. It should be observed, though, that the measured zeta potential values, both positive and negative, are very close to zero. This indicates that subtle changes in the system determine the phase behavior of the PEC.

SF/high MW PAA PEC is positively charged for all SF-PAA mixtures with less than 30% PAA, when the zeta potential reaches zero (Figure 11B). When the PAA% in the mixture is higher than 30, the PEC becomes negatively charged. The net negative charge of the complex decreases up to 60% of PAA in the mixture. Further increases in the PAA% do not result in any additional changes in the zeta potential of the SF/high MW PAA PEC. This suggests that a decrease in turbidity observed for SF/high MW PAA mixtures at a PAA% higher than 60% (Figure 10) is most likely due to a decrease in the amount of complex formed, rather than a change in the solubility of the complex.

A comparison of Figure 11A,B shows that SF-PAA complex formation is influenced by the MW of PAA. The addition of the PAA solution to the SF solution at pH 3 leads to PEC formation, which is positively charged for the mixtures with a low PAA%, for both low and high MW PAA. An increase in the PAA% in the mixture results in a decrease in the net positive charge of the PEC. However, the decrease is more rapid for the high MW PAA, and the complex becomes negatively charged at a lower PAA% compared to the low MW PAA.

3.2.3. Capillary Viscometry of SF/High MW PAA Mixtures

The investigation of PEC formation in 0.5 g/100 mL SF/high MW PAA mixtures of different SF:PAA mass ratios at pH 3 was conducted using capillary viscometry 24 h after the mixture preparation. Figure 12A shows the influence of the PAA concentration on the specific viscosity of the PAA solution. It also shows the specific viscosity of SF-PAA mixtures with different SF:PAA mass ratios (the mixture with 0.5 g/100 mL PAA is considered 100% PAA, while the mixture with 0 g/100 mL PAA is considered 100% SF).

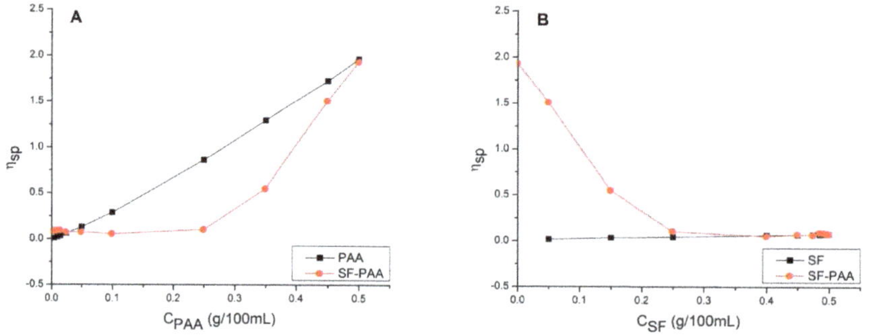

Figure 12. (**A**) Influence of PAA concentration on specific viscosity of PAA solution and SF-PAA mixture with different SF:PAA mass ratios (100% PAA in mixture is at 0.5 g/100 mL PAA), both at pH 3 and 24 h after preparation. (**B**) Influence of SF concentration on specific viscosity of SF solution and SF-PAA mixture with different SF:PAA mass ratios (100% SF in mixture is at 0.5 g/100 mL SF), both at pH 3 and 24 h after preparation.

It can be seen from Figure 12A that the specific viscosity of the PAA solution constantly increases with an increase in PAA concentration. On the other hand, the specific viscosity of the SF-PAA mixture does not change with an increase in the PAA concentration and remains almost constant until the PAA concentration reaches 0.25 g/100 mL (i.e., from 0% to 50% PAA in the mixture). For most of the concentration range, the specific viscosity of the SF-PAA mixture is significantly lower than that of the PAA solutions. Since at this concentration range, SF-PAA complex formation already takes place (Figure 9), the difference in specific viscosity suggests that the SF-PAA PEC's contribution to the specific viscosity of the SF-PAA mixture is significantly lower than the contribution of freely dissolved PAA molecules. Moreover, the constant specific viscosity of SF-PAA mixtures in the 0.00–0.25 g/100 mL PAA concentration range suggests that all of the PAA present in the mixture is entirely used for SF-PAA complex formation and that there are almost no uncomplexed, freely dissolved PAA molecules. A further increase in PAA concentration, above 0.25 g/100 mL, results in a sharp increase in the specific viscosity of SF-PAA mixtures (Figure 12A). At the same time, the difference in the specific viscosity of the PAA solution and SF-PAA mixture, for the same PAA concentration, decreases. The sharp increase and the reduced difference in specific viscosity indicate the presence of a surplus of uncomplexed, freely dissolved PAA in the SF-PAA mixture. This is attributed to the fact that in SF-PAA mixtures with a high PAA%, there is not enough SF to complex all of the dissolved PAA, and therefore some of the PAA remains uncomplexed.

Figure 12B shows the influence of SF concentration on the specific viscosity of the SF solution. It also shows the specific viscosity of SF-PAA mixtures with different SF:PAA mass ratios (where the mixture with 0.5 g/100 mL SF represents 100% SF, and the mixture with 0 g/100 mL SF represents 100% PAA). It can be seen from Figure 12B that the specific viscosity of the SF oligomers solution is low and only slightly increases with an increase in SF concentration. Therefore, the contribution of freely dissolved SF to the specific viscosity of SF-PAA mixtures can be neglected for most of the SF-PAA mixture compositions. The contribution is observed only for the lowest PAA concentrations (i.e., the highest SF concentrations) in the mixtures (Figure 12A) when there is not enough PAA to complex all of the dissolved SF present in the mixture, thereby leaving most of the SF uncomplexed and freely dissolved. The dissolved SF makes the specific viscosity of SF-PAA mixtures with $C_{PAA} < 0.025$ g/100 mL slightly higher than the specific viscosity of the PAA solution (Figure 12A). The viscosimetric study showed that the viscous behavior of SF-PAA mixtures is predominantly determined by the colloidal state of PAA molecules.

3.2.4. Tensiometric Investigation of SF/High MW PAA Mixtures

The influence of SF concentration on the surface tension of SF solutions (0.01–2.00% (w/v)) and the surface tension of SF/high MW PAA mixtures (0.05% (w/v) PAA and 0.01–2.00% (w/v) SF), both at pH 3, is shown in Figure 13. Figure 13 shows that the surface tension of SF solutions decreases with an increase in SF concentration. The decrease is faster at lower SF concentrations (<0.1%); it slows down at the intermediate SF concentrations (0.1–1.0%) and reaches a plateau value for SF concentrations >1%. The obtained dependence is typical for surface active molecules and shows that the decrease in surface tension is due to the adsorption of SF at the air/water interface. The obtained SF surface tension isotherm does not have a sharp break point typical for the critical micelle concentration (CMC), which suggests that the onset of micelle-like SF aggregate formation is not that well defined as is the case in typical surfactant solutions, most likely due to the polydispersity of SF molecules [45]. As suggested by the isotherm, the onset of SF micelle-like aggregate formation most likely takes place in the 0.1–0.5% SF range.

The surface tension of 0.05% (w/v) PAA at pH 3 was c.a. 65 mN/m (Figure 13), which suggests that some of the PAA molecules adsorb at the air/water interface. PAA, a highly charged polyanion, is typically well soluble in water and does not adsorb at the air/water interface at a solution pH > 4. However, at pH 3, the dissociation of carboxylic groups is limited, which leads to a decrease in the PAA charge and consequently makes PAA macromolecules surface-active [46].

The influence of SF concentration on the surface tension of SF/high MW PAA mixtures with 0.05% (w/v) PAA is shown in Figure 13. The figure shows that the surface tension of the mixture sharply decreases with an increase in SF concentration when the SF concentration was less than 0.05% (w/v). In this region, the surface tension of the mixture is significantly lower than the surface tension of corresponding SF solutions, which is attributed to the formation of surface-active SF-PAA electrostatic complexes which form at the air/water interface [47]. A further increase in SF concentration (above 0.05% (*w*/*v*) SF) brings about a further decrease in the surface tension of SF-PAA mixtures, but this time the decrease is significantly slower. The break point of the SF-PAA mixture surface tension isotherm observed at 0.05% (*w*/*v*) SF corresponds to the critical aggregation concentration (CAC), i.e., to the formation of micelle-like aggregates of SF bound to PAA [48]. The CAC was found to be lower than the CMC of pure SF solutions. The onset of SF-PAA complex formation in the bulk (i.e., CAC) was also indicated by the turbidity change in the mixtures, where the SF-PAA mixture at 0.01% (*w*/*v*) SF concentration was clear, while the mixtures with SF concentrations higher than 0.01% (*w*/*v*) were turbid or clear with sediment coacervate phase (not shown). The surface tension of SF-PAA mixtures with SF concentrations >0.05% (*w*/*v*) becomes higher than the surface tension of the corresponding SF solutions and eventually, for the highest SF concentrations studied, approaches the surface tension of the corresponding SF solutions (Figure 13). The surface tension of the mixture becomes equal to the surface tension of SF when the air/water interface in the mixture is entirely covered by SF molecules while all of the PAA is complexed with SF in the bulk, which is expected to take place for the investigated SF-PAA mixtures at SF concentrations higher than 2.0%.

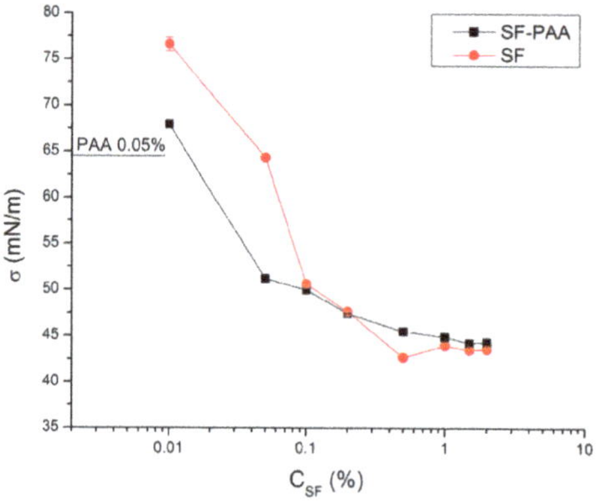

Figure 13. Influence of SF concentration on surface tension of SF solutions and surface tension of SF/high MW PAA mixtures (0.05% (*w*/*v*) PAA and 0.01–2.00% (*w*/*v*) SF), at pH 3.

4. Conclusions

The investigation into the colloidal properties of SF and PAA showed that the two polymers can form PECs through electrostatic interaction when the pH of the SF-PAA mixture is higher than 2.5 and lower than 4.2. In this pH range, SF is in its protonated form and positively charged, while PAA is in its deprotonated form and negatively charged. SF-PAA complex formation occurred at all investigated SF:PAA mass ratios (except for 100:0 and 0:100) when the two polymers were mixed at pH 3 and the total polymer concentration at least 0.1% (*w*/*v*).

The SF complex with the low MW PAA is slightly positively charged for SF:PAA mass ratios ranging from 98:2 up to 20:80 and becomes slightly negatively charged from 10:90 up to 2:98 mass ratios. The positively charged complex remains dispersed even 24 h after preparation when the PAA% in the SF-PAA mixture is 20 or less. The precipitated complex forms sediment when the PAA% in the mixture ranges from 30 to 80.

The SF complex with the high MW PAA is slightly positively charged for SF:PAA mass ratios ranging from 99:1 up to 70:30 and becomes slightly negatively charged from 60:40 up to 5:95 mass ratios. The positively charged complex remains dispersed even 24 h after preparation when the PAA% in the SF-PAA mixture is 5% or less. The precipitated complex forms sediment when the PAA% in the mixture ranges from 8% to 50%. The negatively charged SF-PAA complex remains dispersed in the mixture for at least 24 h when the PAA% ranges from 60 to 95%.

A viscosimetric study on SF/high MW PAA mixtures suggests that practically all of the PAA is bound to SF when the PAA% in the SF-PAA mixture is less than 50%.

The study on SF-PAA electrostatic interactions enables the formation of SF-PAA PECs of tailored properties, which can readily be employed in future studies on designing SF-PAA PEC-based biomaterials, such as biopolymer films and bioadhesives, with advanced properties.

Author Contributions: Conceptualization, J.Š., J.K. and L.S.; methodology, J.Š., J.K., A.B., S.B., L.P. (Lidija Petrović), J.F. and J.M.B.; formal analysis, J.Š. and L.P. (Ljiljana Popović); investigation, J.Š., R.S. and A.S.; resources, J.Š. and J.K.; writing—original draft preparation, J.Š.; writing—review and editing, J.K. and L.S.; visualization, S.A., R.S. and A.S.; supervision, J.K., A.S., S.A., L.P. (Lidija Petrović) and L.P. (Ljiljana Popović). All authors have read and agreed to the published version of the manuscript.

Funding: This research was funded by The Ministry of Science and Higher Education of the Republic of Kazakhstan, Grant No. 14869304 and The Ministry of Science, Technological Development and Innovation of the Republic of Serbia, Grant No. 451-03-66/2024-03/200134 and Grant No. 451-03-65/2024-03/200134.

Institutional Review Board Statement: Not applicable.

Data Availability Statement: The raw data supporting the conclusions of this article will be made available by the authors on request.

Conflicts of Interest: The authors declare no conflicts of interest.

References

1. Gholipourmalekabadi, M.; Sapru, S.; Samadikuchaksaraei, A.; Reis, R.L.; Kaplan, D.L.; Kundu, S.C. Silk fibroin for skin injury repair: Where do things stand? *Adv. Drug Deliv. Rev.* **2020**, *153*, 28–53. [CrossRef] [PubMed]
2. Patel, M.; Dubey, D.K.; Singh, S.P. Phenomenological models of *Bombyx mori* silk fibroin and their mechanical behavior using molecular dynamics simulations. *Mater. Sci. Eng. C* **2020**, *108*, 110414–110434. [CrossRef]
3. Partlow, B.P.; Tabatabai, A.P.; Leisk, G.G.; Cebe, P.; Blair, D.L.; Kaplan, D.L. Silk Fibroin Degradation Related to Rheological and Mechanical Properties. *Macromol. Biosci.* **2016**, *16*, 666–675. [CrossRef] [PubMed]
4. Allardyce, B.J.; Rajkhowa, R.; Dilley, R.J.; Atlas, M.D.; Kaur, J.; Wang, X. The impact of degumming conditions on the properties of silk films for biomedical applications. *Text. Res. J.* **2016**, *86*, 275–287. [CrossRef]
5. Bucciarelli, A.; Greco, G.; Corridori, I.; Pugno, N.M.; Motta, A. A design of experiment rational optimization of the degumming process and its impact on the silk fibroin properties. *ACS Biomater. Sci. Eng.* **2021**, *7*, 1374–1393. [CrossRef] [PubMed]
6. Mottaghitalab, F.; Farokhi, M.; Shokrgozar, M.A.; Atyabi, F.; Hosseinkhani, H. Silk fibroin nanoparticle as a novel drug delivery system. *J. Control. Release* **2015**, *206*, 161–176. [CrossRef] [PubMed]
7. Nguyen, T.P.; Nguyen, Q.V.; Nguyen, V.H.; Le, T.H.; Huynh, V.Q.N.; Vo, D.V.N.; Trinh, Q.T.; Kim, S.Y.; Le, Q.V. Silk Fibroin-Based Biomaterials for Biomedical Applications: A Review. *Polymers* **2019**, *11*, 1933–1958. [CrossRef] [PubMed]
8. Qi, Y.; Wang, H.; Wei, K.; Yang, Y.; Zeng, R.-Y.; Kim, I.S.; Zhang, K.-Q. A review of structure construction of silk fibroin biomaterials from single structures to multi-level structures. *Int. J. Mol. Sci.* **2017**, *18*, 237–258. [CrossRef] [PubMed]
9. Sashina, E.S.; Bochek, A.M.; Novoselov, N.P.; Kirichenko, D.A. Structure and Solubility of Natural Silk Fibroin. *Russ. J. Appl. Chem.* **2006**, *79*, 869–876. [CrossRef]
10. Yao, X.; Zou, S.; Fan, S.; Niu, Q.; Zhang, Y. Bioinspired silk fibroin materials: From silk building blocks extraction and reconstruction to advanced biomedical applications. *Mater. Today Bio* **2022**, *16*, 100381–100408. [CrossRef]
11. Patil, P.P.; Reagan, M.R.; Bohara, R.A. Silk fibroin and silk-based biomaterial derivatives for ideal wound dressings. *Int. J. Biol. Macromol.* **2020**, *164*, 4613–4627. [CrossRef] [PubMed]
12. Liu, X.; Liu, J.; Wang, J.; Wang, T.; Jiang, Y.; Hu, J.; Liu, Z.; Chen, X.; Yu, J. Bioinspired, microstructured silk fibroin adhesives for flexible skin sensors. *ACS Appl. Mater. Interfaces* **2020**, *12*, 5601–5609. [CrossRef] [PubMed]
13. Ruggeri, E.; Kim, D.; Cao, Y.; Fare, S.; De Nardo, L.; Marelli, B. A multilayered edible coating to extend produce shelf life. *ACS Sustain. Chem. Eng.* **2020**, *8*, 14312–14321. [CrossRef]
14. Marelli, B.; Brenckle, M.A.; Kaplan, D.L.; Omenetto, F.G. Silk Fibroin as Edible Coating for Perishable Food Preservation. *Sci. Rep.* **2016**, *6*, 25263–25274. [CrossRef]
15. Grabska-Zielinska, S.; Sionkowska, A. How to improve physico-chemical properties of silk fibroin materials for biomedical applications?—Blending and cross-linking of silk fibroin—A review. *Materials* **2021**, *14*, 1510–1541. [CrossRef] [PubMed]
16. Wani, S.U.D.; Gautama, S.P.; Qadrie, Z.L.; Gangadharappa, H.V. Silk fibroin as a natural polymeric based bio-material for tissue engineering and drug delivery systems-A review. *Int. J. Biol. Macromol.* **2020**, *163*, 2145–2161. [CrossRef] [PubMed]
17. Pandey, V.; Haider, T.; Jain, P.; Gupta, P.N.; Soni, V. Silk as a leading-edge biological macromolecule for improved drug delivery. *J. Drug Deliv. Sci. Technol.* **2020**, *55*, 101294–101314. [CrossRef]

18. Farokhi, M.; Mottaghitalab, F.; Fatahi, Y.; Khademhosseini, A.; Kaplan, D.L. Overview of Silk Fibroin Use in Wound Dressings. *Trends Biotechnol.* **2018**, *36*, 907–922. [CrossRef] [PubMed]
19. Qiao, X.; Miller, R.; Schneck, E.; Sun, K. Influence of pH on the surface and foaming properties of aqueous silk fibroin solutions. *Soft Matter* **2020**, *16*, 3695–3704. [CrossRef]
20. Rao, J.-J.; Chen, Z.-M.; Chen, B.-C. Modulation and Stabilization of Silk Fibroin-Coated Oil-in-Water Emulsions. *Food Technol. Biotechnol.* **2009**, *47*, 413–420.
21. Shang, S.; Zhu, L.; Fan, J. Intermolecular interactions between natural polysaccharides and silk fibroin protein. *Carbohydr. Polym.* **2013**, *93*, 561–573. [CrossRef]
22. Meka, V.S.; Sing, M.K.G.; Pichika, M.R.; Nali, S.R.; Kolapalli, V.R.M.; Kesharwani, P. A comprehensive review on polyelectrolyte complexes. *Drug Discov. Today* **2017**, *22*, 1697–1706. [CrossRef]
23. Kabanov, V.A.; Zezin, A.B. Soluble interpolymeric complexes as a new class of synthetic polyelectrolites. *Pure Appl. Chem.* **1984**, *56*, 343–354. [CrossRef]
24. Kulkarni, A.D.; Vanjari, Y.H.; Sancheti, K.H.; Patel, H.M.; Belgamwar, V.S.; Surama, S.J.; Pardeshi, C.V. Polyelectrolyte complexes: Mechanism, critical experimental aspects, and applications. *Artif. Cells Nanomed. Biotechnol.* **2016**, *44*, 1615–1625. [CrossRef]
25. KayitmazeR, A.B.; Seeman, D.; Minsky, B.B.; Dubin, P.L.; Xu, Y. Protein-polyelectrolyte interactions. *Soft Matter* **2013**, *9*, 2553–2583. [CrossRef]
26. Srinivas, L.; Ramana Murthy, K.V. Preparation and evaluation of polyelectrolyte complex for oral controlled drug delivery. *Asian J. Pharm.* **2010**, *4*, 69–78. [CrossRef]
27. Malay, Ö.; Bayraktar, O.; Batigün, A. Complex coacervation of silk fibroin and hyaluronic acid. *Int. J. Biol. Macromol.* **2007**, *40*, 387–393. [CrossRef]
28. Malay, Ö.; Batigün, A.; Bayraktar, O. pH-and electro-responsive characteristics of silk fibroin-hyaluronic acid polyelectrolyte complex membranes. *Int. J. Pharm.* **2009**, *380*, 120–126. [CrossRef] [PubMed]
29. Ghaeli, I.; De Moraes, M.A.; Bepp, M.M.; Lewandowska, K.; Sionkowska, A.; Ferreira-da-Silva, F.; Ferraz, M.P.; Monteiro, F.J. Phase Behaviour and Miscibility Studies of Collagen/Silk Fibroin Macromolecular System in Dilute Solutions and Solid State. *Molecules* **2017**, *22*, 1368–1385. [CrossRef]
30. Bhardwaj, N.; Kundu, S.C. Silk fibroin protein and chitosan polyelectrolyte complex porous scaffolds for tissue engineering applications. *Carbohydr. Polym.* **2011**, *85*, 325–333. [CrossRef]
31. Kim, E.; Jung, J.-S.; Yoon, S.-G.; Park, W.H. Eco-friendly silk fibroin/tannic acid coacervates for humid and underwater wood adhesives. *J. Colloid Interface Sci.* **2023**, *632*, 151–160. [CrossRef]
32. Wang, F.; Wu, H.; Venkataraman, V.; Hu, X. Silk fibroin-poly(lactic acid) biocomposites: Effect of protein-synthetic polymer interactions and miscibility on material properties and biological responses. *Mater. Sci. Eng. C* **2019**, *104*, 109890–109903. [CrossRef] [PubMed]
33. Gao, Q.; Shao, Z.; Sun, Y.; Lin, H.; Zhou, P.; Yu, T. Complex Formation of Silk Fibroin with Poly(acrylic acid). *Polym. J.* **2000**, *32*, 269–274. [CrossRef]
34. Arakaban, H.; Barani, M.; Akbarizadeh, M.R.; Pal Singh Chauhan, N.; Jadoun, S.; Dehghani Soltani, M.; Zarrintaj, P. Polyacrylic acid Nanoplatforms: Antimicrobial, Tissue Engineering, and Cancer Theranostic Applications. *Polymers* **2022**, *14*, 1259–1292. [CrossRef] [PubMed]
35. Pourmadadi, M.; Farokh, A.; Rahmani, E.; Mahdi Eshaghi, M.; Aslani, A.; Rahdar, A.; Ferreira, L.F.R. Polyacrylic acid mediated targeted drug delivery nano-systems: A review. *J. Drug Deliv. Sci. Technol.* **2023**, *80*, 104169–104181. [CrossRef]
36. Ito, T.; Yamaguchi, S.; Soga, D.; Yoshimoto, T.; Koyama, Y. Preparation of a Bioadhesive Poly(acrylic acid)/Polyvinyl pyrollidone Complex Gel and Its Clinical Effect on Dental Hemostasis. *Gels* **2022**, *8*, 462–475. [CrossRef]
37. Xiong, X.-P.; Ke, Q.-R.; Zhu, S.-Q. Introduction of a reliable method for determination of intrinsic viscosity for any polymer with high precision. *Chin. J. Polym. Sci.* **2014**, *32*, 209–2017. [CrossRef]
38. Brandrup, J.; Immergut, E.H.; Grulke, E.A.; Abe, A.; Bloch, D.R. *Polymer Handbook*, 4th ed.; John Wiley & Sons: Hoboken, NJ, USA, 2005.
39. Sovilj, V. *Fizička Hemija Polimera*, 1st ed.; Tehnološki fakultet Novi Sad: Novi Sad, Serbia, 1996; p. 40.
40. Petrović, L.B.; Milinković, J.R.; Fraj, J.L.; Bučko, S.Đ.; Katona, J.M. An investigation of chitosan and sodium dodecyl sulfate interactions in acetic media. *J. Serb. Chem. Soc.* **2016**, *81*, 575–587. [CrossRef]
41. Laemmli, U.K. Cleavage of structural proteins during the assembly of the head of bacteriophage T4. *Nature* **1970**, *227*, 680–685. [CrossRef]
42. Guo, C.; Li, C.; Kaplan, D.L. Enzymatic Degradation of *Bombyx mori* Silk Materials: A Review. *Biomacromolecules* **2020**, *21*, 1678–1686. [CrossRef]
43. Tirrel, M. Fundamentals of polymer solutions. In *Interactions of Surfactants with Polymers and Proteins*; Goddard, D.E., Ed.; CRC Press: Boca Raton, FL, USA, 1993.
44. Wu, X.; Hou, J.; Li, M.; Wang, J.; Kaplan, D.L.; Lu, S. Sodium dodecyl sulfate-induced rapid gelation of silk fibroin. *Acta Biomater.* **2012**, *8*, 2185–2192. [CrossRef] [PubMed]
45. Yang, Y.; Dicko, C.; Bain, C.D.; Gong, Z.; Jacobs, R.M.J.; Shao, Z.; Terry, A.E.; Vollrath, F. Behavior of silk protein at the air–water interface. *Soft Matter* **2012**, *8*, 9705–9712. [CrossRef]

46. Ishimuro, Y.; Ueberreiter, K. The surface tension of poly(acrylic acid) in aqueous solution. *Colloid Polym. Sci.* **1980**, *258*, 928–931. [CrossRef]
47. Babak, V.G.; Skotnikova, E.A.; Lukina, I.G.; Pelletier, S.; Hubert, P.; Dellacherie, E. Hydrophobically Associating Alginate Derivatives: Surface Tension Properties of Their Mixed Aqueous Solutions with Oppositely Charged Surfactants. *J. Colloid Interface Sci.* **2000**, *225*, 505–510. [CrossRef]
48. Taylor, D.J.F.; Thomas, R.K.; Penfold, J. Polymer/surfactant interactions at the air/water interface. *Adv. Colloid Interface Sci.* **2007**, *132*, 69–110. [CrossRef]

polymers

MDPI

Article

Tuning the Properties of Xylan/Chitosan-Based Films by Temperature and Citric Acid Crosslinking Agent

Martina Camaño Erhardt [1], Yamil Nahún Solier [1,2], María Cristina Inalbon [1,2] and Paulina Mocchiutti [1,2,*]

[1] Instituto de Tecnología Celulósica, Facultad de Ingeniería Química, Universidad Nacional del Litoral, Santiago del Estero 2654, Santa Fe S3000AOJ, Argentina; martinacamano2@gmail.com (M.C.E.); ysolier@fiq.unl.edu.ar (Y.N.S.); cinalbon@fiq.unl.edu.ar (M.C.I.)

[2] Consejo Nacional de Investigaciones Científicas y Técnicas (CONICET), Buenos Aires 1425, Argentina

* Correspondence: paulinam@fiq.unl.edu.ar; Tel.: +54-342-4571160

Abstract: Petroleum-based food packaging causes environmental problems such as waste accumulation and microplastic generation. In this work, biobased films from stable polyelectrolyte complex suspensions (PECs) of xylan and chitosan (70 Xyl/30 Ch wt% mass ratio), at different concentrations of citric acid (CA) (0, 2.5, 5, 7.5 wt%), were prepared and characterized. Films were treated at two temperatures (135 °C, 155 °C) and times (30 min, 60 min) to promote covalent crosslinking. Esterification and amidation reactions were confirmed by Fourier Transform Infrared Spectroscopy and Confocal Raman Microscopy. Water resistance and dry and wet stress–strain results were markedly increased by thermal treatment, mainly at 155 °C. The presence of 5 wt% CA tended to increase dry and wet stress–strain values further, up to 88 MPa—10% (155 °C for 60 min), and 5.6 MPa—40% (155 °C for 30 min), respectively. The UV-blocking performance of the films was improved by all treatments, as was thermal stability (up to T_{onset}: 230 °C). Contact angle values were between 73 and 84°, indicating partly wettable surfaces. Thus, thermal treatment at low CA concentrations represents a good alternative for improving the performance of Xyl/Ch films.

Keywords: esterification; amidation; mechanical properties; thermal properties; water resistance

1. Introduction

At present, the development of new biobased and biodegradable materials for food packaging is a major challenge for researchers. The problem is due to conventional petroleum-based films that produce serious environmental impact such as waste disposal, generation of greenhouse gases, and pollution of oceans and seas. As consumer awareness of sustainability grows, companies are increasingly inclined to adopt environmentally friendly packaging practices. Zhang et al. [1] recently reported that around 40% of the plastic waste worldwide comes from food-packaging films. These statistics highlight the urgent need to explore sustainable alternatives for food packaging, among which polysaccharide-based films excel.

It has been shown that the combination of polysaccharides, at specific mass ratios and medium conditions, can be an alternative for the preparation of films with acceptable mechanical properties [2–5]. Polymers themselves, as well as the intermolecular physical interactions between polysaccharides, such as ionic and secondary interactions (hydrogen bonds and Van der Waals forces), are responsible for the formation of a strong network. Nevertheless, because physical interactions are reversibly linked [6], they weaken in the presence of water, mainly the hydrogen bond interactions [5]. Thus, films show poor resistance to water.

To reduce this effect, permanent networks can be obtained by the production of covalent crosslinking between the functional groups of polysaccharides, either in the absence or presence of a crosslinking agent.

Taking into account our background on xylan/chitosan films [3,7–9] and the results obtained recently by other authors using different compounds to covalently crosslink polysaccharides [1,10,11], it is interesting to evaluate the mode by which a simple thermal treatment, as well as the addition of citric acid followed by a thermal treatment, improves the general properties of xylan/chitosan films due to the formation of covalent crosslinking.

Glucuronoxylan, or just xylan (Xyl), is a hemicellulose present in hardwood trees, such as poplar wood, and represents 20–35% of the lignocellulosic material. It can be isolated from the rest of the matrix (cellulose, lignin, and extractives) by alkaline extraction [8]. Xylan is a weak anionic polyelectrolyte due to the presence of carboxylic acid groups in its chemical structure. On the other hand, chitosan (Ch) is a natural and linear polysaccharide, composed of β-(1,4)-N-acetyl-D-glucosamine units derived from the deacetylation of chitin present in shrimp, crustaceans, and fungal cell walls [12]. It is a weak cationic polyelectrolyte, and its number of amine groups depends on the degree of deacetylation. Thus, the combination of Xyl and Ch solutions at a specific mass ratio and pH allows the formation of stable cationic polyelectrolyte complex (PEC) suspensions.

Crosslinking agents should contain at least two functional groups to react with polymer chains to form a three-dimensional network. Citric acid (CA) has multifunctional groups: three acid groups and one hydroxyl group (Graphical Abstract). At different pH values, the acid groups of citric acid can be ionized or non-ionized. While the ionized groups can form ionic crosslinks, the non-ionized groups, under high temperature, can form covalent crosslinks [1]. This compound is commonly employed in the food industry due to its high safety profile, non-toxic nature, and biodegradability. It is recognized as safe by the FDA (Food and Drug Administration, USA) and EFSA (European Food Safety Authority) [13]. Not only can CA act as a crosslinking agent but it can also contribute to the plasticity of the film. However, note that an increase in plasticity leads to an undesired increase in hydrophilicity [14].

It has been reported that at high temperatures and low pH values, several reactions, such as esterification and amidation, can occur when polysaccharides and citric acid are present. Esterification is the reaction between carboxylic acid groups (or one of their derivatives) and hydroxyl groups to produce esters [1,15] and amidation is the reaction between carboxylic acid groups and cationic amines (NH_3^+). According to McMurry [16] and Tasselli [17], imine formation can also occur between primary amine and aldehyde groups. Carbohydrate hydrolysis can also take place under acidic environmental conditions [1]. Thus, to obtain crosslinks avoiding hydrolysis, the amount of CA added and the conditions for crosslinking reactions should be optimized.

Studies have shown crosslinking reactions between polymers containing different functional groups resulting from thermal treatment [18–20]. Nurkeeva et al. [18] reported that thermally induced intermacromolecular esterification reactions can occur between the carboxylic groups of polyacrylic acid (PAA) and the hydroxyl groups of polyvinyl alcohol (PVA). Blasques et al. [20] reported crosslinked reactions between xanthan chains via esterification reactions at 165 °C in the absence and presence of citric acid. Azeredo et al. [21] reported that citric acid can react with the hydroxyl groups of glucose to form covalent bonds between starch molecules, enhancing the performance of starch films. In addition, it has been reported that sodium hypophosphite can act as a catalyst, accelerating the esterification of polycarboxylic acids by increasing the formation rate of cyclic anhydride intermediate in the CA molecule [10].

In this work, Xyl/Ch polyelectrolyte complex suspensions were prepared containing citric acid at different concentrations (0, 2.5, 5, and 7.5 wt% on a dry mass basis and sodium hypophosphite as the catalyst). Suspensions were characterized by determining zeta-potential and average particle size. Subsequently, films were formed using the casting/evaporation technique. The films were thermally treated (135 °C and 155 °C) for 30 min and 60 min, and the chemical modifications were analyzed by Fourier Transform Infrared Spectroscopy (FTIR) and Confocal Raman Microscopy (CRM). At high temperatures, esterification crosslinking reactions are expected to occur between the polymeric chains of xylan–chitosan and xylan–xylan, or between citric acid–xylan and citric acid–chitosan, while amidation reactions are expected to take place between xylan–chitosan and citric acid–chitosan. Conversely, imine reactions could occur between chitosan and terminal carbonyl groups of xylan.

To the best of our knowledge, this is the first study on the influence of thermal treatment and citric acid addition on the crosslinking of Xyl/Ch films. Dry and wet mechanical properties and swelling capacity were also evaluated. Film morphology was characterized via SEM analysis. Contact angle measurements were performed to assess the surface hydrophobicity of films. Finally, the thermal stability of the films was analyzed by thermogravimetric analysis (TGA).

2. Materials and Methods

2.1. Xylan

Xylan was extracted from twelve-year-old poplar wood by alkaline treatment and bleached using hydrogen peroxide (10% *w*/*w*), as previously explained [8]. Its molecular weight was 56.3 kDa (polydispersity index 1.50), and its glucuronic acid group content was 0.63 meq/g. For film formation, a 4 g/L xylan solution was prepared by dissolving it in distilled water and heating it in a boiling water bath for 15 min.

2.2. Chitosan

Chitosan (Ch) was supplied by Sigma-Aldrich (Darmstadt, Germany). The viscosity average molecular mass was 270 kDa, and the degree of deacetylation weight was 78.1% [7]. A 2.5 g/L solution was prepared by the slow addition of 0.25 g/L acetic acid solution to the Ch powder, maintaining continuous mixing to achieve a homogeneous solution. The preparation was magnetically stirred for 24 h, and filtered with a porous glass filter. The pH of the solution was 4.0.

2.3. Citric Acid and Sodium Hypophosphite

Citric acid and sodium hypophosphite (analytical grade powders) were supplied by Sigma-Aldrich (Darmstadt, Germany) and used to prepare 10 g/L and 5 g/L solutions, respectively.

2.4. Preparation of Xyl/Ch PEC Suspensions in the Presence of Citric Acid and Sodium Hypophosphite

Polyelectrolyte complex (PEC) suspensions (3.55 g/L) were prepared by combining xylan (Xyl) and chitosan (Ch) at pH 4.0. This pH value was selected to achieve the esterification reaction. A Xyl/Ch mass ratio of 70/30 wt% was used for the study, according to previous results [3]. PEC suspensions were prepared in the presence of different concentrations of citric acid as a crosslinking agent (0, 2.5, 5.0, and 7.5 wt% relative to the total dried mass) and sodium hypophosphite as a catalyst (adding 50% of the acid amount: 0, 1.25, 2.5, and 3.75% relative to the total dried mass, respectively, as reported by Zoldners and Kiseleva [22]).

As preliminary trials, PECs containing 10 wt% CA were also prepared. Nevertheless, this high concentration of CA led to the formation of sizeable aggregates, resulting in the formation of non-uniform films with poor mechanical properties.

Citric acid and sodium hypophosphite solutions were added to the xylan solution, and the pH was adjusted. The resulting solution was then added to the chitosan solution at 90 mL/h using a syringe pump. The suspension obtained was later homogenized using an ultrasonic homogenizer (2 min, 500 W, 40% amplitude).

2.5. Determination of Particle Size and Z-Potential Values of PECs

PEC size and z-potential value were estimated using Zetasizer Nano ZS90 (Malvern Instruments, Worcestershire, UK). PEC suspensions were prepared at a lower concentration (1.55 g/L in water at pH 4.0) to be measured in the equipment. At least eight measurements were made for each condition at 25 °C. The refractive index of water (1.334) [23] and a viscosity of 0.8872 mPa·s were used to calculate the properties.

2.6. Preparation of Xyl/Ch Films and Crosslinking

PEC suspensions were cast on silicone molds, and the solvent was evaporated in an air-forced oven at 40 °C until dry. Films were removed from the molds and conditioned for at least 2 h at 50% relative humidity (RH) and 23 °C. Film treatment was conducted as follows: (1) Films were dried in a vacuum oven at 60 °C for 48 h to remove the equilibrium moisture content. According to Wing [24], pre-drying films at a low temperature reduces excess hydrolysis during crosslinking. (2) To induce crosslinking, the pre-dried films were placed between two steel plates (using thin silicone sheets between the film and the plates), and a weight mass exerting a pressure of 30 kgf/m^2 was placed on top to avoid film deformation during treatment. Afterwards, the system was placed in an oven (30 and 60 min) at different temperatures, 135 °C and 155 °C. A high polysaccharide content and a sufficiently high temperature are required to evaporate the water produced in the reaction to be considered irreversible [25]. Untreated films were used as a reference.

2.7. Washing Step

After crosslinking treatment, films were conditioned at 50% RH and 23 °C for 10 h, and then thoroughly washed for 60 min. Each film was placed between two meshes (for protection), immersed in water at pH 4.0, and magnetically stirred. This part of the procedure represents an important step for the removal of salts, the unreacted CA, and the catalyst. Subsequently, the films were dried under tension (pressure of 100 kgf/m^2), as explained in [26], and stored at 50% RH and 23 °C for further characterization.

2.8. FTIR Spectroscopy

The interaction between film components was determined by FTIR analysis. Film samples were reduced to their smallest possible size using scissors and placed under vacuum at 60 °C for 48 h. Potassium bromide was added to the samples and pressed to form pellets for later use. FTIR spectra were obtained using a Fourier transform spectrophotometer (Shimadzu FTIR-8000 spectrometer, (Shimadzu Co., Kyoto, Japan)) and recorded in the 700–4000 cm^{-1} wavelength range with a resolution of 8 cm^{-1}. Each sample was scanned 40 times. Using data analysis and graphics software, the deconvolutions of the interest regions were made to distinguish specific functional groups.

2.9. Confocal Raman Microscopy

Confocal Raman Microscopy was also used to analyze the functional groups in the films treated at 155 °C. The equipment used was a Renishaw Confocal Raman Microscope (Renishaw, Gloucestershire, England), featuring 2 laser excitation lines at 514 and 785 nm. Three scans of 10 s each were conducted for each spectrum. A 785 nm laser at 100% power and a 100× objective were employed. For comparison, the spectra were normalized.

2.10. UV-vis Light Transmittance

The behavior of the films exposed to UV-visible light was studied using a spectrophotometer Cecil 3055 (Cecil Instruments Limited, Cambridge, England) in the range 200 to 800 nm [27]. Films were cut into 12 × 30 mm pieces and placed on the surface of an empty quartz cuvette. The empty cell was used as the blank.

Opacity was measured at 600 nm and calculated with Equation (1), as in [27].

$$Opacity\ value = \frac{Abs_{600}}{t} \quad (1)$$

where Abs_{600} is the absorbance at 600 nm and t is the film thickness (mm).

2.11. Mechanical Film Properties

Dry and wet tensile strengths and elongations at break were measured using an INSTRON 3340 universal testing machine (Instron Corporation, Norwood, MA, USA) with a 1000 N load cell at 23 °C and 50% RH. The trials were conducted following the American Society for Testing and Materials (ASTM D882) standard method [28] with certain modifications: rectangular specimens of 30 mm × 5 mm were prepared, the speed was 3 mm/min, and the initial grip distance was 22 mm. Wet tensile strength was determined after film immersion in distilled water at 23 °C for 1 h, following the International Organization for Standardization ISO 3781:1983 standard method [29]. For each condition, no fewer than ten specimens were used for both dry and wet determinations.

2.12. Swelling Behavior, Solubility, and Equilibrium Moisture Content

Conditioned films were cut into squares of approximately 1 cm^2 each. Then, they were placed in a vacuum oven at 60 °C for 48 h, and the initial dry weight of each sample was recorded ($Weight_{dry\ initial}$). The samples were immersed in distilled water for 4 h. Subsequently, the excess liquid was removed by pressing the samples between two blotters for the weight to be recorded again ($Weight_{wet\ final}$). Finally, the samples were placed in vacuum under the same conditions and then weighted ($Weight_{dry\ final}$). The swelling capacity and solubility were calculated according to Equations (2) and (3), respectively:

$$Swelling\ (\%) = \frac{Weight_{wet\ final} - Weight_{dry\ final}}{Weight_{dry\ final}} \times 100 \quad (2)$$

$$Solubility(\%) = \frac{Weight_{dry\ initial} - Weight_{dry\ final}}{Weight_{dry\ initial}} \times 100 \quad (3)$$

2.13. Thermal Gravimetric Analysis

The thermal stability of films was evaluated by thermogravimetric analysis (TGA). The equipment used was a TGA/SDTA851e Module—Mettler Toledo (TA Instruments, model Q500, New Castle, DE, USA). The films (200 mg) were initially conditioned at 23 °C and 50% RH for 24 h, and subsequently heated from room temperature to 800 °C at a heating rate of 10 °C/min while being exposed to a nitrogen stream. Thermogravimetric curves were built, and the onset (T_{onset}) and endset (T_{endset}) temperatures were determined, as they correspond to temperatures at which significant loss of mass begins and ends, respectively. In addition, the degradation temperature (T_{deg}) obtained using the maximum of the derivative of the TG curve; and the percentage mass residue evaluated at 800 °C ($W_{residual}$) were determined from the curves.

2.14. Scanning Electron Microscopy (SEM)

The surface morphology of films was examined using Scanning Electron Microscopy with Energy-Dispersive Spectroscopy (SEM-EDS, Zeiss Instrument, model CrossBeam 350, Carl Zeiss Microscopy GmbH, Jena, Alemania) in the secondary electron imaging mode using an accelerating voltage of 1 kV. Samples were prepared according to the method described by Schnell et al. [7]. The films were coated with an approximately 10 nm thick gold layer before evaluation.

2.15. Contact Angle

The hydrophobicity/wettability of the film surface was determined by measuring the static contact angle of water on the film surface. Ten distilled water droplets (approximately 18 µL) were deposited onto the dried film surfaces. Contact angles were processed using the LBADSA method implemented in ImageJ 1 software [30].

2.16. Statistical Analysis

The thermal treatments of the films for different time spans and different citric acid concentrations were considered independent experiments. Results were analyzed by one-way analysis of variance (ANOVA), using the RStudio program (version 4.2.3). Differences between the mean values of the film properties were determined using Tukey's test. The significance of the difference was determined at a 95% confidence level ($\alpha = 0.05$). The results are presented as the mean $\pm$ standard deviation (SD), and p-values are reported.

3. Results

3.1. Z-Potential and Average Particle Size

Table 1 shows the z-potential values, average particle sizes, and polydispersity index (PDI) of the Xyl/Ch PEC suspensions. Taking into account Xyl/Ch mass ratio and CA content, the total charge of the PECs was also estimated for each condition (Table 1). For the estimation at pH 4.0: chitosan was completely ionized (75% deacetylated) and xylan (0.1 eq/monomer) was 60% ionized [31]. Also, knowing the dissociation constants of CA (pKa values: 3.13, 4.76, and 6.4 [32]), it was expected that, at pH 4.0, the two acidic groups were partially ionized, while the other acidic group and hydroxyl group remaineed mostly non-ionized (Graphical Abstract). Thus, 1 eq/mol was considered for citric acid. The table shows that all z-potential values were higher than 30 mV, indicating that the suspensions were stable, although the addition of CA gradually reduced the value (from 37 to 30 mV when 7.5 wt% CA was used, $p = 0.00$). It was probable that the anionic groups of the citric acid would crosslink with the free cationic NH_3^+ chitosan groups (ionic crosslink), reducing the z-potential as the total cationic charges of the PECs. However, the system remained cationic and relatively stable.

Table 1. Characterization of Xyl/Ch PEC suspensions in the presence of different concentrations of citric acid and catalyzer.

Identification	Z-Potential (mV)	Average Particle Size (nm)	PDI	meq(+)/g PEC
0% CA	37.15 ± 0.46	1078 ± 24	0.271 ± 0.030	0.99
2.5% CA	35.52 ± 0.76	1037 ± 27	0.286 ± 0.030	0.83
5% CA	34.11 ± 0.53	921 ± 27	0.217 ± 0.030	0.68
7.5% CA	30.05 ± 0.26	876 ± 17	0.239 ± 0.040	0.54

Additionally, the non-ionized acid groups of the CA and the hydroxyl groups could form hydrogen bonds with polysaccharides. Both interactions significantly reduced the average particle size from 1078 to 876 nm when 7.5 wt% CA was used ($p = 0.00$). The PDI value obtained suggests midrange polydispersity suspensions.

Note that due to the increase in concentration during the evaporation of the PEC suspension for film formation, the stability of the suspensions was visually reduced, particularly when CA concentration reached higher values.

3.2. FTIR Analysis

Figure 1 shows the FTIR spectra of the reference film and films containing citric acid thermally treated at 135 °C and 155 °C in the 4000–700 cm^{-1} range.

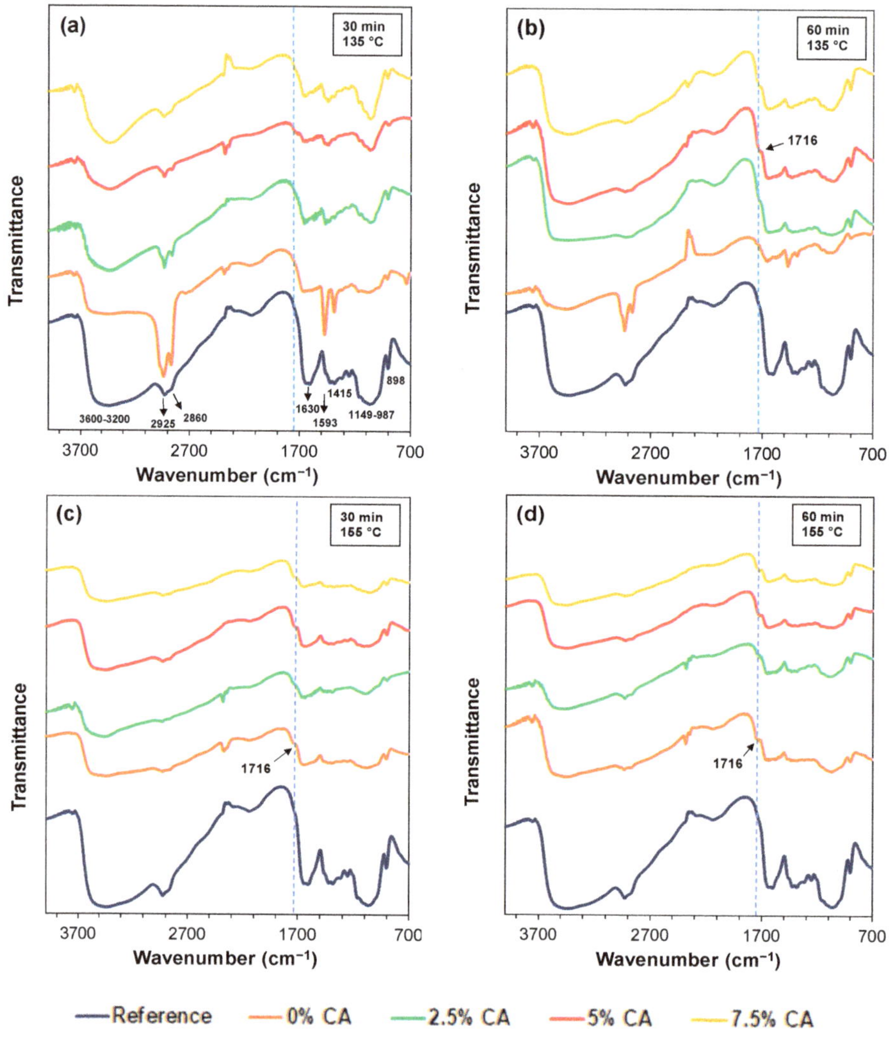

Figure 1. FTIR spectra of the films thermally treated at 135 °C for 30 min (**a**) and 60 min (**b**), and at 155 °C for 30 min (**c**) and 60 min (**d**) at different CA concentrations. The spectrum of the reference film is shown in all figures to allow the comparison.

The peaks in the spectrum of the Xyl/Ch reference film are the same as those in our previously reported work [7], detailed in Figure 1a. A broad band can be observed between 3600 and 3200 cm^{-1}, which is due to the overlaying of the stretching vibration band of the xylan hydroxyl groups (O–H) and the stretching vibration band of the primary amines (N–H) and secondary amides of chitosan. This broad band is also due to hydrogen bonds [33]. The spectrum also shows asymmetric and symmetric aliphatic C–H stretching vibration peaks of CH_3 groups at 2925 cm^{-1} and 2860 cm^{-1}, respectively. The characteristic peaks of chitosan are attributed to amide-stretching vibration (C=O) at around 1630 cm^{-1}, and at around 1593 cm^{-1}, to N–H deformation of primary amines overlapping amide II vibration (1570–1520 cm^{-1}). These peaks are also shown as broad bands [33]. This region also exhibits the vibrations of the asymmetric deformation of NH_3^+ amine salts (1625–1560 cm^{-1}) [33], as well as the stretching vibration of ionized carboxylic groups

(1610–1550 cm^{-1}) of xylan [34,35]. Due to overlap, it is difficult to detect modifications such as those of amidation and imine reactions. The characteristic vibrations of the groups present in polysaccharides are also observed (–CH_2 bending vibration at 1415 cm^{-1} and–CH_3 symmetrical deformation at 1380 cm^{-1}, overlapping with O–H in-plane deformation at 1440–1390 cm^{-1}). A peak can be observed at around 1250 cm^{-1} due to the N–H deformation of secondary amides (amide III band). In addition, a band can be observed between 1170 and 987 cm^{-1}, where several vibrations occur: 1150 cm^{-1} asymmetric stretching of the C–O–C bridge, 1126 cm^{-1} and 1018 cm^{-1} skeletal vibration due to C–O stretching of secondary alcohols found in chitosan and xylan that overlap with 1041 cm^{-1}, C–O bond-stretching of primary alcohols in chitosan [34,36]. The vibration of β-glycosidic linkages between monomers can be observed at 898 cm^{-1} [37]. On the other hand, a peak at around 2400–2300 cm^{-1}, assigned to the vibration of CO_2 in the environment, is observed in all the spectra. The peak may appear upwards or downwards, depending on whether CO_2 absorption in the sample is higher or lower than that in the background.

Figure 1a,b also show the spectra of films treated at 135 °C for 30 and 60 min, respectively. Treatment for 30 min showed no changes in the FTIR spectra when 0 wt% and 2.5 wt% CA were used. However, using 5 wt% and 7.5 wt% CA, a new peak at around 1716 cm^{-1} was observed. When the treatment was for 60 min, this peak appeared for all the cases containing CA. This peak can be attributed to a C=O ester vibrational-stretching band [38–41], indicating the occurrence of esterification reactions. Nevertheless, in the 1730–1710 cm^{-1} region, non-ionized carboxylic acid groups $\nu(C{=}O)_{COOH}$ could also vibrate [34]. Deconvolution of the band in the 1750–1620 cm^{-1} region was made to distinguish carboxylic and ester groups for quantitative evaluation. It was found that, when thermal treatment at 135 °C—30 min was applied on films containing 5 wt% and 7.5 wt% CA, the percentage contribution of ester groups to the total area of the band was 5% and 2%, respectively. On the other hand, when 135 °C—60 min was applied, the contribution of ester groups was 15%, 15%, and 24%, when 2.5 wt%, 5 wt%, and 7.5 wt% CA were used, respectively.

The FTIR spectrum of citric acid powder (Figure S1) reveals two characteristic peaks of similar intensities at 1747 cm^{-1} and 1697 cm^{-1}. Since the peak at 1747 cm^{-1} was not observed in the spectra of all the evaluated films, it can be confirmed that the washing step was effective enough to remove the unreacted citric acid. Conversely, the mono-esters and di-esters resulting from esterification reactions between citric acid and polymers remained in the films anchored to the polysaccharides. These reactions exposed non-ionized and ionized carboxylic acid groups to the solution (Graphical Abstract).

Additionally, Figure 1c,d show the FTIR spectra of films treated at 155 °C for 30 and 60 min, respectively. Both figures already show the peak at 1716 cm^{-1} after the thermal treatment and use of CA, indicating that esterification reactions had indeed occurred. Deconvolution of the band in the 1750–1620 cm^{-1} region was also performed. When thermal treatment—without CA—was applied, the contribution of ester groups to the total area of the band was 11% and 12% for 30 and 60 min, respectively. When CA was used, values increased to 18% and 20% for 5 wt% and 7.5 wt% CA, respectively, for both times studied.

Hence, when 135 °C was used, 5 wt% CA—30 min and 2.5 wt% CA—60 min were enough to generate this crosslinking. On the other hand, at 155 °C, a simple thermal treatment was sufficient to produce ester groups, and the use of CA increased this interaction.

3.3. Mechanical Properties

3.3.1. Dry State

During thermal treatment, esterification and amidation reactions, imine formation, and hydrolysis of carbohydrates occur. The displacement of these reactions to reaction products depends on the conditions of the medium (CA concentration, time, temperature, and pH). More fragile films can be produced by the hydrolysis of carbohydrates, while crosslinking reactions can be favored at high temperatures, thus enhancing film mechanical properties.

Menzel et al. [11] used citric acid for crosslinking starch films and reported that at low pH (1.9, 2.5), the hydrolysis of starch also occurs. In our case, films were prepared at pH 4.0 because, at lower pH values, we found that such films were burned during thermal treatment.

Figure 2a shows the tensile strength of the films after different treatments. Compared with the reference film (53 MPa), when the exclusive thermal treatment was used at 135 °C, it was necessary to extend the time to 60 min to significantly improve this property to 72 MPa ($p = 0.00$). On the other hand, thermal treatment at 155 °C increased this property to 75 MPa for 30 min ($p = 0.00$), and to 79 MPa for 60 min ($p = 0.00$).

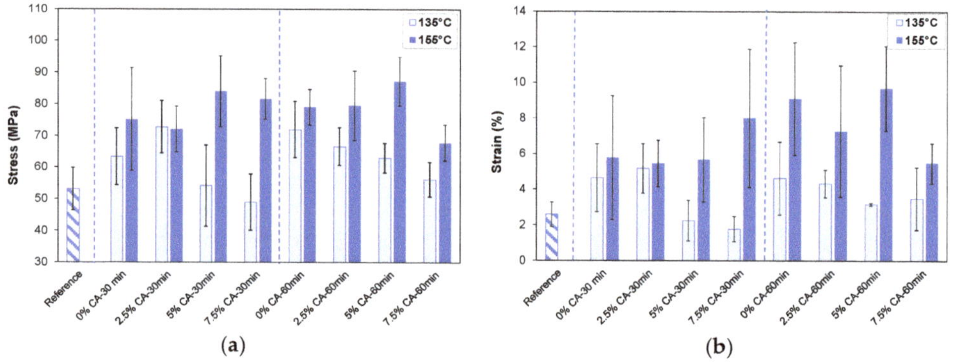

Figure 2. Dry tensile stress (**a**) and dry strain at break (**b**) of reference film (striped pattern), as well as thermally treated films (135 °C, 155 °C) for two times (30 min, 60 min) at different concentrations of CA (0 wt%, 2.5 wt%, 5 wt%, 7.5 wt% CA). Blue dashed lines separate different treatment times.

When CA was added, different behaviors were observed at the two temperatures evaluated. At 135 °C for 30 and 60 min, the films containing citric acid (7.5 wt% CA) showed lower tensile strength compared to that of their respective films which had only been thermally treated (0% CA—30 min ($p = 0.008$) and 0% CA—60 min ($p = 0.005$)). This reduction can be attributed to the effect of the hydrolysis reaction, which was dominant over the crosslinking reaction. No significant difference was observed when 2.5 wt% and 5 wt% CA were used ($p > 0.05$). On the other hand, at 155 °C and independent of time (30 or 60 min), the addition of 5 wt% CA showed a tendency to increase the tensile stress of the thermally treated films (0% CA—30 min and 0% CA—60 min) by 12% and 10%, respectively, although these increases were not statistically different at the 95% confidence level ($p > 0.05$). Particularly, the maximum value was reached when 5 wt% CA—60 min was used (stress at break of 87 MPa). In this case, crosslinking reactions were favored over carbohydrate hydrolysis, improving the strength of films. Higher citric acid concentrations (7.5 wt% CA) reduced this property compared with films containing 5 wt% CA when treatment was performed for 60 min ($p = 0.003$).

Figure 2b shows that the strain at break was significantly increased by thermal treatments, especially when performed at 155 °C for 60 min ($p = 0.00$). When tensile strength and elongation at break improve together, a polymer network with higher homogeneity is obtained [7,14]. With citric acid in the film, different behaviors were observed when thermal treatment was applied at 135 °C or 155 °C. At 135 °C for 30 min, citric acid significantly reduced strain when 7.5 wt% CA was used ($p = 0.00$), making the film more rigid. At 155 °C, no significant differences were observed when CA was used, compared with 0% CA—30 min and 0% CA—60 min ($p > 0.05$). Different authors [1,42] reported that excess CA can increase film strain. In our case, as films were thoroughly washed and unreacted CA was removed, elongations were not expected to be significantly modified.

It can be concluded that the film containing 5 wt% CA thermally treated at 155 °C for 60 min showed the best dry stress–strain mechanical properties. The contribution of thermal treatment to these values was markedly high.

Other authors have shown a significant increase in the tensile strength, using CA as crosslinking agent, of biobased polysaccharides films. Particularly, Wang et al. [14], showed that xylan/poly(vinyl alcohol) (25/75 wt%) films containing 10 wt% CA and treated at 110 °C for 2.5 h increased their tensile strength from 30.3 MPa to 49.3 MPa and decreased strain from 10.7% to 6.8%. Using corn starch-based films, it was shown that the addition of 5 wt% CA and 2.5% sodium hypophosphite, followed by film thermal treatment at 130 °C for 30 min, increased the tensile strength from 10.8 to 18.2 MPa and the elongation from 3.15% to 4.2% [41]. Furthermore, Wu et al. [40], demonstrated that 10 wt% and 15 wt% CA increased the tensile strength of potato starch/chitosan film from 9.7 to 12.5 MPa. In this last case, only ionic crosslinking between the agent and the polysaccharides is expected. The CA crosslinking agent, as well as the catalyst, can also be used for improving the mechanical properties of protein-based films [43]. Thus, the tensile strengths of the films obtained in this work are comparable to those reported in the literature, suggesting good potential for their development as packaging.

The elastic modulus of the reference film was 3500 MPa. Neither thermal treatment nor CA treatment significantly modified this property, which was around 3000–4100 MPa.

3.3.2. Wet State

Figure 3a shows the results of the wet tensile strength. All values were lower than those of the dry tensile strength, indicating that water acted as a plasticizer, and that the structural conformation of the film was modified. It is well known that a plasticizer increases interstitial volume and polymer mobility, resulting in a less dense polymeric network with weaker intermolecular forces [14]. In this case, hydrogen bonds are expected to weaken in the presence of water, while covalent and other physical crosslinks remain present. Chen et al. [5] compared the mechanical properties of dry chitosan/CMC films to those of hydrated films. They reported that hydrogen bonds were disrupted by water molecules; yet, due to some physical crosslinking points resulting from polyelectrolyte complexation, their integrity and dimensional stability were maintained, as also found in our work.

Particularly, the figure shows that thermal treatment at 135 °C had a favorable effect on stress when applied for 60 min ($p = 0.00$). Nevertheless, when CA was used, this property decreased in all cases, except when 2.5 wt% CA was used for 30 min ($p = 0.15$). As previously explained for dry mechanical properties, hydrolysis of carbohydrates in the acid medium occurred, weakening the film. The figure also shows that the application of thermal treatment at 155 °C for 30 min was sufficient to increase the tensile strength of the reference film ($p = 0.00$). On the other hand, at 155 °C for 60 min with 2.5 wt%, 5 wt%, and 7.5 wt% CA, tensile strength significantly increased ($p = 0.038$, $p = 0.0002$, and $p = 0.001$, respectively) compared with 0% CA—60 min, suggesting that certain physical crosslinks between xylan and chitosan, and the covalent crosslinks produced by the reactions, improved the polymeric network.

Figure 3b shows that strain at break also increased when 5 wt% CA and 155 °C were used ($p = 0.006$), compared to 0 wt% CA—30 min. When only thermal treatment was performed for 60 min, the strain at break decreased significantly ($p = 0.00$). However, 5 wt% and 7.5 wt% CA allowed for the flexibility of these last films to be slightly recovered ($p = 0.003$ and $p = 0.00$, respectively).

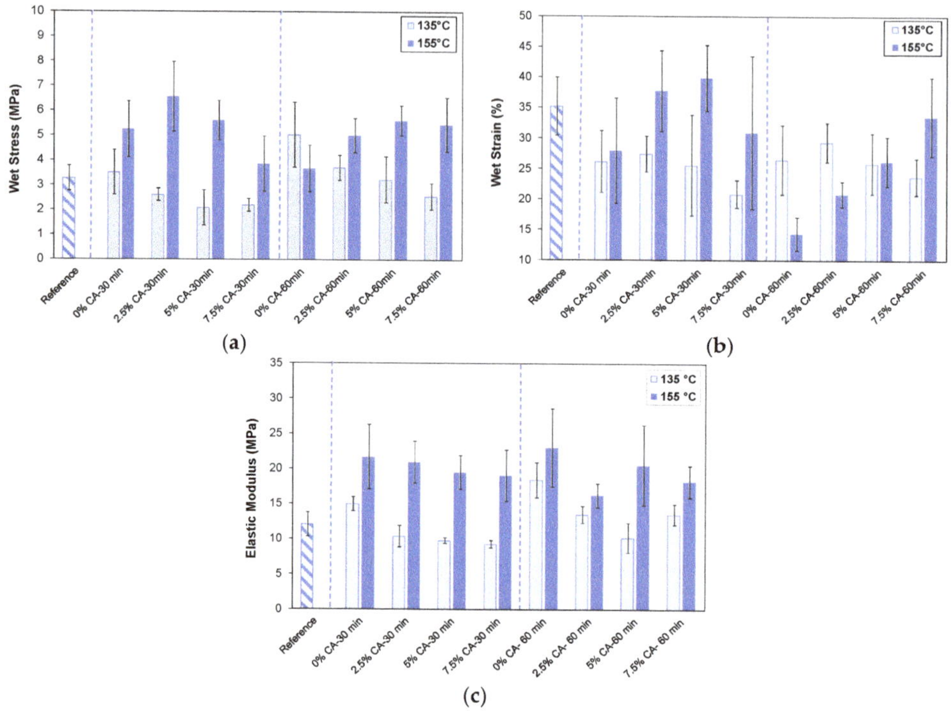

Figure 3. Wet mechanical properties obtained after immersion of films in water for 1 h. Wet tensile stress (**a**), wet strain at break (**b**), and elastic modulus (**c**) of reference film (striped pattern), and thermally treated films (135 °C, 155 °C) for two times (30 min, 60 min) at different concentrations of citric acid (0 wt%, 2.5 wt%, 5 wt%, 7.5 wt% CA). Blue dashed lines separate different treatment times.

It can be concluded that the best wet mechanical properties (tensile strength and strain at break) were obtained when films containing between 2.5 and 5% CA were thermally treated at 155 °C for 30 min.

Similar results were obtained by Liu et al. [44] and Li et al. [45]. Particularly, Liu et al. [44] prepared films using chemically modified soy protein, cured for different times (2–20 min) and temperatures (140–180 °C). Their wet tensile strengths were between 3 and 7 MPa (films conditioned at 90% RH, 48 h). Additionally, Li et al. [45], prepared films using hemicellulose crosslinked with citric acid, and showed wet tensile strengths between 4 and 8 MPa (films conditioned at 95% RH, 24 h). The authors suggested that films with these wet tensile strength values were stable in water and could be used for applications such as packaging for wet food.

Figure 3c shows that the elastic modulus of the reference film is 12 MPa. Although thermal treatments (135 °C and 155 °C), independent of time, significantly increased this property in all cases (p <0.05), CA tended to reduce the elastic modulus. The values obtained after thermal treatment at 155 °C were higher than those at 135 °C, thus intensifying film rigidity due to crosslinking.

3.4. Swelling and Solubility

Swelling is related to the quantity of water absorbed by a film. Low values are required for food packaging applications, as they indicate high resistance to water. Figure 4 shows the swelling capacity of films after immersion in water for 4 h. Not only did all the films maintain their integrity, but they also swelled and resisted dissolution. Crosslinking reactions can enhance the water resistance of the material by chemically and physically reinforcing the network [11,40], which is one of the main objectives of this work. The figure shows that the swelling capacity of the reference film is 269%. Thermal treatment at 135 °C for 30 min and 60 min slightly reduced this property to 244% and 249%, respectively, although it was not statistically significant ($p = 0.27$ and $p = 0.40$, respectively). It was necessary to add 7.5 wt% CA and thermally treat the films at 135 °C for 60 min to significantly reduce the swelling capacity of the film to 207% ($p = 0.028$).

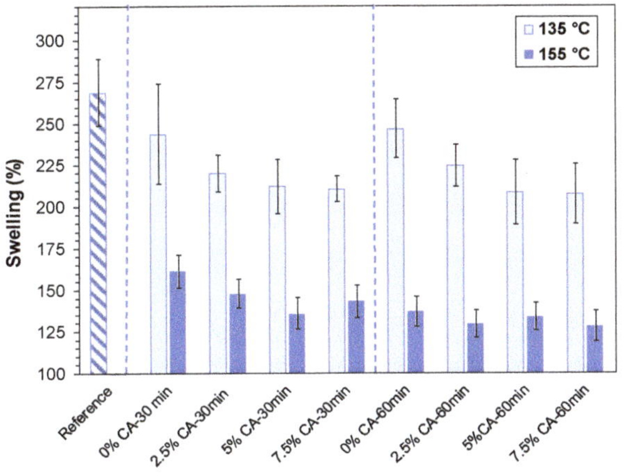

Figure 4. Swelling of the reference film (striped pattern), as well as thermally treated films (135 °C, 155 °C) for two times (30 min, 60 min) at different concentrations of CA (0 wt%, 2.5 wt%, 5 wt%, 7.5 wt% CA). Blue dashed lines separate different treatment times.

At 155 °C, the behavior was markedly different. Thermal treatment without citric acid was significantly effective, decreasing swelling by 39% at 30 min ($p = 0.00$) and 49% at 60 min ($p = 0.00$) to 161% and 137%, respectively. Tukey's test also showed that the difference between them was significant ($p = 0.00$). The occurrence of crosslinking reactions between xylan and chitosan was also confirmed by a decrease in swelling capacity and increase in water resistance. The presence of the crosslinking agent also reduced the swelling capacity of the film. At 155 °C for 30 min, addition of 5 wt% CA reduced the swelling capacity by 16%, compared to the same thermal treatment with no citric acid ($p = 0.00$), to 136%, in agreement with the improvement seen in wet tensile strength (Figure 3a). Several authors have also shown that the swelling capacity of biobased films decreased as esterification and crosslinking degrees increased [14,46].

No significant differences were observed at higher concentrations of CA. This phenomenon may be attributed to the excess unreacted carboxylic acid groups of the citric acid molecule anchored to the carbohydrates. Lipatova and Yusova [41] prepared starch films crosslinked with citric acid at high temperature. Using complexometric titration, they concluded that esterification reactions produced mono-, di-, and tri-esters of CA. Conversely, it has been reported that the probability of tri-esters is uncertain [11]. It is plausible that the favorable effect of reduced swelling due to intermolecular covalent bonds may be offset by the presence of exposed carboxylic acid groups, as well as OH groups, contributing to increased swelling.

The low swelling value obtained at 155 °C—60 min (137 g of water/100 g of film) could not be significantly reduced with citric acid ($p > 0.05$).

The solubility values of all the films were lower than 3% with no observable trend related to treatment conditions or CA content.

3.5. Characterization of Films Thermally Treated at 155 °C

Taking into account the results of the FTIR spectra, mechanical properties, and swelling capacity, it is necessary to reach a temperature of 155 °C to markedly improve film properties. Thus, films treated at 155 °C at different times and CA concentrations were characterized by Confocal Raman Microscopy, their microstructure, surface wettability, UV-visible light barrier, and thermal behavior. Results are shown in the following Sections 3.5.1–3.5.5.

3.5.1. Confocal Raman Microscopy Analysis

Infrared and Raman spectroscopic methods are complementary for the analysis of chemical structures. Figure S2 shows the Raman spectrum of the reference film in the range of 3600–400 cm^{-1}. Between 3580 and 3350 cm^{-1}, the O–H stretching vibration of alcohols present in carbohydrates overlaps with the O–H stretching vibration due to hydrogen bonds. In addition, in the region of 3450–3180 cm^{-1}, the stretching vibration of the N–H bond of primary amines NH_2 and secondary amides of chitosan (hydrogen bonded) also vibrates [34]. At around 2900 cm^{-1}, a sharp peak can be observed due to the symmetric stretching vibrations of $-CH_3$ and $-OCH_3$ from chitosan and xylan, respectively. The region of the 1800–1500 cm^{-1} spectra was analyzed in more detail (Figure 5). Figure S2 also shows a medium peak at 1467 cm^{-1} due to the deformation vibration (scissor) of $-CH_2$, and at 1422 cm^{-1} and 1388 cm^{-1} due to the asymmetric and symmetric vibration bands of the methyl group, respectively. Moreover, $-CH_2$ deformation vibration (wagging) can also be seen at 1322 cm^{-1}. In the 1330–1215 cm^{-1} range, a band can also be observed due to the new amide III vibration [34]. Furthermore, a sharp peak at 1096 cm^{-1} can be observed due to the C–N stretching of amides [34]. The C–O–C vibration of carbohydrates at 902 cm^{-1} can also be identified, as well as the C–O deformation vibration at 494 cm^{-1} [34].

Figure 5 shows the Raman spectra in the region of 1800–1500 cm^{-1} where the functional groups of the reaction products, such as those from esterification and amidation, can be identified. Figure 5a–d, on the left side, show those films thermally treated for 30 min, while Figure 5e–h, on the right side, show the spectra corresponding to 60 min of treatment. Weak bands in the range of 1750–1725 cm^{-1} can be assigned to the C=O stretching vibration of ester groups that can partially overlap with the weak C=O stretching vibration of non-ionized carboxylic groups (1735–1700 cm^{-1}) [34]. In addition, in the region between 1650 and 1550 cm^{-1}, bands can be assigned to the vibration of amide I groups that can only partially overlap with the C=O asymmetric stretching vibration of ionized carboxylic groups (around 1600 cm^{-1}). Figure 5a shows that thermal treatment for 30 min at 155 °C barely increased the intensity of the bands corresponding to the ester groups, amides, and ionized carboxylic acids, indicating that chemical reactions occurred. The areas of the bands in the region of 1670–1620 cm^{-1} associated with the amide I group's vibrations were estimated. Compared with 0% CA—30 min, when 2.5 wt% CA was used (Figure 5b), a large increase was observed in this region (the area increased by 20%). Yet, when 5 wt% CA was used, the amide band increased even more (the area increased by 37% compared with 0% CA—30 min). The amide groups were more easily detected using this technique than via FTIR analysis. Bands attributed to esters and ionized carboxylic groups also increased when 5 wt% CA was used. These results agree with those obtained from FTIR analysis and confirm that citric acid was anchored through esterification reactions, producing crosslinks between polymers, as well as mono- and di- esters and amidation reactions. The intensities of the bands of the film containing 7.5 wt% CA thermally treated for 30 min (Figure 5d) were only slightly higher than the reference film.

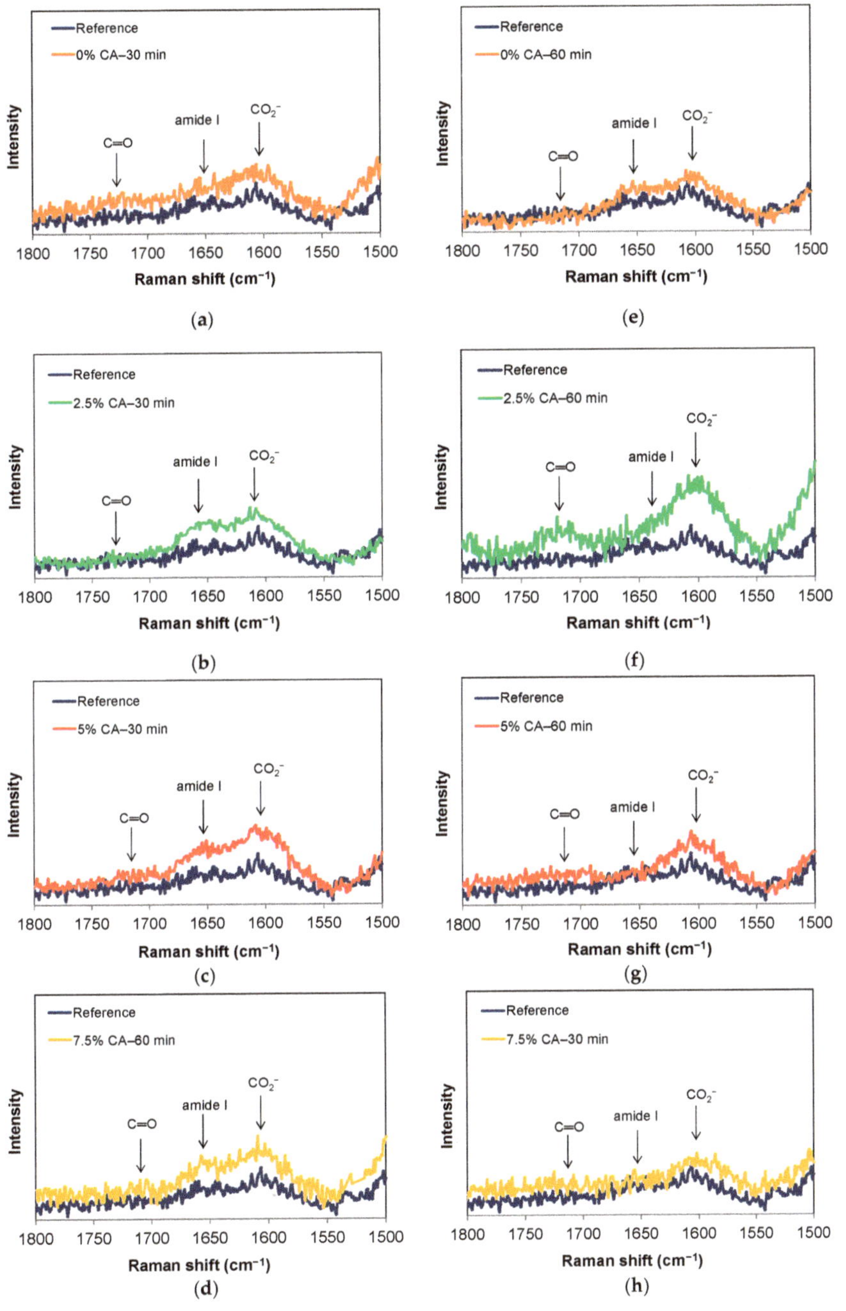

Figure 5. Confocal Raman Microscopy normalized spectra of the reference film and films treated at 155 °C for 30 min (**a**–**d**) and 60 min (**e**–**h**) at different concentrations of citric acid (0 wt%, 2.5 wt%, 5 wt%, 7.5 wt% CA). The spectrum for the reference film is shown in all figures to allow comparison.

When the films containing 5 wt% CA were thermally treated for 60 min, the intensities of the band of the amide I groups decreased (the area was reduced by 39% compared with 0% CA—60 min). On the other hand, the spectra of the film with 2.5% CA—60 min showed the most intense peak of esters and ionized carboxylic acids (the areas were increased by 150% and 36%, respectively). The ester peak was difficult to observe in the FTIR spectra for the last condition (Figure 1d).

Vandenabeele et al. [47] also reported the vibration of C=O of esters at around 1710 cm^{-1} and C=O of amide I at around 1650 cm^{-1} when Raman spectroscopy was used for proteins. In addition, the appearance of a weak asymmetric stretching vibration at 1605 cm^{-1} was attributed to the presence of ionized carboxylic groups [35,48].

3.5.2. UV Barrier and Opacity Properties

It is well known that UV-light (200–400 nm) activates oxidation reactions in fatty and polyunsaturated food and can modify their organoleptic characteristics. Figure 6 shows the UV-visible absorbance spectra of the films (values also consider the light dispersed). Film thickness was 45–55 μm. All spectra showed saturation of absorbance between 200 and 230 nm. Compared to the reference film, the absorbance of the treated films was markedly higher, up to 350 nm, indicating better UV-barrier properties. The peak at 280 nm shows that thermal treatment increased the absorbance value, while the presence of citric acid slightly reduced this effect. Nevertheless, absorbance values were always higher than those of the reference film.

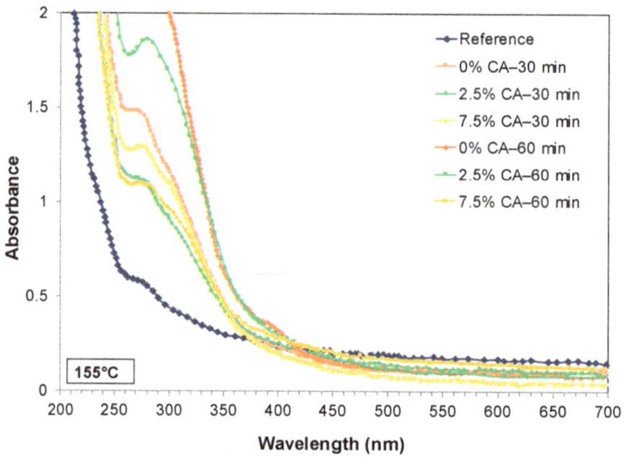

Figure 6. Absorbance spectra as a function of wavelength of the different films. Except for the reference film, all other films were treated at 155 °C.

Table S1 shows the opacity values of the films (measured at 600 nm). Compared to the reference film (opacity 3.5), thermal treatment and CA reduced opacity (values around 2), probably due to crosslinking, producing a denser film with less light dispersion. Several authors [27,40,41,49] produced highly transparent films with opacity values lower than 2. Thus, it suggests that treatments improved film transparency. This feature is of major importance since the food contained in the packaging is clearly visible to the consumer [27].

3.5.3. SEM Microscopy Analysis

Film microstructure was analyzed using SEM. Figure S3 shows the surface of the reference film, as well as the films thermally treated (155 °C) or thermally treated in the presence of the crosslinking agent. All films showed a smooth and continuous structure, without cracks or pores, indicating that components were homogeneously mixed in the matrices. In our previous works [7,31] we had found that when films are homogeneous and dense, differences in SEM cross-section images due to treatments are difficult to observe. It is well known that cracks and pores can affect the barrier properties of films [1].

Compared to the film thermally treated for 30 min without CA, the surface roughness slightly increased when 2.5% CA—30 min was used. This increase in roughness could be attributed to the introduction of a new component in the matrix. Although roughness can affect the hydrophobicity of the surface of films, a higher content of CA and longer reaction times decreased roughness to some extent, probably due to an increase in the compatibility between all components.

Films treated with citric acid showed larger number of white-colored spots on the surface, which could be ascribed to the presence of residual salts or PEC aggregation. Several authors observed PEC aggregation on the surface of chitosan-based films incorporated with different biobased extracts [50–52]. Energy-Dispersive Spectroscopy (EDS) analysis was conducted on the aggregates and on the matrix of films (7.5 wt% CA, 155 °C, 30 and 60 min). It was found that certain aggregates had a similar composition to that of the matrix (mainly C, O, N), and others showed the presence of salts (low concentrations of Ca, Mg, Na, S).

3.5.4. Surface Wettability

Figure 7 shows the contact angles of all the films. This property allows estimating the ability of the material to repel or absorb water, which is important for food packaging. A high water-contact angle indicates hydrophobicity (>90°), while a low contact angle indicates hydrophilicity (<90°). The figure shows that films of Xyl/Ch without treatment had a contact angle of 84°. The thermal treatment for 30 or 60 min did not markedly modify this value. Nevertheless, citric acid produced a decreasing tendency in this property compared to the thermally treated film, suggesting higher surface hydrophilicity. This behavior could be attributed to the mono and di-esters anchored to the carbohydrates, which left hydroxyl and carboxylic acid groups exposed to the medium.

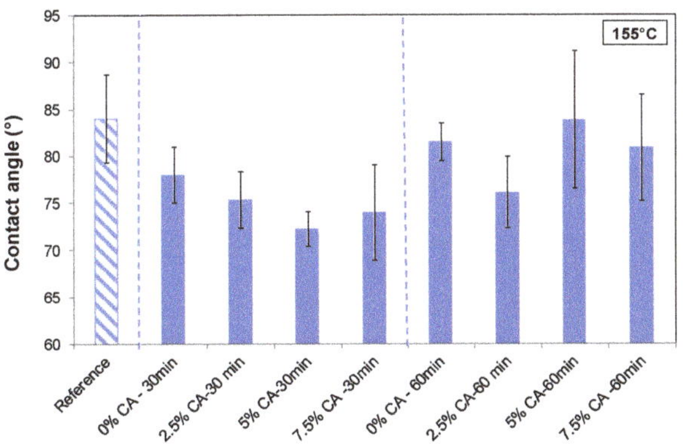

Figure 7. Contact angle results of the reference film (striped pattern), as well as thermally treated films (155 °C) for two times (30 min, 60 min) at different concentrations of CA (0 wt%, 2.5 wt%, 5 wt%, 7.5 wt% CA). Blue dashed lines separate different treatment times.

Ryan and Poduska [53] reported that a rough surface can reduce the contact angle for a droplet when the surface is hydrophilic, in agreement with the results of our work. Taking into account SEM results (Figure S3), rougher films (2.5% CA—30 min; 7.5% CA—30 min) showed lower contact angles. In contrast, films thermally treated for a longer time (60 min) containing citric acid showed both smoother surfaces and higher contact angles, reaching the values of the reference film and confirming acceptable compatibility between the components in the matrix.

All obtained values indicate that films are slightly hydrophilic. They also indicate that films are of the same order of hydrophilicity or higher than other polysaccharide-based films reported in the literature. For example, Wu et al. [40] obtained potato starch/chitosan composite films by crosslinking with citric acid (5 to 20% at 55° C) whose contact angles were between 37° and 56°. Gordobil et al. [2] prepared xylan and acetylated xylan films containing different concentrations of cellulose or acetylated cellulose, respectively, observing values ranging from 57° (xylan film) to 77° (acetylated xylan/acetylated cellulose film). On the other hand, using chitosan/acorn starch/eugenol films, Zheng et al. [4] obtained values ranging from 74°, for chitosan/acorn starch films, to 89° when there was eugenol.

Wu et al. [40] and Jiang et al. [54] reported that the hydrophobicity of biobased films depends on the structure of the network, which can be improved by crosslinking. They also used citric acid, suggesting that this phenomenon may be linked to a reduction in polar groups due to the formation of ester groups between citric acid and polymers. Wu et al. [40] found that, when a high citric acid concentration was used (20%), the contact angle decreased, probably due to the free hydrophilic groups (–COOH and –OH) from the excessive addition of citric acid.

3.5.5. Thermal Stability

Figure S4 shows the thermogravimetric curves of the different films; Table 2 summarizes characteristic temperatures (T_{onset}, T_{deg}, and T_{endset}), and the percentage of mass residue at 800 °C ($W_{residual}$). The figure shows a first thermal event in which mass loss was seen in all films. This event occurred in a temperature range of 54–110 °C and can be attributed to the loss of absorbed moisture from the films. It was found that, in this temperature range, the reference film lost 11% of the total mass, while for thermally treated films (30 or 60 min), a minor mass loss of around 7.5% was found, indicating that films absorbed less water. However, although thermal treatment was applied, the presence of citric acid in the films increased the percentage of water mass absorbed to 10–15%, probably because the mono- and di-esters anchored to the carbohydrates left hydrophilic groups exposed to the medium.

Table 2. Characteristic temperatures.

Identification	T_{onset} (°C)	T_{deg} (°C)	T_{endset} (°C)	$W_{residual}$ (%)
Reference	224	246	280	4.6
0% CA—30 min	224	247	293	28.4
2.5% CA—30 min	229	250	285	11.0
5% CA—30 min	230	254	289	27.2
7.5% CA—30 min	230	250	280	13.9
0% CA—60 min	231	255	295	29.8
2.5% CA—60 min	231	255	302	31.0
5% CA—60 min	229	246	289	31.7
7.5% CA—60 min	231	252	282	9.7

Table 2 shows that the thermal properties of the reference film were modified after the different treatments. The lowest T_{onset} value was obtained for the reference film (224 °C). For the rest of the treatments, except for the film treated at 155 °C for 30 min, this value increased up to around 230 °C, indicating higher thermal stability. This increase is expected, as more crosslinks were formed within the network. In general, T_{deg} was also increased by the treatments.

The solid residue of the reference film at 800 °C was around 5%, while the solid residue for films treated at 155 °C, regardless of whether citric acid was present, was around 10–31%. This increase also indicates that thermal resistance was markedly improved. Guevara et al. [55] reported a similar behavior when chitosan/pectin films were crosslinked with calcium chloride, thus concluding that the increase in thermal stability when the crosslinking agent was present was attributed to a more complex structure that delayed thermal decomposition.

4. Conclusions

Highly transparent and homogeneous films with good UV-blocking performance were prepared from xylan and chitosan stable polyelectrolyte complex suspensions containing different concentrations of citric acid (0 wt%, 2.5 wt%, 5 wt%, 7.5 wt% CA). They also exhibited smooth, continuous, and partly wettable surfaces.

Amidation and esterification reactions between polysaccharides and a citric acid crosslinking agent substantially contribute to the enhancement of the dry and wet mechanical properties of films, as well as their water resistance. Major favorable changes were observed when thermal treatment was applied at 155 °C. However, the citric acid slightly improved these properties mainly when 2.5 wt% CA and 5 wt% CA were used. The reactions also produced mono- and di-esters anchored to the carbohydrates, which could limit improvement in the properties, mainly the water resistance of the films. The results also indicate that crosslinking reactions and carbohydrate hydrolysis are concomitant processes that depend on temperature and citric acid concentration.

The thermal stability of the films increased due to thermal treatments in the presence of citric acid, reaching an onset temperature of 230 °C for all films.

These results suggest that biobased films with interesting mechanical properties can be obtained, contributing to environmental protection and preservation.

Supplementary Materials: The following supporting information can be downloaded at https://www.mdpi.com/article/10.3390/polym16172407/s1, Figure S1: FTIR spectrum of the citric acid powder used in this work; Figure S2: Confocal Raman spectrum of the Xyl/Ch reference film (the region between 1800 and 1200 cm^{-1} is explained in detail in Figure 2); Figure S3: Surface Scanning Electron Micrographs of 70 Xyl/30 Ch wt% films prepared using different concentrations of citric acid (0, 2.5, 5, and 7.5 wt%) and thermally treated at 155 °C for different times (30 and 60 min). The magnification was 8000×; Figure S4. Thermogravimetric curves of the different films; Table S1: Opacity values of the films calculated using Equation (1) at 660 nm.

Author Contributions: Conceptualization, P.M. and M.C.I.; methodology, M.C.E. and Y.N.S.; formal analysis, M.C.I. and P.M.; investigation, M.C.E. and P.M.; resources, M.C.I. and P.M.; data curation, M.C.E. and M.C.I.; writing—original draft preparation, M.C.E. and P.M.; writing—review and editing, M.C.I. and P.M.; visualization, Y.N.S. and P.M.; supervision, P.M.; project administration, M.C.I. and P.M.; funding acquisition, M.C.I. and P.M. All authors have read and agreed to the published version of the manuscript.

Funding: This research was funded by Agencia Nacional de Promoción de la Investigación, el Desarrollo Tecnológico y la Innovación—PICT 2019 N°03246; Universidad Nacional del Litoral—CAI+D 2020 PI, Tipo II-N° 50620190100135LI and Consejo Nacional de Investigaciones Científicas y Técnica—PIP 2022–2024 GI 11220210100627CO.

Data Availability Statement: The original contributions presented in the study are included in the article/Supplementary Materials, further inquiries can be directed to the corresponding author.

Acknowledgments: The authors acknowledge Carolina Mosconi, certified translator: professional license No 649 Translator's Association of the Province of Córdoba, Argentina (CTPPC); and M. S. Moro for her help in the laboratory.

Conflicts of Interest: The authors declare no conflicts of interest.

References

1. Zhang, W.; Roy, S.; Assadpour, E.; Cong, X.; Jafari, S.M. Cross-linked biopolymeric films by citric acid for food packaging and preservation. *Adv. Colloid Interface Sci.* **2023**, *314*, 102886. [CrossRef]
2. Gordobil, O.; Egüés, I.; Urruzola, I.; Labidi, J. Xylan-cellulose films: Improvement of hydrophobicity, thermal and mechanical properties. *Carbohydr. Polym.* **2014**, *112*, 56–62. [CrossRef]
3. Schnell, C.N.; Galván, M.V.; Peresin, M.S.; Inalbon, M.C.; Vartiainen, J.; Zanuttini, M.A.; Mocchiutt, P. Films from xylan/chitosan complexes: Preparation and characterization. *Cellulose* **2017**, *24*, 4393–4403. [CrossRef]

4. Zheng, K.; Xiao, S.; Li, W.; Wang, W.; Chen, H.; Yang, F.; Qin, C. Chitosan-acorn starch-eugenol edible film: Physico-chemical, barrier, antimicrobial, antioxidant and structural properties. *Int. J. Biol. Macromol.* **2019**, *135*, 344–352. [CrossRef] [PubMed]
5. Chen, P.; Xie, F.; Tang, F.; McNally, T. Thermomechanical-induced polyelectrolyte complexation between chitosan and carboxymethyl cellulose enabling unexpected hydrolytic stability. *Compos. Sci. Technol.* **2020**, *189*, 108031. [CrossRef]
6. Berger, J.; Reista, M.; Mayer, J.M.; Felt, O.; Peppas, N.A.; Gurny, R. Structure and interactions in covalent and ionically crosslinked chitosan hydrogels for biomedical applications. *Eur. J. Pharm. Biopharm.* **2004**, *57*, 19–34. [CrossRef] [PubMed]
7. Schnell, C.N.; Galván, M.V.; Solier, Y.N.; Inalbon, M.C.; Zanuttini, M.A.; Mocchiutti, P. High strength biobased films prepared from xylan/chitosan polyelectrolyte complexes in the presence of ethanol. *Carbohydr. Polym.* **2021**, *273*, 118602. [CrossRef] [PubMed]
8. Schnell, C.N.; Inalbon, M.C.; Minari, R.J.; Mocchiutti, P. Use of micro/nano- and nanofibrillated cellulose to improve the mechanical properties and wet performance of xylan/chitosan films: A comparison. *Appl. Polym. Mater.* **2023**, *5*, 7867–7877. [CrossRef]
9. Solier, Y.N.; Schnell, C.N.; Galván, M.V.; Mocchiutti, P.; Zanuttini, M.A.; Inalbon, M.C. Fast preparation of flexible wet-resistant and biodegradable films from a stable suspension of xylan/chitosan polyelectrolyte complexes. *J. Polym. Environ.* **2022**, *30*, 114–124. [CrossRef]
10. Dastidar, T.G.; Netravali, A.N. 'Green' crosslinking of native starches with malonic acid and their properties. *Carbohydr. Polym.* **2012**, *90*, 1620–1628. [CrossRef]
11. Menzel, C.; Olsson, E.; Plivelic, T.S.; Andersson, R.; Johansson, C.; Kuktaite, R.; Järnström, L.; Koch, K. Molecular structure of citric acid cross-linked starch films. *Carbohydr. Polym.* **2013**, *96*, 270–276. [CrossRef]
12. Mansilla, A.Y.; Albertengo, L.; Rodríguez, M.S.; Debbaudt, A.; Zúñiga, A.; Casalongué, C.A. Evidence on antimicrobial properties and mode of action of a chitosan obtained from crustacean exoskeletons on *Pseudomonas syringae* pv. tomato DC3000. *Appl. Microbiol. Biotechnol.* **2013**, *97*, 6957–6966. [CrossRef] [PubMed]
13. Olsson, E.; Menzel, C.; Johansson, C.; Andersson, R.; Koch, K.; Järnström, L. The effect of pH on hydrolysis, cross-linking and barrier properties of starch barriers containing citric acid. *Carbohydr. Polym.* **2013**, *98*, 1505–1513. [CrossRef] [PubMed]
14. Wang, S.; Rena, J.; Li, W.; Sun, R.; Liu, S. Properties of polyvinyl alcohol/xylan composite films with citric acid. *Carbohydr. Polym.* **2014**, *103*, 94–99. [CrossRef] [PubMed]
15. Zhao, Y.; Sun, H.; Yang, B.; Weng, Y. Hemicellulose-based films: Potential green films for food packaging. *Polymers* **2020**, *12*, 1775. [CrossRef]
16. McMurry, J. Aldehydes and Ketones: Nucleophilic Addition Reactions. In *Organic Chemistry*, 9th ed.; Cengage Learning: Boston, MA, USA, 2016; pp. 605–640.
17. Tasselli, F.; Mirmohseni, A.; Seyed Dorraji, M.S.; Figoli, A. Mechanical, swelling and adsorptive properties of dry-wet spun chitosan hollow fibers crosslinked with glutaraldehyde. *Reactive Funct. Polym.* **2013**, *73*, 218–223. [CrossRef]
18. Nurkeeva, Z.S.; Mun, G.A.; Dubolazov, A.V.; Khutoryanskiy, V.V. pH effects on the complexation, miscibility and radiation-induced crosslinking in poly(acrylic acid)-poly(vinyl alcohol) blends. *Macromol. Biosci.* **2005**, *5*, 424–432. [CrossRef]
19. Rhim, J.W.; Sohn, M.Y.; Joo, H.J.; Lee, K.H. Pervaporation separation of binary organic-aqueous liquid mixtures using crosslinked PVA membranes. I. Characterization of the reaction between PVA and PAA. *J. Appl. Polym. Sci.* **1993**, *50*, 679–684. [CrossRef]
20. Blasques Bueno, V.; Bentini, R.; Catalani, L.H.; Freitas Siqueira Petri, D. Synthesis and swelling behavior of xanthan-based hydrogels. *Carbohydr. Polym.* **2013**, *92*, 1091–1099. [CrossRef]
21. Azeredo, H.M.C.; Waldron, K.W. Crosslinking in polysaccharide and protein films and coatings for food contact—A review. *Trends Food Sci. Technol.* **2016**, *52*, 109–122. [CrossRef]
22. Zoldners, J.; Kiseleva, T. Modification of hemicelluloses with polycarboxylic acids. *Holzforschung* **2013**, *67*, 567–571. [CrossRef]
23. Cegnar, M.; Kerč, J. Self-assembled polyelectrolyte nanocomplexes of alginate, chitosan and ovalbumin. *Acta Chim. Slov.* **2010**, *57*, 431–441. [PubMed]
24. Wing, R.E. Starch citrate: Preparation and ion exchange properties. *Starch/Stärke* **1996**, *48*, 275–279. [CrossRef]
25. Shen, L.; Xu, H.; Kong, L.; Yang, Y. Non-toxic crosslinking of starch using polycarboxylic acids: Kinetic study and quantitative correlation of mechanical properties and crosslinking degrees. *J. Polym. Environ.* **2015**, *23*, 588–594. [CrossRef]
26. Solier, Y.N.; Sznaider, F.; Navarro, D.A.; Mocchiutti, P.; Inalbon, M.C. Sulfation of poplar wood xylan using a deep eutectic solvent: Effects of the degree of substitution on the properties of biobased hydrogels. *Cellulose* **2024**, *31*, 1189–1203. [CrossRef]
27. Haghighi, H.; Gullo, M.; La China, S.; Pfeifer, F.; Siesler, H.W.; Licciardello, F.; Pulvirenti, A. Characterization of bio-nanocomposite films based on gelatin/polyvinyl alcohol blend reinforced with bacterial cellulose nanowhiskers for food packaging applications. *Food Hydrocoll.* **2021**, *113*, 106454. [CrossRef]
28. *ASTM D 882-02*; Standard Test Method for Tensile Properties of Thin Plastic Sheeting. ASTM: West Conshohocken, PA, USA, 2002.
29. *ISO 3781*; Paper and Board–Determination of Tensile Strength after Immersion in Water. ISO: Genève, Switzerland, 1983.
30. Allasia, M.; Aguirre, M.; Gugliotta, L.M.; Minari, R.J.; Leiza, J.R. High biobased content waterborne latexes stabilized with casein. *Prog. Org. Coat.* **2022**, *168*, 106870. [CrossRef]
31. Schnell, C.N.; Galván, M.V.; Zanuttini, M.A.; Mocchiutti, P. Hydrogels from xylan/chitosan complexes for the controlled release of diclofenac sodium. *Cellulose* **2020**, *27*, 1465–1481. [CrossRef]

32. Krukowski, S.; Karasiewicz, M.; Kolodziejski, W. Convenient UV-spectrophotometric determination of citrates in aqueous solutions with applications in the pharmaceutical analysis of oral electrolyte formulations. *J. Food Drug Anal.* **2017**, *25*, 717–722. [CrossRef]
33. Yuyu, E.; Qi, Q.; Chang, Z.; Jiang, J.; Yao, X.; Li, P.; Lei, F.; Wang, K. Fabrication of pH-sensitive galactomannan/glycerol bio-composite films for food packaging applications. *React. Funct. Polym.* **2022**, *181*, 105465. [CrossRef]
34. Socrates, G. *Infrared and Raman Characteristic Group Frequencies: Tables and Charts*, 3rd ed.; Wiley: New York, NY, USA, 2001.
35. Synytsya, A.; Copíková, J.; Matejka, P.; Machovic, V. Fourier transform Raman and infrared spectroscopy of pectins. *Carbohydr. Polym.* **2003**, *54*, 97–106. [CrossRef]
36. Xu, J.; Xia, R.; Yuan, T.; Sun, R. Use of xylooligosaccharides (XOS) in hemicelluloses/chitosan-based films reinforced by cellulose nanofiber: Effect on physicochemical properties. *Food Chem.* **2019**, *298*, 125041. [CrossRef] [PubMed]
37. Sun, R.C.; Tomkinson, J.; Ma, P.L.; Liang, S.F. Comparative study of hemicelluloses from rice straw by alkali and hydrogen peroxide treatments. *Carbohydr. Polym.* **2000**, *42*, 111–122. [CrossRef]
38. Jia, C.; Chen, L.; Shao, Z.; Agarwal, U.P.; Hu, L.; Zhu, J.Y. Using a fully recyclable dicarboxylic acid for producing dispersible and thermally stable cellulose nanomaterials from different cellulosic sources. *Cellulose* **2017**, *24*, 2483–2498. [CrossRef]
39. Luo, J.; Huang, K.; Xu, Y.; Fan, Y. A comparative study of lignocellulosic nanofibrils isolated from celery using oxalic acid hydrolysis followed by sonication and mechanical fibrillation. *Cellulose* **2019**, *26*, 5237–5246. [CrossRef]
40. Wu, H.; Lei, Y.; Lu, J.; Zhu, R.; Xiao, D.; Jiao, C.; Xia, R.; Zhang, Z.; Shen, G.; Liu, Y.; et al. Effect of citric acid induced crosslinking on the structure and properties of potato starch/chitosan composite films. *Food Hydrocoll.* **2019**, *97*, 105208. [CrossRef]
41. Lipatova, I.M.; Yusova, A.A. Effect of mechanical activation on starch crosslinking with citric acid. *Int. J. Biol. Macromol.* **2021**, *185*, 688–695. [CrossRef]
42. Hernandez, M.S.; Luduena, L.N.; Flores, S.K. Citric acid, chitosan and oregano essential oil impact on physical and antimicrobial properties of cassava starch films. *Carbohydr. Polym. Technol. Appl.* **2023**, *5*, 100307. [CrossRef]
43. Sakkara, S.; Venkatesh, K.; Reddy, R.; Nagananda, G.S.; Meghwal, M.; Patil, J.H.; Reddy, N. Characterization of crosslinked macrotyloma Uniflorum (Horsegram) protein films for packaging and medical applications. *Polym. Test.* **2020**, *91*, 106794. [CrossRef]
44. Liu, L.; Chen, Z.; Wang, B.; Yang, Y. Improving wet strength of soy protein films using oxidized sucrose. *J. Appl. Polym. Sci.* **2015**, *132*, 41473. [CrossRef]
45. Li, J.; Liu, Y.; Sun, B.; Zhang, R. Improving the wet strength of hemicelluloses based composite films by citric acid crosslinking. *J. Wood Chem. Technol.* **2020**, *41*, 1–9. [CrossRef]
46. Zou, G.; Qu, J.; Zou, X. Optimization of water absorption of starch/PVA composites. *Polym. Compos.* **2007**, *28*, 674–679. [CrossRef]
47. Vandenabeele, P.; Wehling, B.; Moens, L.; Edwards, H.; De Reu, M.; Van Hooydonk, G. Analysis with micro-Raman spectroscopy of natural organic binding media and varnishes used in art. *Anal. Chim. Acta* **2000**, *407*, 261–274. [CrossRef]
48. Hernández, R.; Sacristán, J.; Mijangos, C. Sol/Gel transition of aqueous alginate solutions induced by Fe^{2+} cations. *Macromol. Chem. Phys.* **2010**, *211*, 1254–1260. [CrossRef]
49. Wang, Y.; Chen, S.; Yao, Y.; Wu, N.; Xu, M.; Yin, Z.; Zhao, Y.; Tu, Y. Effects of citric acid crosslinking on the structure and properties of ovotransferrin and chitosan composite films. *Int. J. Biol. Macromol.* **2023**, *229*, 268–281. [CrossRef] [PubMed]
50. Yong, H.; Wang, X.; Bai, R.; Miao, Z.; Zhang, X.; Liu, J. Development of antioxidant and intelligent pH-sensing packaging films by incorporating purple-fleshed sweet potato extract into chitosan matrix. *Food Hydrocoll.* **2019**, *90*, 216–224. [CrossRef]
51. Zhang, W.; Li, X.; Jiang, W. Development of antioxidant chitosan film with banana peels extract and its application as coating in maintaining the storage quality of apple. *Int. J. Biol. Macromol.* **2020**, *154*, 1205–1214. [CrossRef]
52. Kadam, A.A.; Sing, S.; Gaikwad, K.K. Chitosan based antioxidant films incorporated with pine needles (*Cedrus deodara*) extract for active food packaging applications. *Food Control* **2021**, *124*, 107877. [CrossRef]
53. Ryan, B.J.; Poduska, K.M. Roughness effects on contact angle measurements. *Am. J. Phys.* **2008**, *76*, 1074–1078. [CrossRef]
54. Jiang, S.; Qiao, C.; Liu, R.; Liu, Q.; Xu, J.; Yao, J. Structure and properties of citric acid cross-linked chitosan/poly(vinyl alcohol) composite films for food packaging applications. *Carbohydr. Polym.* **2023**, *312*, 120842. [CrossRef]
55. Guevara, M.M.S.; Saucedo-Rivalcoba, V.; Rivera-Armenta, J.L.; Elvira-Torales, L.I. Evaluation of a cross-linking agent in the preparation of films based on chitosan and pectin for food packaging applications. *Cellulose Chem. Technol.* **2022**, *56*, 1061–1070. [CrossRef]

polymers

MDPI

Review

Characterization of Polyallylamine/Polystyrene Sulfonate Polyelectrolyte Microcapsules Formed on Solid Cores: Morphology

Aleksandr L. Kim, Egor V. Musin, Yuri S. Chebykin and Sergey A. Tikhonenko *

Institute of Theoretical and Experimental Biophysics, Russian Academy of Science, Institutskaya St., 3, 142290 Puschino, Moscow Region, Russia; kimerzent@gmail.com (A.L.K.); eglork@gmail.com (E.V.M.); kobepoftruth@gmail.com (Y.S.C.)
* Correspondence: tikhonenkosa@gmail.com

Abstract: Polyelectrolyte microcapsules (PMC) based on polyallylamine and polystyrene sulfonate are utilized in various fields of human activity, including medicine, textiles, and the food industry, among others. However, characteristics such as microcapsule size, shell thickness, and pore size are not sufficiently studied and systematized, even though they determine the possibility of using microcapsules in applied tasks. The aim of this review is to identify general patterns and gaps in the study of the morphology of polyelectrolyte microcapsules obtained by the alternate adsorption of polystyrene sulfonate and polyallylamine on different solid cores. First and foremost, it was found that the morphological change in polyelectrolyte microcapsules formed on different cores exhibits a significant difference in response to varying stimuli. Factors such as ionic strength, the acidity of the medium, and temperature have different effects on the size of the microcapsules, the thickness of their shells, and the number and size of their pores. At present, the morphology of the microcapsules formed on the melamine formaldehyde core has been most studied, while the morphology of microcapsules formed on other types of cores is scarcely studied. In addition, modern methods of nanoscale system analysis will allow for an objective assessment of PMC characteristics and provide a fresh perspective on the subject of research.

Keywords: polyelectrolyte microcapsules; morphology; size of microcapsules; thickness of shell; pores of microcapsules; pH; ionic strength; temperature; PSS; PAH; solid core

1. Introduction

A polyelectrolyte microcapsule (PMC) is a modular, spherical-shaped structure that was first introduced in 1998 as a new type of nanoengineered multifunctional material [1,2]. They are prepared by the layer-by-layer deposition of oppositely charged polyelectrolytes on the surface of spherical templates (cores) ranging in size from 60 nm to several micrometers, followed by the dissolution of the core [3,4]. A schematic representation of this process is presented in Figure 1.

The step-by-step formation of the microcapsule shell allows its properties such as thickness, permeability, elasticity, stiffness, functionality, and others to be configured to meet specific requirements [5–9]. One of the most important features of capsules is their multifunctionality due to the possibility of modifying the shell with organic molecules [10], polymers [11], inorganic nanoparticles [12], carbon nanotubes [13], antibodies [14], etc. In addition, these capsules can be filled with a wide range of different molecules such as proteins/enzymes, drugs, nanoparticles, polyelectrolytes, fluorescent tags, and even bacterial spores [15–21].

The relative simplicity of their preparation and the possibility of controlling various properties will, in the long term, allow their use in many fields of human activity such as catalysis, agriculture, biotechnology, medicine, etc. [22,23]. In fact, they have great potential as containers for targeted drug delivery [24,25], the intracellular transport of small peptides [26], corrosion protection [27], biosensors [28], bioreactors [29], etc. As can be seen from the above, there are numerous studies devoted to the application of PMCs in various fields. However, the study of their physicochemical and mechanical properties, as well as the morphology of polyelectrolyte microcapsules, is necessary for the successful encapsulation of substances and the prediction of the influence of environmental conditions on the structure of microcapsules and on the encapsulated substances.

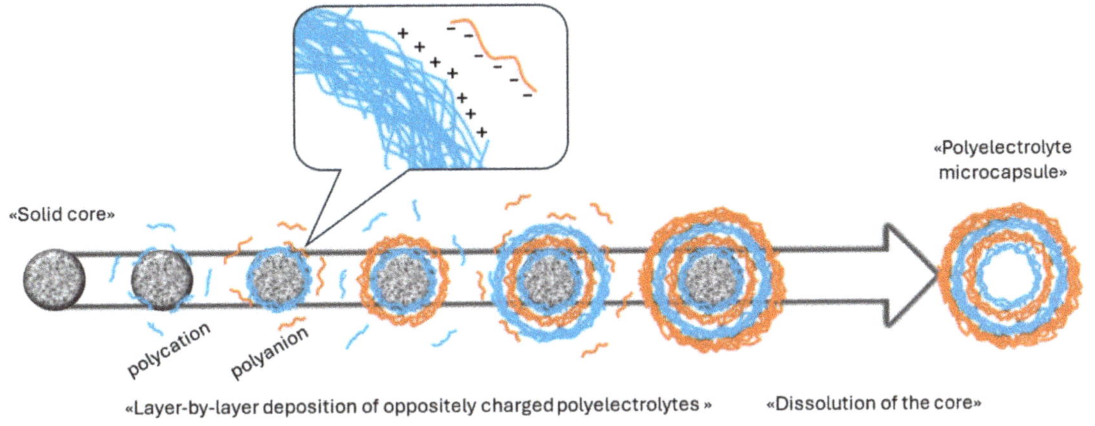

Figure 1. Scheme of the preparation of polyelectrolyte microcapsules.

In particular, the size of the microcapsules, the thickness of their shell, and the size of the pores are the characteristics that determine the possibility of using microcapsules in the given tasks. The size/diameter of the microcapsules should not exceed the minimum diameter of the channel of the system in which the microcapsules are to be used.

For example, the minimum diameter of mammalian blood vessels can vary significantly. Therefore, the size of the PMCs should not exceed the diameter of the vessels to avoid occlusion [30]. The thickness of the shell affects the deformation and rupture of the capsule shell under an external load, which is a critical parameter for ensuring the protection and/or release of encapsulated materials in delivery systems, diagnostic systems, and smart materials [31]. The size of the pores determines the ability of substances of different sizes to pass through the shell of a polyelectrolyte microcapsule, both from the external environment into the microcapsule and vice versa [32]. Furthermore, it has been demonstrated that various physicochemical factors, including ionic strength, pH, temperature, and solvent polarity, can influence the morphological parameters of polyelectrolyte microcapsules [3,33,34]. The underlying molecular mechanisms of these effects remain poorly understood, despite their critical importance, given the necessity of considering a range of incubation conditions for different applications.

At present, the morphological characteristics of polyelectrolyte microcapsules obtained by the alternate adsorption of polystyrene sulfonate and polyallylamine on solid templates are the most extensively studied, but the results of these studies are fragmented and lack coherence. Therefore, the objective of this review is to collate and organize existing studies that investigate the morphology of polyelectrolyte microcapsules obtained by the alternating adsorption of polystyrene sulfonate and polyallylamine on solid templates.

2. Size and Shape of Polyelectrolyte Microcapsules

The morphology of polyelectrolyte microcapsules, including the size and shape of the microcapsules, as well as the thickness, porosity, and density of their shell, is a key characteristic that determines the degree of capsule loading, release rate, and stability under different environmental conditions. These factors are crucial to consider when developing targeted delivery systems, diagnostic systems, sorption systems, and other applications.

It is essential to consider the size and shape of the PMC when designing polyelectrolyte microcapsules for certain applications. First of all, it is necessary to take into account the minimum channel diameter of the system in which the microcapsules will be utilized because the size/diameter of the microcapsules should not exceed this minimum diameter. For example, the minimum diameter of a human blood vessel is 8 μm, while that of a rat is 4 μm [30]. Therefore, depending on the object of study, it is necessary to create certain conditions for the formation of microcapsules of the required size. In addition, the size of PMCs significantly determines the adhesion behavior and the interaction of the capsule with environmental objects. In particular, their uptake by cells [35], efficiency of drug delivery [36], rate of removal from the bloodstream, and interaction with immune cells depend on the size of the microcapsules [37]. The shape of polyelectrolyte microcapsules is also a critical parameter as it affects the cellular processing of polyelectrolyte microcapsules [38] on the adhesion of capsules [39] and stability [40].

Therefore, depending on the object of study, specific conditions must be created for the formation of microcapsules of the required size. Moreover, the size of PMCs significantly influences the adhesion behavior and the interaction of the capsule with environmental objects. Specifically, factors such as their uptake by cells [35], the efficiency of drug delivery [31], the rate of removal from the bloodstream, and their interaction with immune cells are dependent on the size of the microcapsules [36]. The shape of polyelectrolyte microcapsules is also a crucial parameter as it affects the cellular processing of the capsules [37], their adhesion [38], and stability [39].

One of the key factors determining the size and shape of polyelectrolyte microcapsules is the type of core/template used for microcapsule formation. Meanwhile, various external factors can also influence their size, which also needs to be taken into account for their application across various fields.

2.1. Influence of the Core Type on the Size and Shape of the PMC

The size and shape of polyelectrolyte microcapsules obtained by the alternate adsorption of polystyrene sulfonate and polyallylamine are determined by the core used, as it sets the initial arrangement of the polyelectrolyte layers of PMC. Broadly speaking, cores can be classified into two types: hard and soft [8]. Soft cores are formed mainly on emulsion drops. However, capsules derived from such cores are characterized by high polydispersity in their size and low stability upon surface contact [41]. In this regard, solid cores are more commonly used for the creation of polyelectrolyte microcapsules.

The most studied and widespread PMCs are microcapsules formed on melamine formaldehyde (MF) cores [1]. Polystyrene (PS) is also frequently used [2]. However, the complete removal of these cores requires solvents that are aggressive to biomolecules (pH below 1, THF, DMF, etc.). As a result, $MnCO_3$ [6] and $CaCO_3$ [42] cores which can be completely removed from PMC under relatively mild conditions (pH $\approx$ 6, EDTA, ethylene glycol tetraacetic acid, etc.) have been widely used [34]. Based on a literature analysis, the aforementioned cores are most commonly used in PMC preparation, making microcapsules based on these cores the most studied. In particular, the influence of MF, PS, $MnCO_3$, and $CaCO_3$ cores on the shape and size of polyelectrolyte microcapsules has been described in detail. Their images, obtained by scanning electron microscopy, are presented in Figure 2 [3,34,43,44].

As observed in Figure 1, MF, PS, $MnCO_3$, and $CaCO_3$ particles exhibit a regular spherical shape, or close to it in the case of $CaCO_3$. Figure 2 presents a view of these particles, which significantly differ in surface roughness. The size of the polyelectrolyte microcapsules is dependent on the size and shape of the particle templates, which fall within the following ranges: melamine formaldehyde cores, 0.06 to 10 μm [4,45]; polystyrene cores, 0.1 to 16 μm [45,46]; $MnCO_3$, 1.85 μm to 300 μm [47]; and $CaCO_3$, 0.4 μm to 10 μm [47]. In contrast to PS and MF, $CaCO_3$ and $MnCO_3$ particles do not possess a perfect spherical shape with a smooth surface, but rather exhibit a rougher and more complex topography, which may influence the eventual shape and properties of the polyelectrolyte microcapsule. Subsequently, we will provide a detailed discussion on PMCs formed on different types of cores.

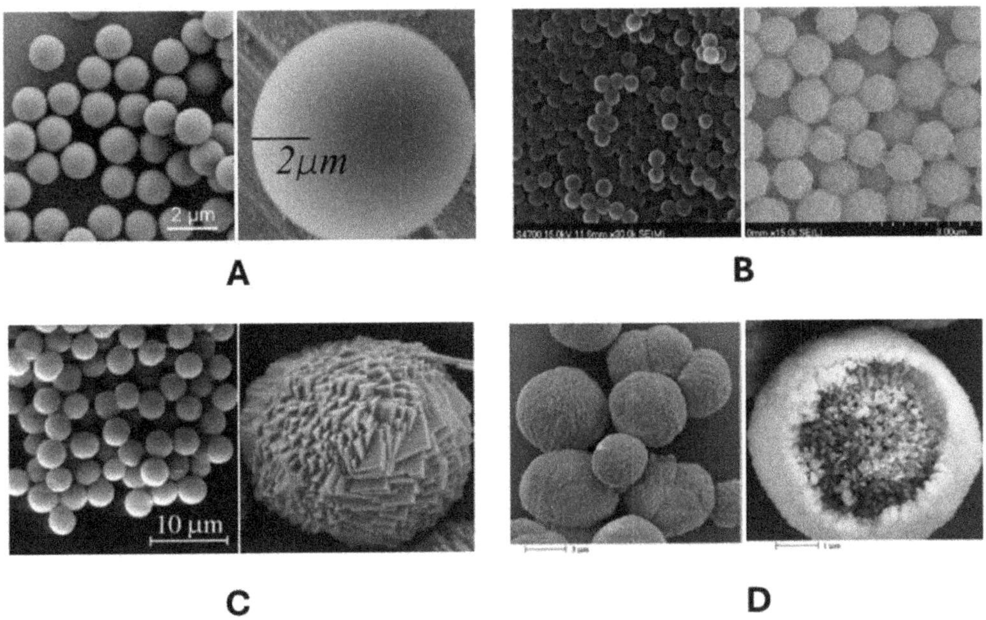

Figure 2. SEM images of (**A**) melamine formaldehyde [3,43]; (**B**) polystyrene [44]; (**C**) $MnCO_3$ [34]; and (**D**) $CaCO_3$ [34] particles. Adapted with permission from Refs. [3,34,43,44].

The size of polyelectrolyte microcapsules, formed on MF particles, can range from 60 nm to 10 μm [4]. According to Ljuljevic et al. [22], the size of the PMCs matches the size of the MF particles themselves (5.6 ± 0.2 μm), which is also confirmed in this study [1]. However, in the work of C.Gao et al. [23], the process of MF core dissolution was observed in real time using confocal laser scanning microscopy. It was shown that the capsules swell as a result of the MF core's dissolution. The initial size of the PMC was 6.4 μm, which subsequently increased to 9.1 μm. Over time, the size of the capsule returned to its original state. The authors found that MF oligomers are too large to easily diffuse out of the capsules. The temporarily high concentration of dissolved MF oligomers inside the capsules creates osmotic pressure. This situation leads to the swelling of the capsules and exerts great mechanical stress on the polyelectrolyte film of the PMC shell. Moreover, this effect significantly increases with the increase in the core diameter [45].

PMCs formed on polystyrene cores exhibit similar characteristics. In the works of Christophe De'jugnat*, Gleb B. Sukhorukov, Heuvingh, and J. et al., it was observed that the size of PMCs did not differ from the size of the core after its dissolution [5,46,48]. Similar to the case with MF cores, the size of the PMC briefly increased before returning to its previous size. The reason for this increase is the same as with MF cores—the dissolution products of the polystyrene core are long polymer chains (Mw about 106 Da), which results in increased osmotic pressure inside the PMC.

In the case of $MnCO_3$ serving as the core, the size and shape of the PMC coincide with the size and shape of the template [6]. However, similar to the previously described studies, no research has been conducted on the change in the microcapsule size during the dissolution of the core. Furthermore, comparative studies of polyelectrolyte microcapsules formed on melamine formaldehyde (MF) and manganese carbonate ($MnCO_3$) particles, and on polystyrene (PS) and $MnCO_3$ particles, were carried out in the works of Sukhorukov et al. [6] and Déjugnat, C. and Sukhorukov, G.B. [48]. According to the results of these studies, no significant difference was found in the morphology of the shells based on MF, PS, and $MnCO_3$ particles. The size and shape of the prepared polyelectrolyte microcapsules correspond to the size and shape of the original templates.

However, the PMCs formed on a $CaCO_3$ core exhibit significant differences. In the work of Volodkin et al. [42], $CaCO_3$ particles of an approximately spherical shape, ranging from 4 to 6 μm in size and covered with a polyelectrolyte shell, were presented. After the core's removal, the size and shape of the PMCs were preserved without a short-term increase in size. The authors suggested that the polyelectrolyte complex formed on the surface/inside of the capsule creates a strong yet porous network, protecting the capsule from high osmotic pressure and allowing Ca^{2+} ions to rapidly penetrate through the pores without causing destruction. The size of these pores can reach several tens of nanometers, aiding in maintaining an osmotic equilibrium. Additionally, the study [49] presented cross-sections of polyelectrolyte microcapsules formed on $CaCO_3$ particles containing protein and on protein-free $CaCO_3$ particles. The protein-free polyelectrolyte particles are complex internal structures consisting of multiple filamentous and enclosed nanoelements of a polyelectrolyte nature with a thickness of 20–30 nm. PMCs containing six to eight polyelectrolyte layers lack an outer shell. This structure is presumably due to the surface morphology of $CaCO_3$ microparticles being very rough and comprising a large number of combined carbonate nanoparticles. The internal structure of $CaCO_3$ microparticles is channel-like, with pore sizes in the range of tens of nanometers [42]. Polyelectrolyte microcapsules, which have more than nine polyelectrolyte layers, possess a formed shell [42], which confines the internal space of the microcapsule. As in the case of PMCs with fewer layers, this internal space is a polyelectrolyte complex with a complex channel-like structure.

In instances where the $CaCO_3$ core contains protein molecules [49], polyelectrolytes are located only in the near-surface layer. The outer, spatially organized shell confines the inner volume, which is filled with a protein solution. This effect arises due to the "channels" of $CaCO_3$ particles being filled with protein molecules [50], resulting in the adsorption of the polyelectrolyte on the outer surface of the spherulite.

Less commonly used templates for polyelectrolyte microcapsule formation include $CdCO_3$ [34], SiO_2 [51], and crystalline forms of substances that require encapsulation, such as fluorescein [52] and ibuprofen [19], among others.

The morphology of $CdCO_3$ differs significantly from $CaCO_3$ and $MnCO_3$. The surface of $CdCO_3$ is close to smooth while the shape of the core is close to cubic [34]. At the same time, no increase in the size of $CdCO_3$ is observed when the core dissolves, while the cubic shape is preserved after its removal.

The size of PMCs formed on SiO_2 can vary from 0.03 to 100 μm [45], depending on the particle size. It can be presumed that the PMCs obtained on this template may increase in size after the removal of the core. In particular, Yang et al. demonstrated an almost twofold increase in the size of microcapsules after core removal [51]. However, due to the extremely small size of the SiO_2 particles (200 nm) used in this study, limitations are imposed by the diffraction limit of resolution for optical microscopy, making this method unsuitable for the direct size estimation of PMCs. Therefore, transmission electron microscopy, which involves heating the sample during sample preparation, was used for size determination in the work of Yang et al. This resulted in the loss of about 65% of the polyelectrolyte mass.

From what has been written above, it can be concluded that the type of core used and its size are fundamental factors which the size and shape of polyelectrolyte microcapsules depend on. However, the size and shape of PMCs can be modified by external factors such as the pH of the medium, the ionic strength of the solution, and the incubation temperature, as well as by increasing the number of shell layers.

2.2. Influence of the Number of Layers on the Size and Shape of the PMC

The most obvious way to change the size of polyelectrolyte microcapsules is to increase the number of polyelectrolyte layers in the shell, as each subsequent layer of polyelectrolyte increases the thickness of the shell and, consequently, the size of the PMC. However, the addition of one layer increases the size of the PMC by a maximum of 20 nm. In several studies, it has been confirmed that the size of PMCs formed on PS and MF cores remains largely unchanged by the number of layers of the PMC shell [45,48]. A similar pattern is observed in the case of $MnCO_3$ as a core [48,53].

2.3. Influence of the pH of the Medium on the Size and Shape of the PMC

Another factor that can influence the size and shape of polyelectrolyte microcapsules is the pH of the solution.

In the work of Christophe De'jugnat* and Gleb B. Sukhorukov, it was shown that PMCs on a PS template increase in size in an alkaline medium, while the capsule size decreases when PMCs are incubated at low pH values [48]. The increase in size is presumably due to the deprotonation of the ammonium groups of PAH under alkaline conditions, which results in an increase in the size of PMCs due to the mutual repulsion of PSS. The decrease in the microcapsule size is presumably due to a greater number of PAH amino groups becoming positively charged, which enhances the interaction with the anionic sulfonate groups of PSS. As a result, the polymer chains shrink, leading to smaller capsules.

However, the change in the size of PMCs at different pH values depends on the number of shell layers. In an acidic environment (pH $\approx$ 1), PMCs with fewer layers become significantly smaller than microcapsules with more layers. This is presumably because capsules with more layers have a stiffer shell, which prevents them from shrinking [46]. However, in an alkaline medium, the authors observed a different result with a similar nature of the phenomenon. Microcapsules containing 10 and 12 layers decrease in size in an alkaline medium by almost twofold (from pH 11.5), while if there are 14 and 16 layers, they conversely increase in size by twofold [48]. Capsules with fewer layers lack sufficient stiffness to resist shrinkage, while PMCs with more layers have a rigid shell [46]. However, this does not explain the effect where the PMC shell becomes smaller than the original core. Nevertheless, the authors conducted an experiment in which PMCs in an alkaline medium (pH > 11) with varying numbers of layers were incubated in 0.5 M NaCl. The authors hypothesized that the addition of salt would increase the shell's flexibility, causing the PMCs to decrease in size. Indeed, according to the work of Heuvingh et al. [46], the capsule shell exhibits a pronounced softening with the increasing ionic strength of the solution, but only at salt concentrations from 3 M NaCl. Thus, it can be inferred that the increase in the softness and size of PMCs at 0.5 M NaCl may be due to the synergy of the effects of ionic strength and an alkaline medium. As a result, it can be inferred that the PMC shell becomes smaller than the original core due to the enhanced effect of ionic force on the PMC shell under alkaline conditions.

In the same paper, Christophe De'jugnat* and Gleb B. Sukhorukov [48] showed that PMCs formed on the $MnCO_3$ core (3.6 μm) increase in size by a factor of three in an alkaline medium (pH 12.5). Moreover, this effect was observed regardless of the number of polyelectrolyte layers. The authors attributed this effect to the deprotonation of ammonium groups of PAH at an increasing pH. As a result, the polyelectrolyte shell is dominated by a negative charge due to PSS, which increases the electrostatic repulsion of polyelectrolytes and increases the mutual distance, leading to an increase in the size of the PMC and eventually to the complete destruction of the shell.

PMCs formed on MF or $CaCO_3$ particles do not change their size at different pH values. This effect was demonstrated in the work of Gleb B. Sukhorukov et al. [54] where confocal microscopy images show that the size of PMCs at pH 10 and 3 does not differ from the size of the MF particles on which the microcapsules are formed. In the work of Changyou Gao et al. [23], the dissolution of the MF core at pH 1.1 was studied, and it was demonstrated that the particle size does not change during the entire time of core dissolution and the removal of its degradation products. A similar effect was demonstrated using $CaCO_3$ as a template. In the work of Tong et al. [55], it was shown that the size of PMCs remains the same at both pH 2.5 and 7, which was also confirmed in Ref. [56]. However, at a pH above 12.5, capsule dissolution was observed within 3 s [46]. Meanwhile, at low pH values (below 1), the dissolution of capsules was also observed over time, over a period of 30 min [57].

Summarizing the above, it can be asserted that the size of polyelectrolyte microcapsules changes depending on the type of core and the pH of the solution: specifically, PMCs formed on MF and $CaCO_3$ cores do not change their size at different pH values of the medium, and PMCs formed on PS cores decrease in size in an acidic medium and increase in size in an alkaline medium. In the case of PMCs formed on $MnCO_3$, the size of the capsules increases in an alkaline medium; however, there is currently no data on the impact of low pH values on the size of such PMCs, which requires additional research in this area.

2.4. *Effect of Ionic Strength on the Size and Shape of the PMC*

In the work of Heuvingh et al., it was demonstrated that with an increasing NaCl concentration in the solution, a decrease in the size of PMCs formed on the PS core was observed [46]. A significant difference in size was noted at salt concentrations of 3 M and higher, with a maximum reduction of 20%. It was also shown that the incubation time of PMCs in a medium with a high ionic strength does not affect the size change. After incubating the PMCs in a salt solution for 30 min, the size stabilizes, and even if the PMCs are incubated in a 5 M NaCl solution for 3 weeks, the size remains the same as when incubated for 30 min. Additionally, this process is irreversible. Re-incubating PMCs in pure water maintains the size that was present when incubated in salt. The authors suggested that the addition of salt leads to a softening of the ionic interactions between the polyelectrolytes forming the multilayer structure, thereby allowing the molecular rearrangement and annealing of the capsule structure. Furthermore, the authors have shown that the salt anion, rather than the cation, has a greater impact on the size change of the PMC. While the $SO_4{}^{2-}$ ion has a much smaller effect (11% reduction) than Cl^-, NO^{3-} has a stronger effect on the morphology change (38% reduction). This effect of reducing the size of PMCs is consistent with the position of anions in the Hofmeister series, in which ions are classified according to their lyotropic properties, i.e., their ability to salt in or salt out proteins. Given that some amino acids are characterized by electrostatic interactions in proteins similar to the inter/intra polyelectrolyte interaction, it can be assumed that the structural influence of anions of the Hofmeister series on the polyelectrolyte complex of the PMC shell will have a similar character. This also supports the assumption of a decrease in the size of the PMC with an increasing salt concentration due to changes in the structure of the multilayer shell.

Similar observations on the effect of different salts on the morphology of polyelectrolyte capsules were independently made by Georgieva et al. for the PAH/PSS system formed on the MF core [58]. However, the concentration of salts used was 1.7 M, which does not allow us to draw a definite conclusion about the influence of NaCl and ionic strength on the size of PMCs since at this concentration, the decrease in the size of PMCs is only 2%. At the same time, in the work of Lebedeva et al. [59], it was shown that the size of PMCs formed on the MF core does not change at NaCl concentrations from 0 to 5 M. Thus, we can conclude that the effect of salt on the size of PMCs formed on the MF and PS core is different.

When considering PMCs formed on the MF core containing PSS, there was also a decrease in the size of the PMCs observed with the increasing ionic strength, but this decrease differed from that of microcapsules formed on the PS core. In particular, Lebedeva et al. showed [59] that PMCs formed on the MF core, filled with PSS, significantly increase in size compared to "empty" PMCs. This is due to the high osmotic pressure caused by the encapsulated PSS [60]. Increasing the concentration of NaCl in the solution leads to a decrease in the diameter of these PMCs up to a concentration of 3 M, after which the size of the microcapsules does not change. However, even in a 5 M NaCl solution, the size of the "filled" PMCs does not reach the size of the "empty" capsules. Presumably, this effect is due to the excessive osmotic pressure caused by the counterions of the encapsulated PSS.

The behavior of PMCs formed on $CaCO_3$ cores in solutions with different ionic strengths is significantly different compared to the PMCs described above. In the work of Pechenkin et al., it was shown that PMCs morphologically do not change in the presence of NaCl salt in solution, even at a concentration of 6.1 M NaCl [57]. Thus, it can be assumed that the decrease in the size of PMCs obtained on an MF or PS core is due not only to the rearrangement of the shell, but also to the influence of external osmotic pressure on the PMC shell, which leads to its reduction. Meanwhile, in PMCs formed on a $CaCO_3$ core, this shell is absent and as a result, there is no change in size in the presence of salt. Another possible explanation could be differences in the shape of the shell; the matrix structure of the capsules formed on $CaCO_3$ may have sufficient stiffness to resist shrinkage.

2.5. Influence of the Temperature of the Medium on the Size and Shape of the PMC

Kim et al. suggested that the PMCs of a PSS/PAH composition formed on MF cores decreased slightly (less than 10%) when incubated at 80 °C for 5 h, which is comparable to the measurement error [61]. However, Leporatti et al. showed that polyelectrolyte capsules of the composition (PSS/PAH) decrease in size upon heating [62]. In addition, the authors studied the effect of temperature on the size of PMCs at different MF particle sizes, with different outer layers of PSS or PAH, and in the presence or absence of heating the acid solution used to remove the core. As a result, it was shown that the parameters described above did not significantly affect the reduction of the PMCs, finding that, on average, the PMCs were reduced by approximately 15% at 70 °C. This reduction caused by heating was attributed to the transition to a more coiled arrangement of polymers in the multilayers due to the breaking and reforming of ion pairs, resulting in a more compact arrangement of polyelectrolyte molecules. Prior to heating, the size and shape of the capsule were maintained by charge bonding, and the breaking of this bond upon heating caused shrinkage. A similar effect was observed by Kohler et al. at a higher temperature of 120 °C [3], whereby the PMCs decreased by more than 50 percent.

At present, there are no studies investigating the size of PMCs formed on the PS cores at different incubation temperatures. However, a number of studies have applied thermal effects on this type of microcapsule in order to achieve applications [63,64]. Based on the results of these studies, it can be inferred that changing the temperature does not significantly affect the size of the microcapsules. However, these studies do not directly address the effect of temperature, which limits the ability to draw specific conclusions regarding this aspect. Consequently, further studies specifically aimed at investigating the effect of temperature on the microcapsule size are needed for a more complete understanding.

In the case of PMCs formed on $CaCO_3$, the work of A. V. Dubrovskii et al. found that the capsules decrease in size with an increasing temperature and heating time, regardless of whether they consist of an even or odd number of layers [65]. When microcapsules were incubated for 20 min at 90 °C, a decrease in size was observed. The PMCs with an even number of layers ($(PSS/PAH)_{6/8/10}$) decreased by 28% and those with an odd number of ($(PSS/PAH)_{7/9}$) layers decreased by 24%, respectively. In the case of polyelectrolyte microcapsules formed on $CaCO_3$ + protein spherulites, a decrease in capsule size was also observed. However, in this case, the influence of the parity/oddity of the polymer microcapsule layers had an opposite effect. Microcapsules with an even number of layers ($(PSS/PAH)_{6/8/10}$) reduced by 26% and those with an odd number of layers ($(PSS/PAH)_{7/9}$) reduced by 34%, respectively. It should be noted that in these types of PMCs, unlike microcapsules formed on $CaCO_3$ spherulites without protein, polyelectrolytes are located only in the capsule shell, limiting its internal volume filled with protein solution, although some of the protein may be bound to the shell itself. This difference is probably due to the presence of the shell in these capsules, in contrast to the matrix arrangement of polyelectrolytes throughout the internal volume of capsules formed on a $CaCO_3$ core.

According to the above description, it can be concluded that polyelectrolyte microcapsules formed on different types of cores significantly differ in size changes in response to different physicochemical stimuli. When MF and $CaCO_3$ were used as cores, polyelectrolyte microcapsules did not change their size at different pH values of the medium, whereas PMCs formed on PS cores decreased in an acidic medium and increased in an alkaline medium with subsequent disintegration (which is also characteristic of PMCs on $MnCO_3$). Given that, morphologically, PMCs formed on MF and PS cores do not differ, we can hypothesize that the reason for the difference in the size change of PMCs and the pH of the medium lies in the process associated with the removal of the cores. In particular, the work of Christophe De'jugnat* and Gleb B. Sukhorukov [48] has already demonstrated the influence of tetrahydrofuran, a solvent used to remove the templates of PMCs formed on PS cores, on the stability of PMCs formed on $MnCO_3$. After 15 h of incubation in this solvent, the shell stability of PMCs formed on $MnCO_3$ was significantly increased in an alkaline medium. In addition, oligomers formed as a result of the incomplete dissolution of the MF cores may remain in the PMC shell, which may affect its mechanical properties and, as a consequence, the size change under the influence of the solution's pH [23].

Also, the ionic strength of the solution does not affect the size of PMCs formed on MF and $CaCO_3$ cores, but most likely the mechanisms of the lack of influence may be different. In the case of using PS cores, the PMCs decrease at ionic strengths above 3 M. The differences in PMCs obtained on MF and PS cores can be explained in the same way, as in the case of the influence of the solution's pH. However, the nature of this phenomenon may be different and requires separate study.

In the case of temperature treatment, the polyelectrolyte microcapsules formed on MF cores and $CaCO_3$ cores are characterized by a decrease in size, which is also characteristic of microcapsules obtained on PS cores; however, due to the lack of work aimed at studying this particular effect, we cannot state this unequivocally. Nevertheless, we can observe a general tendency to decrease the size of all presented types of capsules during temperature treatment, which may be directly related to the interpolyelectrolyte interaction between polyallylamine and polystyrene sulfonate, in which polyelectrolytes are arranged more compactly relative to each other.

3. Thickness of the Shell of Polyelectrolyte Microcapsules

The thickness of the PMC shell is one of the most important parameters of microcapsule morphology because the thickness of the shell determines the deformation and rupture of the capsule shell under the action of external load, which is extremely important to ensure the protection and/or release of encapsulated materials. In particular, the thickness of the shell of the PMC primarily affects the stability of the microcapsule during core disruption. If the shell is thinner than a certain value, it will be destroyed by osmotic pressure. In addition, depending on the application and the physicochemical or mechanical conditions, capsule disruption may be desirable or preventable. Therefore, an understanding of the conditions that affect the thickness of the PMC shell is essential in the design of PMCs.

3.1. *Influence of the Number of Layers on the Thickness of the PMC's Shell*

The number of layers of the PMC shell that adsorb onto the core directly affects its thickness. The literature data are presented in Table 1.

Table 1 shows that the shell thickness at the same number of layers does not differ significantly in the case of using MF, $MnCO_3$, or PS as a core. Moreover, the shell thickness of the PMC increases linearly with the increasing number of polyelectrolyte layers, with a thickness increment of 3–3.8 nm per layer. It is noted in Dong et al. that the minimum thickness of the PMC shell (formed on the MF core) should not be less than 10 nm because it will lead to its destruction under the action of osmotic pressure in the process of removing the core from the microcapsule [60]. Considering that PMCs formed on MF, $MnCO_3$, and PS cores have an extremely similar morphology and level of shell deformation under the action of osmotic pressure as a result of the removal of these cores [5,6,48,66], it can be assumed that the behavior of these capsules will be similar for shell thicknesses less than 10 nm.

Table 1. PMC's shell thickness depending on the type of core.

Number of Layers	Types of Core									
	MF [60]		PS [48]		$CaCO_3$ [42,50]		$CaCO_3$ + Protein [42,50]		$MnCO_3$ [48]	
	Shell Thickness, nm	Estimated Thickness of One Layer, nm	Shell Thickness, nm	Estimated Thickness of One Layer, nm	Shell Thickness, nm	Estimated Thickness of One Layer, nm	Shell Thickness, nm	Estimated Thickness of One Layer, nm	Shell Thickness, nm	Estimated Thickness of One Layer, nm
6	-	-	-	-	No shell		36.92± 3.28	6.15 ± 0.5	-	-
8	20 ± 2	2.5 ± 0.25	-	-	No shell		58.43 ± 2.82	7.3 ± 0.3	-	-
10	32 ± 3	3.2 ± 0.4	33 ± 2	3.3 ± 0.2	41	4.1	61.54 ± 3.21	6.15 ± 0.3	-	-
12	42 ± 1	3.5 ± 0.1	44.4 ± 2.4	3.7 ± 0.2	-	-	-	-	45.8 ± 4.7	3.8 ± 0.4
14	47 ± 2	3.35 ± 0.15	47.6 ± 2.7	3.4 ± 0.2	-	-	-	-	-	-
16	59 ± 3	3.69 ± 0.2	62.4 ± 3.1	3.9 ± 0.2	-	-	-	-	47.3 ± 2.9	3.5 ± 0.3
18	66 ± 1	3.67 ± 0.05	-	-	-	-	-	-	-	-

Polyelectrolyte microcapsules formed on $CaCO_3$ cores differ significantly. PMCs containing six to eight polyelectrolyte layers are devoid of an outer shell. Such a structure was presumably formed due to the complex internal channel-like organization of $CaCO_3$ microparticles [42]. PMCs that have more than nine polyelectrolyte layers possess a shell that is about 46 nm thick [49]. However, Volodkin et al., in their study, measured the thickness of the polyelectrolyte complex that is formed on the $CaCO_3$ core (without removing the core). The thickness of the PMC shell on $CaCO_3$ is (six layers) 95 nm, with a single layer thickness of 8 nm, which is 3–5 times larger than in the case of PMCs on smooth surfaces [42]. The authors suggested that polyelectrolyte adsorption on the very rough surface of the $CaCO_3$ microcore favors the formation of such a thick and porous polyelectrolyte network for the capsule wall. Obviously, the polyelectrolyte complex inside the capsule also has a significant share in the increase in thickness, but it is not possible to assess this. However, it is worth noting that when the cores are removed, there are many processes that affect the thickness of the shell, which does not allow us to compare PMCs with a removed core and with a non-removed core.

In the case of PMC formation on $CaCO_3$ particles containing protein, the microcapsules form a pronounced shell. The authors have shown that the thickness of the shell increases with the increasing number of polyelectrolyte layers [49]. Six-layer microcapsules possess a fragmented outer shell with a thickness of 31–46 nm, while in seven-layer microcapsules, the thickness ranges from 43 to 49 nm. Starting from eight polyelectrolyte layers, the shell has a constant thickness of 60–70 nm.

3.2. *Influence of pH, Ionic Strength, and Temperature of the Medium on the Thickness of the PMC's Shell*

As described above in Section 2.1, the thickness of the PMC shell, as well as its size, can be influenced by the pH environment, ionic strength, and temperature of the medium. However, there are currently no studies on the influence of the pH environment, ionic strength, and temperature of the medium on the thickness of the PMC shell formed exclusively on the MF core.

Georgieva et al. conducted a study on the impact of the solution's ionic strength on the shell thickness of PMCs that were formed on the MF core. The findings demonstrated a 30–40% increase in shell thickness, as a function of salt (LiCl, NaCl, KCl, and CsCl), given a constant ionic strength (I = 1.67) [58]. In contrast, research by Lebedeva et al. indicated that the shell thickness of PMC increases by 10–15% with an increase in the NaCl concentration up to 2.5 M [59]. However, any further increase in the salt concentration results in a 40% reduction in the thickness of the PMC shell. It is postulated that these effects occur because, at a NaCl concentration up to 2.5 M, PSS and PAH are shielded from each other by the electrolyte's counterions. Moreover, a substantial increase in the salt concentration leads to a significant rise in the dissociation of polyelectrolytes from the shell.

The ionic strength is one of the main factors of the shell thickness not only after core removal, but also in the process of shell formation around the template. In particular, Dong et al. showed that the ionic strength at which the PMC shell is formed, prior to core dissolution, also affects the thickness of the wall of the formed capsule [60]. In particular, at the lower salt concentration of 0.25 M, the shell thickness is 1.5–2 times thinner compared to the standard concentration of 0.5 M. However, it is worth noting that the minimum thickness at which the PMC shell is stable is the same in both conditions and is 10 nm.

Studies devoted directly to the influence of the pH of the medium on the shell thickness of PMCs formed on MF cores are absent in the literature. However, Georgieva et al. studied the effect of 1.67 M Na_2HPO_4 (I = 10 M) at pH 8.72 and 1.67 M NaH_2PO_4 (I = 10 M) at pH 3.92 on the shell thickness of such PMCs. As can be seen, the concentrations of the salts correspond to the same ionic strength, but the pH of the solution differs. As a result, an increase of 30% was shown at pH 8.72, which corresponds to an increase in shell thickness from the ionic strength of the solution. However, at the same ionic strength but at pH 3.92, the shell thickness of PMCs increases by only 13% [58]. Accordingly, it can be assumed that the smaller increase in the shell thickness can be attributed to the low pH of the solution, which is associated with an increase in the $[H^+]$ concentration, which leads to an increase in the number of charged amino groups of PAH and, consequently, increases the interaction with PSS.

The influence of temperature on the thickness of the PMC shell, as presented in the literature, is exclusively based on the example of microcapsules formed on MF cores. Estrela-Lopis et al., for instance, demonstrated that maintaining a temperature of 70 °C for 4 h resulted in an approximately 38% increase in the thickness of the capsule wall [67]. This finding was corroborated by Leporetti et al., who reported a 32% increase in shell thickness [62]. An even more significant effect was demonstrated by Kohler et al., whereby PMCs incubated at 120 °C showed an initial shell thickness of 14.4 ± 1.4 nm, which increased more than tenfold to 160 ± 12 nm after heating [3]. The authors proposed that the thickening of the capsule wall was due to a rearrangement of the macromolecular layer constituents. This process would require the disruption and recombination of ionic pairs formed between charged groups of polyanions and polycations, which are crucial for layer stability. Consequently, the polymers that form the layer would adopt a conformation that is more entropically favorable.

4. Porosity of the Shell of Polyelectrolyte Microcapsules

Nanoscale pores/holes primarily determine the permeability of the shell of polyelectrolyte microcapsules. Currently, there are no definitive data confirming the existence of pores or other openings in the PMC shell. Typically, electron microscopy methods, which require PMC preparation through heating to temperatures above 300 °C or the use of organic solvents, are utilized to measure their sizes. This preparation process may lead to the formation of breakage. However, it is widely accepted that a decrease in the polyelectrolyte interactions of the PMC shell due to exposure to varying environmental conditions could result in the formation of pores/holes. These pores/holes could enable the penetration of high-molecular-weight substances.

The shell structure of polyelectrolyte microcapsules formed on an MF core varies with the number of layers and shell thickness. For a thickness of 12 to 16 nm, a mesoporous PMC shell structure with a large pore size was observed by Dong et al. Conversely, for shell thicknesses ranging from 16 to 28 nm, the shell structure is microporous [60]. 'Mesoporous' capsules have an average pore size of 2 to 50 nm, presumably formed by the breakthrough release of MF oligomers with an average size of 2–3 nm. The 'microporous' PMCs have an average pore size of less than 2 nm, which is likely the result of the trapping of MF oligomers by the multilayer shell obtained during core removal. According to previous studies, up to 20% (by weight) of the released MF remains in the shell [68].

For PMCs formed on a $CaCO_3$ core, the lack of a distinct shell complicates the description of pore presence in this structure. Porous $CaCO_3$ microparticles, measuring 4.5–5.0 µm, have a surface area of 8.8 m^2/g and an average pore size of 35 nm [42]. Approximately 59% of the particle volume consists of solid $CaCO_3$, while around 41% comprises the internal pore volume. This structure allows small molecules of PSS and PAH, as small as a few nanometres in size, to penetrate the interior through the pores during successive polyelectrolyte treatment, thereby forming a polyelectrolyte complex. It is suggested that the channel-like structure of the $CaCO_3$ microcore, with its substantial internal volume, promotes the formation of a dense and porous polyelectrolyte network in the capsule wall, featuring numerous holes larger than 100 nm.

Macroscopic properties such as permeability, elasticity, etc., are strongly dependent on the environment: the temperature, solvent, ionic strength, and pH of the solution. Among others, the pore size of the shell of polyelectrolyte microcapsules depends on these parameters.

4.1. *Influence of pH of the Medium on the Porosity of the PMC's Shell*

In particular, Antipov et al. studied the influence of the acidity of the medium on the permeability of the shell of PMCs formed on the MF core [15]. However, the authors also demonstrated the influence of the acidity of the medium on the pore size of the microcapsule shell. The obtained images are presented in Figure 3 [15].

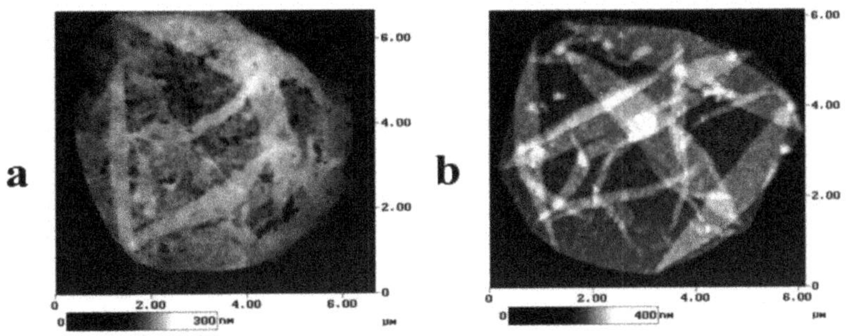

Figure 3. SFM images of polyelectrolyte microcapsules: (**a**)—pH 3.5; (**b**)—pH 10.5 [15]. Adapted with permission from Ref. [15].

Figure 3 shows that when the microcapsules are incubated in the pH 3.5 solution, holes as small as 100 nm are formed. In addition, the authors proved the reversibility of this process. When the same microcapsules are subsequently incubated in pH 10.5 (as well as at pH 12), these holes are absent.

However, the authors suggest that the presence of holes in the SFM preparation (dried during preparation) does not necessarily imply their presence in the capsules in the solution. It is likely that at a low pH, damage to the shell walls occurs, potentially leading to the formation of holes during drying due to layer shrinkage. This clearly indicates a weakening of the polyelectrolyte bond, which could increase permeability [15]. Moreover, numerous studies have explored the encapsulation of substances into this type of capsules by altering the solution's acidity. For instance, Sukhorukov et al. incorporated FITC-dextran 75 kDa into PMC at pH 3. Above pH 8, FITC-dextran could not penetrate the capsule shell [54]. Similarly, Tiurina et al. encapsulated the proteolytic enzyme, achymotrypsin, in hollow PSS/PAH shells. The encapsulation was conducted at pH 4.0, where, according to the authors, pores formed in the capsule walls. Subsequent washing at pH 8.0 sealed the shell walls [69].

Halozan et al. demonstrated that $(PAH/PSS)_8$ microcapsules, prepared on PS templates, were sensitive to a pH shift to the alkaline region [70]. Above pH 11.2, the polyelectrolyte network in the microcapsule wall undergoes changes in its structure and porosity.

This pH region may correspond to the formation of pores in the wall due to the segregation of the polymer network, which is associated with the significant deprotonation of PAH. It is well established that the degree of protonation of a weak polyelectrolyte is highly dependent on its environment, particularly on the nature of the polyelectrolyte with which it is bound. Above pH 11, the degree of protonation of the PAH-PSS complex is virtually zero [33]. The fully ionized and negatively charged PSS causes the capsule wall to swell due to repulsion between its charged monomeric groups. This mechanism differs from the encapsulation mechanism at an acidic pH in (PSS/PAH) microcapsules templated on melamine formaldehyde particles [15]. As the pH decreases, the charge density on PAH increases, leading to partial polymer degradation and increased shell permeability [71]. Oligomers resulting from the incomplete dissolution of the MF core can also protonate at a low pH, contributing to the increase in permeability [23].

The most unusual and unexpected is the response to acidity of the PMC medium formed on the $CaCO_3$ core. Pechenkin et al. demonstrated that the remaining $CaCO_3$ in the capsules, after core removal, dissolves in a strong acid (0.1 M HCl or more concentrated), forming CO_2, which is expected to diffuse out of the capsule. However, lowering the pH of the solution leads to the annealing of the polyelectrolytes (reducing their mutual interaction), resulting in the rearrangement of the polyelectrolyte complex by carbon dioxide bubbles. Consequently, this process decreases the CO_2 diffusion rate and leads to the formation of CO_2 bubbles inside the capsule. The bubble grows, causing the capsule to swell and its structure to change from matrix-like (completely filled with polymers) to shell-like (with a well-defined shell and hollow inner lumen). As the capsule expands, the shell thins, and when it reaches a critical size, shell permeability increases due to the formation of defects and the escape of CO_2. The internal pressure within the capsule then decreases, causing the capsule to shrink back to its original size. Once the defects in the shell are restored, the capsule begins to inflate again. These oscillating inflate–deflate cycles repeat 4–5 times with a decreasing amplitude until most of the $CaCO_3$ has dissolved or the capsule shell's defects can no longer be repaired. Eventually, the capsule collapses [57].

From the above, it can be inferred that the acidity of the medium affects the pore size of polyelectrolyte microcapsules formed on different templates in various ways. In an acidic medium, the pores of the PMC shell on an MF core increase in size, but close in an alkaline medium. Conversely, for PMC on a PS core, the pores increase in size or number in an alkaline medium. Given that the microcapsules formed on the MF matrix have the same polyelectrolyte composition as those derived from polystyrene, it is possible that the increase in porosity under alkaline pH conditions is due to factors other than the chemical properties of the polyelectrolyte shell. This could include the spatial organization of the polyelectrolytes, the effect of an incompletely removed core, or its solvent. Meanwhile, microcapsules formed on a $CaCO_3$ core are not directly affected by the pH environment.

4.2. Influence of Ionic Strength and Temperature on the Porosity of the PMC's Shell

At present, there are no studies devoted to exploring the impact of ionic strength and solution temperature on the pore size of PMCs. However, if we hypothesize that the primary mechanism influencing the permeability of the microcapsule shell is the alteration in pore size, we can extrapolate from the results that focus on the effect of ionic strength and solution temperature on the permeability of PMCs.

For instance, Ibarz et al. investigated the effect of the solution's ionic strength on the penetration of fluorescently labeled PAH through the shell of PMCs formed on the MF core. They demonstrated that at 10^{-3} M NaCl, M PAH $\pm$ rho(rhodamine) (70 kDa) could not penetrate the shell [72]. However, increasing the salt concentration above 10^{-2} M NaCl enabled the labeled polymer to pass through the shell. The authors proposed that this effect is due to an increase in pore size resulting from the higher ionic strength of the solution. On the other hand, Georgieva et al. suggested that PMC permeability can be attributed both to the presence of pores or defects in the capsule wall, formed during the preparation process, and to electrostatic interactions between the penetrating molecules and the charged surface of the capsule [73].

In the case of PMCs formed on a PS particle, Halozan et al. [70] demonstrated that the addition of 0.2 M NaCl more than doubled the amount of substance (PAH) incorporated into the PMC. However, the authors suggest that in this instance, the ionic strength of the solution did not result in an increase in the pore size of the shell. Instead, it led to a decrease in the hydrodynamic radius of the polyelectrolyte (PAH) from 8 nm in pure water to 5 nm with the addition of 0.1 M NaCl, which facilitated the penetration of more polyelectrolyte.

In the study conducted by Ibarz et al., a significant decrease (three-fold) in substance release from the PMCs formed on the MF core was observed after heating them at 80 °C for 30 min. This is presumably associated with a decrease in the pore size in the PMC shell [74]. However, it should be noted that the effect of the temperature alters the permeability of PMCs on a consistent basis.

4.3. Influence of Solvent Type on the Porosity of the PMC's Shell

The influence of solvent remains a relatively underexplored area. In their work, Kim et al. studied the effect of solvent on the shell of PMCs formed on an MF core. They demonstrated that both the total shell pore area and pore size are dependent on the solvent used. For example, the total pore area was estimated to be approximately 1% for capsules immersed in water, and 1% and 4% for capsules immersed in ethanol/water and acetone/water mixtures, respectively. The analysis of pore size distribution in water indicates that the majority of the pores have a radius of 5–10 nm. No significant changes in pore size were found for capsules treated with ethanol/water mixtures. However, capsules treated with acetone/water mixtures exhibited a distinct change in pore size distribution, with the relative number of pores with a radius less than 10 nm decreasing by a factor of two. This was accompanied by the emergence of large pores with extremely irregular, non-circular shapes, with the average radius of large pores exceeding 40 nm [75]. This effect was utilized in the work of Lvov et al. to encapsulate urease in polyelectrolyte microcapsules [20].

5. Density of the Shell of Polyelectrolyte Microcapsules

Currently, the densities of the PMC shell are determined by mathematical calculations. In their studies, authors determine the shell's thickness and volume using methods such as small-angle neutron scattering [67,76], transmission electron microscopy (TEM), and scanning force microscopy (SFM) [3], among others. However, to date, there is no method that allows for the direct determination of shell density, necessitating the use of several assumptions. This has been acknowledged by the authors in their respective studies.

One such assumption is the belief that the properties of the PMC shell are independent of the type of core used. As mentioned earlier, several studies use data (such as shell thickness) derived solely from PMCs formed on the MF core [5,61,76–79], even though the authors themselves may use polystyrene or other cores.

Another assumption is that the calculation of the PMC shell's density does not account for the influence of the sample preparation procedure on the shell's thickness. However, the use of high temperatures (up to 60 °C), organic solvents, and other environmental conditions can significantly impact the PMC shell, its thickness, and consequently, the calculation of its density.

In these studies, the effect of various environmental conditions such as the temperature, ionic strength, and pH of the solution on shell density does not consider the change in the mass of the PMC shell due to these factors. However, there are several studies [80–82] that have demonstrated the loss of polyelectrolytes by the PMC shell depending on different environmental conditions. Therefore, given that mass is one of the key parameters of density, such an assumption may not be entirely accurate.

Also, under various environmental conditions, as mentioned earlier [15,20,54,57,69,70, 72–74], the number and size of pores in the shell of polyelectrolyte microcapsules change.

However, these parameters are not taken into account by the authors in their studies. Nonetheless, the formation of a hole in the shell of a PMC can lead to an increase in shell density or in its thickness (which also affects the density calculation), provided that the mass of the shell remains constant.

In summary, given the issues described above, it can be concluded that there is a current need for new methods to determine the density of the PMC shell that are free from these limitations. Additionally, for a more accurate determination of the PMC shell density under various environmental conditions using standard methods, it is essential to consider the impact of these conditions on shell mass loss and pore formation.

6. Conclusions

Based on this review, it can be concluded that the morphology of PMCs is most extensively studied on microcapsules that are formed on MF cores. There are considerably fewer studies on the morphology of microcapsules formed on other types of cores. However, most applied research often uses the results of studies obtained for polyelectrolyte microcapsules formed on a melamine formaldehyde core to explain the influence of environmental conditions or to calculate the physicochemical characteristics of PMCs formed on different cores. This approach is not entirely correct, as this review has shown that significant differences were found in the changes in microcapsule size, shell thickness, and pore size in response to physicochemical stimuli.

When considering the influence of the medium's pH on the size of PMCs, it is observed that PMCs formed on MF and $CaCO_3$ cores do not change their size at different pH values of the medium. However, PMCs formed on PS cores decrease in size in an acidic medium and increase in size in an alkaline medium, followed by subsequent destruction (a characteristic also observed in PMCs on $MnCO_3$). In the case of exposure to ionic solution strength, a similar lack of influence on the size of PMCs formed on MF and $CaCO_3$ cores was observed, though the mechanisms of the phenomenon may differ. PMCs formed on PS cores are characterized by a decrease in size at ionic strengths above 3 M. Temperature effects lead to a reduction in the size of all types of capsules presented in this work, regardless of the template used.

In the context of studying the influence of the core type on the shell thickness of polyelectrolyte microcapsules (PMCs), it can be noted that no significant differences are observed for the same number of layers (this applies to MF, $MnCO_3$, and PS cores). However, the thickness of the PMC shell increases linearly with the increasing number of polyelectrolyte layers. The polyelectrolyte microcapsules formed on a $CaCO_3$ core, whose shell is formed after nine polyelectrolyte layers, are significantly different, with a thickness of about 46 nm. In the case of PMCs formed on $CaCO_3$ particles containing protein, the microcapsules form a pronounced shell already after the sixth layer. However, the calculated thickness of one layer is twice that of the thickness of one layer of the shells of PMCs formed on MF, PS, or $MnCO_3$ cores.

It has also been shown that the acidity of the medium affects the pore size of polyelectrolyte microcapsules formed on different templates in varying ways. In an acidic medium, the pores of the PMC shell on the MF core become larger, and in an alkaline medium, they close. In contrast, for PMCs on the PS core, the pores increase in size or number in an alkaline medium. Microcapsules formed on a $CaCO_3$ core, however, are not directly affected by the pH of the medium. Despite these observations, it is not currently possible to make any assumptions about the mechanisms of this phenomenon, as there are no studies on the influence of other stimuli (ionic strength and temperature). This highlights the need for further research in this area, which may elucidate the impact of pore formation in the PMC shell on its permeability.

Separately, we would like to note that some studies conducted earlier require the use of more advanced methods for analyzing nanoscale systems, which will allow a more accurate and objective assessment of the characteristics of PMCs without the harsh conditions of sample preparation: sample heating, the use of organic solvents, and other conditions. Modern methods, such as Tunable Resistive Pulse Sensing (TRPS) technology [83] nano-impact electrochemistry [84], single-particle inductively coupled plasma mass spectrometry [85], focused ion beam nanoslice tomography [86], and others, allow us to avoid these drawbacks and adequately estimate the size of polyelectrolyte microcapsules (PMCs), the thickness of the shell, and the size of their pores. These approaches will not only allow researchers to clarify the already obtained data, but will provide an opportunity to evaluate new aspects of PMC behavior under different conditions to develop more effective methods of PMC application in various fields, ranging from medicine to electronics.

Author Contributions: Conceptualization, A.L.K. and S.A.T.; data curation, A.L.K., E.V.M. and Y.S.C.; formal analysis, A.L.K.; project administration, S.A.T. and A.L.K.; supervision, S.A.T.; visualization, A.L.K. and E.V.M.; writing—original draft preparation, A.L.K.; writing—review and editing, A.L.K., S.A.T., E.V.M. and Y.S.C. All authors have read and agreed to the published version of the manuscript.

Funding: This research was funded by the State Assignment of the Russian Federation No. 075-00224-24-01.

Institutional Review Board Statement: Not applicable.

Data Availability Statement: Not applicable.

Conflicts of Interest: The authors declare no conflicts of interest.

References

1. Sukhorukov, G.B.; Donath, E.; Davis, S.; Lichtenfeld, H.; Caruso, F.; Popov, V.I.; Möhwald, H. Stepwise polyelectrolyte assembly on particle surfaces: A novel approach to colloid design. *Polym. Adv. Technol.* **1998**, *9*, 759–767. [CrossRef]
2. Sukhorukov, G.B.; Donath, E.; Lichtenfeld, H.; Knippel, E.; Knippel, M.; Budde, A.; Möhwald, H. Layer-by-layer self assembly of polyelectrolytes on colloidal particles. *Colloids Surf. A Physicochem. Eng. Asp.* **1998**, *137*, 253–266. [CrossRef]
3. Köhler, K.; Shchukin, D.G.; Sukhorukov, G.B.; Möhwald, H. Drastic Morphological Modification of Polyelectrolyte Microcapsules Induced by High Temperature. *Macromolecules* **2004**, *37*, 9546–9550. [CrossRef]
4. Gao, C.; Donath, E.; Möhwald, H.; Shen, J. Spontaneous Deposition of Water-Soluble Substances into Microcapsules: Phenomenon, Mechanism, and Application. *Angew. Chem. Int. Ed.* **2002**, *41*, 3789–3793. [CrossRef]
5. Dubreuil, F.; Elsner, N.; Fery, A. Elastic properties of polyelectrolyte capsules studied by atomic-force microscopy and RICM. *Eur. Phys. J. E* **2003**, *12*, 215–221. [CrossRef] [PubMed]
6. Sukhorukov, G.B.; Shchukin, D.G.; Dong, W.; Möhwald, H.; Lulevich, V.V.; Vinogradova, O.I. Comparative Analysis of Hollow and Filled Polyelectrolyte Microcapsules Templated on Melamine Formaldehyde and Carbonate Cores. *Macromol. Chem. Phys.* **2004**, *205*, 530–535. [CrossRef]
7. Köhler, K.; Shchukin, D.G.; Möhwald, H.; Sukhorukov, G.B. Thermal Behavior of Polyelectrolyte Multilayer Microcapsules. 1. The Effect of Odd and Even Layer Number. *J. Phys. Chem. B* **2005**, *109*, 18250–18259. [CrossRef]
8. Skirtach, A.G.; Yashchenok, A.M.; Möhwald, H. Encapsulation, release and applications of LbL polyelectrolyte multilayer capsules. *Chem. Commun.* **2011**, *47*, 12736. [CrossRef] [PubMed]
9. De Geest, B.G.; Skirtach, A.G.; Mamedov, A.A.; Antipov, A.A.; Kotov, N.A.; De Smedt, S.C.; Sukhorukov, G.B. Ultrasound-Triggered Release from Multilayered Capsules. *Small* **2007**, *3*, 804–808. [CrossRef]
10. Marchenko, I.V.; Plotnikov, G.S.; Baranov, A.N.; Saletskii, A.M.; Bukreeva, T.V. Formation and destruction of polyelectrolyte microcapsules modified by rhodamine 6G. *J. Surf. Investig. X-ray Synchrotron Neutron Tech.* **2010**, *4*, 95–98. [CrossRef]
11. Amin, M.K.; Boateng, J.S. Enhancing Stability and Mucoadhesive Properties of Chitosan Nanoparticles by Surface Modification with Sodium Alginate and Polyethylene Glycol for Potential Oral Mucosa Vaccine Delivery. *Mar. Drugs* **2022**, *20*, 156. [CrossRef]
12. Bukreeva, T.; Orlova, O.A.; Sulyanov, S.; Grigoriev, Y.V.; Dorovatovskii, P. A New Approach to Modification of Polyelectrolyte Capsule Shells by Magnetite Nanoparticles. *Crystallogr. Rep.* **2011**, *56*, 880–883. [CrossRef]
13. Reshetilov, A.; Plekhanova, Y.; Tarasov, S.; Tikhonenko, S.; Dubrovsky, A.; Kim, A.; Kashin, V.; Machulin, A.; Wang, G.J.; Kolesov, V.; et al. Bioelectrochemical properties of enzyme-containing multilayer polyelectrolyte microcapsules modified with multiwalled carbon nanotubes. *Membranes* **2019**, *9*, 53. [CrossRef]
14. Johnston, A.P.R.; Kamphuis, M.M.J.; Such, G.K.; Scott, A.M.; Nice, E.C.; Heath, J.K.; Caruso, F. Targeting cancer cells: Controlling the binding and internalization of antibody-functionalized capsules. *ACS Nano* **2012**, *6*, 6667–6674. [CrossRef] [PubMed]
15. Antipov, A.A.; Sukhorukov, G.B. Polyelectrolyte multilayer capsules as vehicles with tunable permeability. *Adv. Colloid Interface Sci.* **2004**, *111*, 49–61. [CrossRef]
16. Kazakova, L.I.; Shabarchina, L.I.; Anastasova, S.; Pavlov, A.M.; Vadgama, P.; Skirtach, A.G.; Sukhorukov, G.B. Chemosensors and biosensors based on polyelectrolyte microcapsules containing fluorescent dyes and enzymes. *Anal. Bioanal. Chem.* **2013**, *405*, 1559–1568. [CrossRef]

17. Balkundi, S.S.; Veerabadran, N.G.; Eby, D.M.; Johnson, G.R.; Lvov, Y.M. Encapsulation of Bacterial Spores in Nanoorganized Polyelectrolyte Shells. *Langmuir* **2009**, *25*, 14011–14016. [CrossRef] [PubMed]
18. Musin, E.V.; Kim, A.L.; Dubrovskii, A.V.; Kudryashova, E.B.; Tikhonenko, S.A. Decapsulation of dextran by destruction of polyelectrolyte microcapsule nanoscale shell by bacillus subtilis bacteria. *Nanomaterials* **2020**, *10*, 12. [CrossRef]
19. Qiu, X.; Donath, E.; Möhwald, H. Permeability of Ibuprofen in Various Polyelectrolyte Multilayers. *Macromol. Mater. Eng.* **2001**, *286*, 591. [CrossRef]
20. Lvov, Y.; Antipov, A.A.; Mamedov, A.; Möhwald, H.; Sukhorukov, G.B. Urease Encapsulation in Nanoorganized Microshells. *Nano Lett.* **2001**, *1*, 125–128. [CrossRef]
21. Balabushevitch, N.G.; Sukhorukov, G.B.; Moroz, N.A.; Volodkin, D.V.; Larionova, N.I.; Donath, E.; Mohwald, H. Encapsulation of proteins by layer-by-layer adsorption of polyelectrolytes onto protein aggregates: Factors regulating the protein release. *Biotechnol. Bioeng.* **2001**, *76*, 207–213. [CrossRef] [PubMed]
22. Lulevich, V.V.; Radtchenko, I.L.; Sukhorukov, G.B.; Vinogradova, O.I. Deformation Properties of Nonadhesive Polyelectrolyte Microcapsules Studied with the Atomic Force Microscope. *J. Phys. Chem. B* **2003**, *107*, 2735–2740. [CrossRef]
23. Gao, C.; Moya, S.; Lichtenfeld, H.; Casoli, A.; Fiedler, H.; Donath, E.; Möhwald, H. The Decomposition Process of Melamine Formaldehyde Cores: The Key Step in the Fabrication of Ultrathin Polyelectrolyte Multilayer Capsules. *Macromol. Mater. Eng.* **2001**, *286*, 355–361. [CrossRef]
24. Kim, A.L.; Musin, E.V.; Oripova, M.J.; Oshchepkova, Y.I.; Salikhov, S.I.; Tikhonenko, S.A. Polyelectrolyte Microcapsules—A Promising Target Delivery System of Amiodarone with the Possibility of Prolonged Release. *Int. J. Mol. Sci.* **2023**, *24*, 3348. [CrossRef]
25. Borodina, T.N.; Rumsh, L.D.; Kunizhev, S.M.; Sukhorukov, G.B.; Vorozhtsov, G.N.; Feldman, B.M.; Markvicheva, E.A. Polyelectrolyte microcapsules as the systems for delivery of biologically active substances. *Biochem. Suppl. Ser. B Biomed. Chem.* **2008**, *2*, 88–93.
26. Palankar, R.; Skirtach, A.G.; Kreft, O.; Bédard, M.; Garstka, M.; Gould, K.; Möhwald, H.; Sukhorukov, G.B.; Winterhalter, M.; Springer, S. Controlled Intracellular Release of Peptides from Microcapsules Enhances Antigen Presentation on MHC Class I Molecules. *Small* **2009**, *5*, 2168–2176. [CrossRef]
27. Kartsonakis, I.A.; Danilidis, I.L.; Pappas, G.S.; Kordas, G.C. Encapsulation and Release of Corrosion Inhibitors into Titania Nanocontainers. *J. Nanosci. Nanotechnol.* **2010**, *10*, 5912–5920. [CrossRef]
28. Plekhanova, Y.V.; Tikhonenko, S.A.; Dubrovsky, A.V.; Kim, A.L.; Musin, E.V.; Wang, G.-J.; Kuznetsova, I.E.; Kolesov, V.V.; Reshetilov, A.N. Comparative study of electrochemical sensors based on enzyme immobilized into polyelectrolyte microcapsules and into chitosan gel. *Anal. Sci.* **2019**, *35*, 1037–1043. [CrossRef]
29. Pastorino, L.; Dellacasa, E.; Noor, M.R.; Soulimane, T.; Bianchini, P.; D'Autilia, F.; Antipov, A.; Diaspro, A.; Tofail, S.A.M.; Ruggiero, C. Multilayered Polyelectrolyte Microcapsules: Interaction with the Enzyme Cytochrome C Oxidase. *PLoS ONE* **2014**, *9*, e112192. [CrossRef]
30. Müller, B.; Lang, S.; Dominietto, M.; Rudin, M.; Schulz, G.; Deyhle, H.; Germann, M.; Pfeiffer, F.; David, C.; Weitkamp, T. High-resolution tomographic imaging of microvessels. In *Developments in X-ray Tomography VI*; Stock, S.R., Ed.; SPIE: San Diego, CA, USA, 2008; Volume 7078, p. 70780B.
31. Vinogradova, O.I.; Lebedeva, O.V.; Kim, B.-S. Mechanical Behavior and Characterization of Microcapsules. *Annu. Rev. Mater. Res.* **2006**, *36*, 143–178. [CrossRef]
32. Tong, W.; Song, H.; Gao, C.; Möhwald, H. Equilibrium Distribution of Permeants in Polyelectrolyte Microcapsules Filled with Negatively Charged Polyelectrolyte: The Influence of Ionic Strength and Solvent Polarity. *J. Phys. Chem. B* **2006**, *110*, 12905–12909. [CrossRef] [PubMed]
33. Petrov, A.I.; Antipov, A.A.; Sukhorukov, G.B. Base–Acid Equilibria in Polyelectrolyte Systems: From Weak Polyelectrolytes to Interpolyelectrolyte Complexes and Multilayered Polyelectrolyte Shells. *Macromolecules* **2003**, *36*, 10079–10086. [CrossRef]
34. Antipov, A.A.; Shchukin, D.; Fedutik, Y.; Petrov, A.I.; Sukhorukov, G.B.; Möhwald, H. Carbonate microparticles for hollow polyelectrolyte capsules fabrication. *Colloids Surf. A Physicochem. Eng. Asp.* **2003**, *224*, 175–183. [CrossRef]
35. An, Z.; Kavanoor, K.; Choy, M.L.; Kaufman, L.J. Polyelectrolyte microcapsule interactions with cells in two- and three-dimensional culture. *Colloids Surf. B Biointerfaces* **2009**, *70*, 114–123. [CrossRef] [PubMed]
36. De Koker, S.; De Geest, B.G.; Cuvelier, C.; Ferdinande, L.; Deckers, W.; Hennink, W.E.; De Smedt, S.; Mertens, N. In vivo cellular uptake, degradation, and biocompatibility of polyelectrolyte microcapsules. *Adv. Funct. Mater.* **2007**, *17*, 3754–3763. [CrossRef]
37. Verkhovskii, R.; Ermakov, A.; Sindeeva, O.; Prikhozhdenko, E.; Kozlova, A.; Grishin, O.; Makarkin, M.; Gorin, D.; Bratashov, D. Effect of Size on Magnetic Polyelectrolyte Microcapsules Behavior: Biodistribution, Circulation Time, Interactions with Blood Cells and Immune System. *Pharmaceutics* **2021**, *13*, 2147. [CrossRef] [PubMed]
38. Shimoni, O.; Yan, Y.; Wang, Y.; Caruso, F. Shape-Dependent Cellular Processing of Polyelectrolyte Capsules. *ACS Nano* **2013**, *7*, 522–530. [CrossRef]
39. Hatami Boura, S.; Peikari, M.; Ashrafi, A.; Samadzadeh, M. Self-healing ability and adhesion strength of capsule embedded coatings—Micro and nano sized capsules containing linseed oil. *Prog. Org. Coat.* **2012**, *75*, 292–300. [CrossRef]
40. Kalenichenko, D.; Nifontova, G.; Karaulov, A.; Sukhanova, A.; Nabiev, I. Designing Functionalized Polyelectrolyte Microcapsules for Cancer Treatment. *Nanomaterials* **2021**, *11*, 3055. [CrossRef]

41. Grigoriev, D.O.; Bukreeva, T.; Möhwald, H.; Shchukin, D.G. New Method for Fabrication of Loaded Micro- and Nanocontainers: Emulsion Encapsulation by Polyelectrolyte Layer-by-Layer Deposition on the Liquid Core. *Langmuir* **2008**, *24*, 999–1004. [CrossRef]
42. Volodkin, D.V.; Petrov, A.I.; Prevot, M.; Sukhorukov, G.B. Matrix Polyelectrolyte Microcapsules: New System for Macromolecule Encapsulation. *Langmuir* **2004**, *20*, 3398–3406. [CrossRef]
43. Karasev, V.Y.; Polishchuk, V.A.; Gorbenko, A.P.; Dzlieva, E.S.; Mironova, I.I.; Pavlov, S.I. Changes in the surface structure of melamine-formaldehyde particles in complex plasma. *J. Phys. Conf. Ser.* **2018**, *946*, 012156. [CrossRef]
44. Cho, Y.-S.; Shin, C.H.; Han, S. Dispersion Polymerization of Polystyrene Particles Using Alcohol as Reaction Medium. *Nanoscale Res. Lett.* **2016**, *11*, 46. [CrossRef]
45. Peyratout, C.S.; Dähne, L. Tailor-Made Polyelectrolyte Microcapsules: From Multilayers to Smart Containers. *Angew. Chem. Int. Ed.* **2004**, *43*, 3762–3783. [CrossRef] [PubMed]
46. Heuvingh, J.; Zappa, M.; Fery, A. Salt Softening of Polyelectrolyte Multilayer Capsules. *Langmuir* **2005**, *21*, 3165–3171. [CrossRef]
47. Parakhonskiy, B.V.; Yashchenok, A.M.; Konrad, M.; Skirtach, A.G. Colloidal micro- and nano-particles as templates for polyelectrolyte multilayer capsules. *Adv. Colloid Interface Sci.* **2014**, *207*, 253–264. [CrossRef]
48. Déjugnat, C.; Sukhorukov, G.B. pH-Responsive Properties of Hollow Polyelectrolyte Microcapsules Templated on Various Cores. *Langmuir* **2004**, *20*, 7265–7269. [CrossRef]
49. Kazakova, L.I.; Dubrovskiĭ, A.V.; Moshkov, D.A.; Shabarchina, L.I.; Sukhorukov, B.I. An electron microscopy study of the structure of polyelectrolyte microcapsules containing protein and containing no protein. *Biofizika* **2007**, *52*, 850–854. [PubMed]
50. Sukhorukov, G.B.; Volodkin, D.V.; Günther, A.M.; Petrov, A.I.; Shenoy, D.B.; Möhwald, H. Porous calcium carbonate microparticles as templates for encapsulation of bioactive compounds. *J. Mater. Chem.* **2004**, *14*, 2073–2081. [CrossRef]
51. Yang, X.; Han, X.; Zhu, Y. (PAH/PSS)5 microcapsules templated on silica core: Encapsulation of anticancer drug DOX and controlled release study. *Colloids Surf. A Physicochem. Eng. Asp.* **2005**, *264*, 49–54. [CrossRef]
52. Antipov, A.A.; Sukhorukov, G.B.; Donath, E.; Möhwald, H. Sustained release properties of polyelectrolyte multilayer capsules. *J. Phys. Chem. B* **2001**, *105*, 2281–2284. [CrossRef]
53. Vinogradova, O.I.; Andrienko, D.; Lulevich, V.V.; Nordschild, S.; Sukhorukov, G.B. Young's Modulus of Polyelectrolyte Multilayers from Microcapsule Swelling. *Macromolecules* **2004**, *37*, 1113–1117. [CrossRef]
54. Sukhorukov, G.B.; Antipov, A.A.; Voigt, A.; Donath, E.; Möhwald, H. pH-controlled macromolecule encapsulation in and release from polyelectrolyte multilayer nanocapsules. *Macromol. Rapid Commun.* **2001**, *22*, 44–46. [CrossRef]
55. Tong, W.; Dong, W.; Gao, C.; Möhwald, H. Charge-Controlled Permeability of Polyelectrolyte Microcapsules. *J. Phys. Chem. B* **2005**, *109*, 13159–13165. [CrossRef]
56. Haložan, D.; Riebentanz, U.; Brumen, M.; Donath, E. Polyelectrolyte microcapsules and coated $CaCO_3$ particles as fluorescence activated sensors in flowmetry. *Colloids Surf. A Physicochem. Eng. Asp.* **2009**, *342*, 115–121. [CrossRef]
57. Pechenkin, M.A.; Möhwald, H.; Volodkin, D.V. pH- and salt-mediated response of layer-by-layer assembled PSS/PAH microcapsules: Fusion and polymer exchange. *Soft Matter* **2012**, *8*, 8659. [CrossRef]
58. Georgieva, R.; Dimova, R.; Sukhorukov, G.; Ibarz, G.; Möhwald, H. Influence of different salts on micro-sized polyelectrolyte hollow capsules. *J. Mater. Chem.* **2005**, *15*, 4301. [CrossRef]
59. Lebedeva, O.V.; Kim, B.-S.; Vasilev, K.; Vinogradova, O.I. Salt softening of polyelectrolyte multilayer microcapsules. *J. Colloid Interface Sci.* **2005**, *284*, 455–462. [CrossRef]
60. Dong, W.-F.; Ferri, J.K.; Adalsteinsson, T.; Schönhoff, M.; Sukhorukov, G.B.; Möhwald, H. Influence of Shell Structure on Stability, Integrity, and Mesh Size of Polyelectrolyte Capsules: Mechanism and Strategy for Improved Preparation. *Chem. Mater.* **2005**, *17*, 2603–2611. [CrossRef]
61. Kim, B.-S.; Fan, T.-H.; Lebedeva, O.V.; Vinogradova, O.I. Superswollen Ultrasoft Polyelectrolyte Microcapsules. *Macromolecules* **2005**, *38*, 8066–8070. [CrossRef]
62. Leporatti, S.; Gao, C.; Voigt, A.; Donath, E.; Möhwald, H. Shrinking of ultrathin polyelectrolyte multilayer capsules upon annealing: A confocal laser scanning microscopy and scanning force microscopy study. *Eur. Phys. J. E* **2001**, *5*, 13–20. [CrossRef]
63. Prevot, M.; Déjugnat, C.; Möhwald, H.; Sukhorukov, G.B. Behavior of Temperature-Sensitive PNIPAM Confined in Polyelectrolyte Capsules. *ChemPhysChem* **2006**, *7*, 2497–2502. [CrossRef]
64. Park, M.-K.; Deng, S.; Advincula, R.C. Sustained Release Control via Photo-Cross-Linking of Polyelectrolyte Layer-by-Layer Hollow Capsules. *Langmuir* **2005**, *21*, 5272–5277. [CrossRef]
65. Dubrovskii, A.V.; Shabarchina, L.I.; Kim, Y.A.; Sukhorukov, B.I. Influence of the temperature on polyelectrolyte microcapsules: Light scattering and confocal microscopy data. *Russ. J. Phys. Chem. A* **2006**, *80*, 1703–1707. [CrossRef]
66. Gao, C.; Donath, E.; Moya, S.; Dudnik, V.; Möhwald, H. Elasticity of hollow polyelectrolyte capsules prepared by the layer-by-layer technique. *Eur. Phys. J. E* **2001**, *5*, 21–27. [CrossRef]
67. Estrela-Lopis, I.; Leporatti, S.; Clemens, D.; Donath, E. Polyelectrolyte multilayer hollow capsules studied by small-angle neutron scattering (SANS). *Soft Matter* **2009**, *5*, 214–219. [CrossRef]
68. Moya, S.; Dähne, L.; Voigt, A.; Leporatti, S.; Donath, E.; Möhwald, H. Polyelectrolyte multilayer capsules templated on biological cells: Core oxidation influences layer chemistry. *Colloids Surf. A Physicochem. Eng. Asp.* **2001**, *183–185*, 27–40. [CrossRef]
69. Tiourina, O.P.; Antipov, A.A.; Sukhorukov, G.B.; Larionova, N.I.; Lvov, Y.; Möhwald, H. Entrapment ofα-Chymotrypsin into Hollow Polyelectrolyte Microcapsules. *Macromol. Biosci.* **2001**, *1*, 209–214. [CrossRef]

70. Haložan, D.; Déjugnat, C.; Brumen, M.; Sukhorukov, G.B. Entrapment of a Weak Polyanion and H^+/Na^+ Exchange in Confined Polyelectrolyte Microcapsules. *J. Chem. Inf. Model.* **2005**, *45*, 1589–1592. [CrossRef]
71. Han, B.; Chery, D.R.; Yin, J.; Lu, X.L.; Lee, D.; Han, L. Nanomechanics of layer-by-layer polyelectrolyte complexes: A manifestation of ionic cross-links and fixed charges. *Soft Matter* **2016**, *12*, 1158–1169. [CrossRef]
72. Ibarz, G.; Dähne, L.; Donath, E.; Möhwald, H. Smart Micro- and Nanocontainers for Storage, Transport, and Release. *Adv. Mater.* **2001**, *13*, 1324. [CrossRef]
73. Georgieva, R.; Moya, S.; Hin, M.; Mitlöhner, R.; Donath, E.; Kiesewetter, H.; Möhwald, H.; Bäumler, H. Permeation of Macromolecules into Polyelectrolyte Microcapsules. *Biomacromolecules* **2002**, *3*, 517–524. [CrossRef]
74. Ibarz, G.; Dähne, L.; Donath, E.; Möhwald, H. Controlled Permeability of Polyelectrolyte Capsules via Defined Annealing. *Chem. Mater.* **2002**, *14*, 4059–4062. [CrossRef]
75. Kim, B.-S.; Lebedeva, O.V.; Koynov, K.; Gong, H.; Glasser, G.; Lieberwith, I.; Vinogradova, O.I. Effect of Organic Solvent on the Permeability and Stiffness of Polyelectrolyte Multilayer Microcapsules. *Macromolecules* **2005**, *38*, 5214–5222. [CrossRef]
76. Estrela-Lopis, I.; Leporatti, S.; Moya, S.; Brandt, A.; Donath, E.; Möhwald, H. SANS Studies of Polyelectrolyte Multilayers on Colloidal Templates. *Langmuir* **2002**, *18*, 7861–7866. [CrossRef]
77. Kim, B.; Vinogradova, O.I. pH-Controlled Swelling of Polyelectrolyte Multilayer Microcapsules. *J. Phys. Chem. B* **2004**, *108*, 8161–8165. [CrossRef]
78. Kim, B.-S.; Fan, T.-H.; Vinogradova, O.I. Thermal softening of superswollen polyelectrolyte microcapsules. *Soft Matter* **2011**, *7*, 2705. [CrossRef]
79. Glinel, K.; Sukhorukov, G.B.; Möhwald, H.; Khrenov, V.; Tauer, K. Thermosensitive Hollow Capsules Based on Thermoresponsive Polyelectrolytes. *Macromol. Chem. Phys.* **2003**, *204*, 1784–1790. [CrossRef]
80. Dubrovskii, A.V.; Berezhnov, A.V.; Kim, A.L.; Tikhonenko, S.A. Behaviour of FITC-Labeled Polyallylamine in Polyelectrolyte Microcapsules. *Polymers* **2023**, *15*, 3330. [CrossRef]
81. Dubrovskii, A.V.; Kochetkova, O.Y.; Kim, A.L.; Musin, E.V.; Seraya, O.Y.; Tikhonenko, S.A. Destruction of shells and release of a protein from microcapsules consisting of non-biodegradable polyelectrolytes. *Int. J. Polym. Mater. Polym. Biomater.* **2019**, *68*, 160–164. [CrossRef]
82. Musin, E.V.; Kim, A.L.; Tikhonenko, S.A. Destruction of polyelectrolyte microcapsules formed on CaCO3 microparticles and the release of a protein included by the adsorption method. *Polymers* **2020**, *12*, 520. [CrossRef]
83. Ejjigu, N.; Abdelgadir, K.; Flaten, Z.; Hoff, C.; Li, C.-Z.; Sun, D. Environmental noise reduction for tunable resistive pulse sensing of extracellular vesicles. *Sens. Actuators A Phys.* **2022**, *346*, 113832. [CrossRef]
84. Kirk, K.A.; Luitel, T.; Narouei, F.H.; Andreescu, S. Nanoparticle Characterization through Nano-Impact Electrochemistry: Tools and Methodology Development. In *Nanoparticles in Biology and Medicine. Methods in Molecular Biology*; Humana: New York, NY, USA, 2020; pp. 327–342.
85. Kéri, A.; Sápi, A.; Ungor, D.; Sebők, D.; Csapó, E.; Kónya, Z.; Galbács, G. Porosity determination of nano- and sub-micron particles by single particle inductively coupled plasma mass spectrometry. *J. Anal. At. Spectrom.* **2020**, *35*, 1139–1147. [CrossRef]
86. Glasscott, M.W.; Pendergast, A.D.; Choudhury, M.H.; Dick, J.E. Advanced Characterization Techniques for Evaluating Porosity, Nanopore Tortuosity, and Electrical Connectivity at the Single-Nanoparticle Level. *ACS Appl. Nano Mater.* **2019**, *2*, 819–830. [CrossRef]

polymers

MDPI

Article

A Novel Fractional Brownian Dynamics Method for Simulating the Dynamics of Confined Bottle-Brush Polymers in Viscoelastic Solution

Shi Yu [1,*], Ruizhi Chu [1,2], Guoguang Wu [1,2] and Xianliang Meng [1,2]

1 Department of Chemical Engineering, China University of Mining & Technology, Xuzhou 221116, China; rzchu@cumt.edu.cn (R.C.); ggwu@cumt.edu.cn (G.W.); meng27@cumt.edu.cn (X.M.)

2 Key Laboratory of Coal-Based CO_2 Capture and Geological Storage, China University of Mining & Technology, Xuzhou 221116, China

* Correspondence: 5804@cumt.edu.cn or shiyu_cumt@163.com; Tel.: +86-1825-211-9142

Abstract: In crowded fluids, polymer segments can exhibit anomalous subdiffusion due to the viscoelasticity of the surrounding environment. Previous single-particle tracking experiments revealed that such anomalous diffusion in complex fluids (e.g., in bacterial cytoplasm) can be described by fractional Brownian motion (fBm). To investigate how the viscoelastic media affects the diffusive behaviors of polymer segments without resolving single crowders, we developed a novel fractional Brownian dynamics method to simulate the dynamics of polymers under confinement. In this work, instead of using Gaussian random numbers ("white Gaussian noise") to model the Brownian force as in the standard Brownian dynamics simulations, we introduce fractional Gaussian noise (fGn) in our homemade fractional Brownian dynamics simulation code to investigate the anomalous diffusion of polymer segments by using a simple "bottle-brush"-type polymer model. The experimental results of the velocity autocorrelation function and the exponent that characterizes the subdiffusion of the confined polymer segments can be reproduced by this simple polymer model in combination with fractional Gaussian noise (fGn), which mimics the viscoelastic media.

Keywords: fractional Brownian motion; Brownian dynamics (BD) simulation; bottle-brush polymers

1. Introduction

Macromolecular diffusion in crowded and heterogeneous cytoplasm is essential in many intracellular processes, such as metabolism, signal transduction, transcription, and translation [1,2]. Many imaging experiments demonstrate that chromosomal loci and cytoplasmic particles exhibit anomalous subdiffusion in living cells [3–6]. Thus, it is of great interest to investigate the dynamics of the genome and cytosol in bacterial cells both experimentally and theoretically to elucidate the origin of this anomalous diffusion. Due to the temporal and spatial resolution limit of single-molecule tracking, studying the interactions among various biological macromolecules in real time is extremely challenging. Therefore, many researchers examined how the dynamics of bacterial genomes and cytosol are altered by different treatments to bypass the resolution limit. Parry et al. [7] revealed that the bacterial cytoplasm is fluidized by ATP-dependent cellular metabolic activities, which agrees with Weber's observation [3] that the motions of genomic loci can be suppressed by the depletion of ATP in *E. coli* bacterial cells. Okumus et al. [8] showed that the diffusion of cytoplasmic proteins can be slowed down significantly by compressive forces exerted onto *E. coli* cells. Moreover, Wlodarski et al. [9,10] investigated the effects of sublethal antibiotic treatments on the dynamics of both genome and cytosol in *E. coli*. Although bacterial genome and cytosol responses to external perturbation can provide insight into intracellular dynamics, a detailed molecular model that resolves the heterogeneous and crowded cytoplasm is still necessary to fully understand the subdiffusive behaviors of various biological macromolecules.

All-atom molecular models of cytoplasm [11], DNA [12], and cell membranes [13] can characterize the molecular-level interactions in cells with an unprecedented resolution. However, the all-atom molecular dynamics method is computationally expensive, which drives the development of widely used coarse-grained models of biological macromolecules, such as Martini force field [14], the oxDNA model [15], and Shinoda's model of lipids [16]. Although molecular dynamics simulations can be sped up dramatically with those coarse-grained force fields due to the reduction in the number of degrees of freedom, it is still difficult to reproduce the anomalous diffusion by using those coarse-grained models since cytoplasmic particles/proteins and chromosomal loci can scan a large fraction of bacterial cells by performing subdiffusion [3,7,8]. Therefore, extensive research has focused on the simulation of cytoplasmic anomalous diffusion by using mesoscopic scale models [17]. Ando et al. [18] and Hasnain et al. [19] used spherical particles or several connected beads of different sizes to simulate bacterial cytoplasm and demonstrated that the anomalous diffusion can be attributed largely to the crowding effect. Słyk et al. [20] compared the effects of hard crowders on cytoplasmic diffusion to that of soft crowders, describing both hard and soft crowders by spherical particles. Swain et al. [21] studied the effects of crowding and confinement on the dynamics of bacterial chromosomes by using a "bottle-brush"-type polymer model [22,23]. Although the dynamics of the loop formation of a nucleoid can be very complex due to topological entanglements as well as polydispersity in loop sizes, the helical shape with a $\sim$1 μm pitch of bacterial chromosomes in rod-shaped bacteria like *E. coli* can be predicted by a minimal structured polymer model [22–24]. In addition, the subdiffusion of chromosomal loci due to DNA bridging was also simulated by S. Subramanian and S. M. Murray [25] by using the bond fluctuation method.

To further reduce the degrees of freedom of the system, different types of random walks (e.g., obstructed diffusion, fractional Brownian motion, and continuous time random walks) in which cytosol is modeled implicitly were proposed to explain the anomalous diffusion of tracers in cells [26]. Although anomalous subdiffusion can be generated with obstructed diffusion (OD) [27] as well as a continuous time random walk (CTRW) [28,29], the subdiffusion in a cell is likely a result of the intrinsic viscoelasticity of cytoplasm since the cytoplasmic particle velocity is anticorrelated for short time lags, which is contradicting to OD or CTRW models [3,29]. Such an anticorrelation of particle displacements can be successfully reproduced by fractional Brownian motion (fBm) [28–30] because fractional Brownian motion is a highly correlated/anticorrelated process [31]. On the other hand, cytoplasmic subdiffusion has a non-Gaussian distribution of increments [6,7], which is similar to the so-called "Brownian yet non-Gaussian" dynamics found in a growing number of biological, soft, and active matter systems [32]. Many researchers [32–34] adopted the diffusing or random diffusivity models to explain this non-Gaussian behavior. The predictions of the diffusing diffusivity model qualitatively agree with the experimental observations of the non-Gaussian behaviors of cytoplasmic tracers since the effects of the heterogeneous environment on subdiffusive particles have been included in such models [35]. Baldovin et al. [36] argued that such diffusing diffusivity models work at a mesoscopic scale, which is not directly related to the underlying molecular processes. So, they [36] proposed a specific example of the dynamical behaviors of macromolecules during simple polymerization processes, which can bridge the gap between diffusing diffusivity models and molecular-scale processes. In addition, metabolic activities that drive the system out of the equilibrium state might also contribute to the long tails of the displacement distributions of the particles performing non-Gaussian diffusion. Such nonequilibrium activities were simulated by active forces in Fodor's work [37].

As discussed above, the crowding effects, the viscoelasticity of the cytoplasm, the DNA bridging, the heterogeneous environment causing diffusing diffusivity, and even the nonequilibrium activity have been considered separately in previous theoretical and numerical works. However, the non-Gaussian subdiffusion of chromosomal loci and cytoplasmic particles has yet to be fully reproduced since Markovian processes (as in simple diffusing diffusivity models) cannot describe the anticorrelation of particle displacements and fBm cannot simulate non-Gaussian behaviors as fBm is a mean-zero Gaussian process [30,31]. Moreover, the application of the fractional Brownian motion model to viscoelastic-type subdiffusion was often limited to single tracers [29]. Therefore, in this manuscript, we consider a particular model that combines a "bottle-brush"-type polymer model [22] and fractional Brownian motion to explore to what extent the anomalous diffusion of polymer segments is affected by the dynamical structure of the bottle-brush polymers and the viscoelasticity of the surrounding environment. The effects of viscoelastic crowded media on the polymer segments were included in the fractional Brownian motion increments for each polymer bead, allowing us to consider the crowders implicitly to reduce the degrees of freedom of the whole system. Another advantage of introducing fractional Brownian motion to bottle-brush polymers simulations is that the nonequilibrium activities that take place in vivo can be modeled as accelerated Brownian dynamics [38] with a large Hurst parameter ($H \in (0.5, 1)$) for fractional Brownian motion. Although the heterogeneity of the surrounding environment might be introduced to this model by random variations in the polymer bead size in an overdamped regime, here, in this work, we focus on the simulation of the "non-Brownian yet Gaussian" diffusion of bottle-brush polymers by using a polymer model composed of fractional diffusive monomers.

2. Models and Simulation Methods

2.1. Force Fields

In this manuscript, the confined bottle-brush polymer was modeled by a linear backbone chain of 200 beads connected by springs [22]. Moreover, 40-bead side loops were attached to every bead of the backbone chain as described by D. Chaudhuri and B. M. Mulder [22,23]. As discussed as follows, to study to what extent the diffusive behaviors of polymer segments are affected by polymer stiffness, both FENE (finitely extensible nonlinear elastic) bond energy [21] and harmonic bond energy [22] were used to model the springs connecting the polymer beads. The FENE bond energy is given by

$$U_{FENE} = -\frac{1}{2}KR^2 ln\left[1 - \left(\frac{d_i - \sigma}{R}\right)^2\right] \tag{1}$$

where $d_i = |\mathbf{d_i}|$ is the distance between consecutive beads, i.e., the bond length ($\mathbf{d}_i = \mathbf{r}_{i+1} - \mathbf{r}_i$ is the bond vector); $K = 30\epsilon/\sigma^2$; $R = 1.5\sigma$; and $\epsilon = 1.0$ and $\sigma = 1.0$ are the unit of energy and unit of length in our simulations, respectively. The harmonic bond energy is as follows:

$$U_{harmonic} = \frac{1}{2}k_{bond}(d_i - \sigma)^2 \tag{2}$$

where $k_{bond} = 100\epsilon/\sigma^2$ is the harmonic bond energy coefficient, d_i is the distance between consecutive beads, and σ is the equilibrium bond length and the unit of length. To avoid the overlapping of polymer beads, the excluded volume interactions between the nonbonded beads are modeled by the Weeks–Chandler–Andersen (WCA) potential [22], which is given by

$$U_{WCA}(r_{ij}) = \begin{cases} 4\epsilon\left[\left(\frac{\sigma}{r_{ij}}\right)^{12} - \left(\frac{\sigma}{r_{ij}}\right)^{6} + \frac{1}{4}\right], & r_{ij} \leq 2^{1/6}\sigma \\ 0, & r_{ij} > 2^{1/6}\sigma \end{cases} \tag{3}$$

where $r_{ij} = |\mathbf{r}_{ij}|$ is the distance between the ith bead and j^{th} bead; the cutoff distance for the WCA potential is $2^{1/6}\sigma$. The confining walls (grey surfaces, as shown in Figure 1a) for all polymer beads are modeled through the U_{wall} [22], which is given by

$$U_{wall}(r_{iw}) = \begin{cases} 2\pi\epsilon\left[\frac{2}{5}\left(\frac{\sigma}{r_{iw}}\right)^{10} - \left(\frac{\sigma}{r_{iw}}\right)^{4} + \frac{3}{5}\right], & r_{iw} \le \sigma \\ 0, & r_{iw} > \sigma \end{cases} \tag{4}$$

where r_{iw} is the distance between the ith bead and the wall. Some polymer beads may be close to (i.e., $r_{iw} < \sigma$) both the cylindrical side surface and the top (or bottom) surface, and then U_{wall} potentials for these two surfaces are computed separately. As shown in Figure 1a, the length of the cylindrical confinement in the z-axis direction is 50.75 σ, and the radius of the cylinder is 14.75 σ.

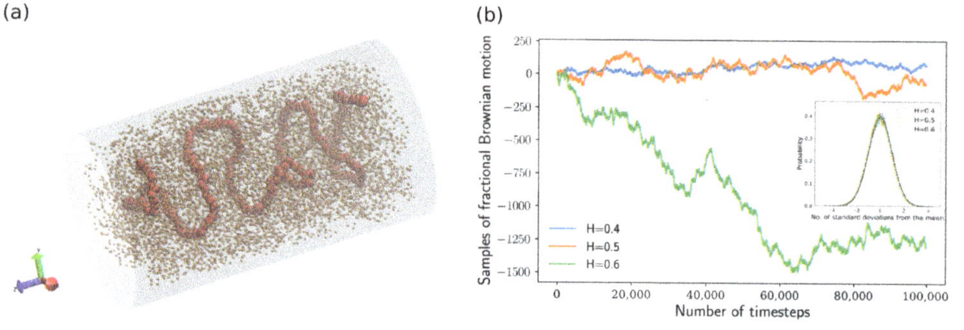

Figure 1. (**a**) Snapshot of "bottle-brush" polymer model under cylindrical confinement. The backbone chain and side loops are represented by red and orange beads, respectively. (**b**) Samples of fractional Brownian motion (fBm) for $H = 0.4$, $H = 0.5$, and $H = 0.6$. (Inset) histogram of fractional Gaussian noise (fGn), i.e., fBm increments for $H = 0.4$, $H = 0.5$, and $H = 0.6$. The black dashed curve represents the normal distribution.

2.2. Fractional Brownian Dynamics Simulation

We update the bead positions based on the following equation [39]:

$$\mathbf{r}^{(n+1)} = \mathbf{r}^{(n)} + \frac{\Delta t}{6\pi\eta a_{bead}}\mathbf{F}^{(n)} + \sqrt{\frac{2k_B T \Delta t}{6\pi\eta a_{bead}}}\,\xi^{(n)} \tag{5}$$

where $\mathbf{r}^{(n+1)}$ and $\mathbf{r}^{(n)}$ are the bead positions at time step $n+1$ and n, respectively; Δt is the time step; $\mathbf{F}^{(n)}$ is the total non-Brownian force on the bead at time step n; and $a_{bead} = \frac{1}{2}\sigma$ is the radius of the beads. Here we take $\eta = \frac{200}{3\pi\sigma}$ to be the cytoplasm viscosity. We use reduced units for our simulations where $k_B T = 1$. We assume that all the beads of the polymer model perform fractional Brownian motion; that is, bead positions can be defined by the stochastic representation [40] described as follows if the total non-Brownian force is zero:

$$B_H(t) := \frac{1}{\Gamma\left(H+\frac{1}{2}\right)}\left(\int_{-\infty}^{0}\left[(t-s)^{H-\frac{1}{2}} - (-s)^{H-\frac{1}{2}}\right]dB(s) + \int_{0}^{t}(t-s)^{H-\frac{1}{2}}dB(s)\right) \tag{6}$$

where Γ is the Gamma function $\Gamma(\alpha) := \int_0^\infty x^{\alpha-1}e^{-x}dx$, $H \in (0,1)$ is the Hurst parameter, and the integrator B is an ordinary Brownian motion. Note that the fractional Brownian motion B_H is reduced to an ordinary Brownian motion by taking $H = \frac{1}{2}$ in Equation (6). Three example tracks of one-dimensional (1D) fractional Brownian motion for different Hurst parameters are depicted in Figure 1b. To ensure that polymer beads perform fractional Brownian motion if the total non-Brownian force is absent, $\xi^{(n)}$ in Equation (5) was chosen as a vector of fractional Brownian motion increments, i.e., fractional Gaussian noise (fGn), at time step n. The three components of the vector $\xi^{(n)}$ in the three-dimensional Cartesian coordinate system are denoted by $X_1^{(n)}$, $X_2^{(n)}$, and $X_3^{(n)}$:

$$X_i^{(n)} := B_H(n+1) - B_H(n) \quad (i = 1,2,3) \tag{7}$$

Each $X_i^{(n)}$ has a standard normal distribution [41], as shown in the inset of Figure 1b. The autocovariance function $\gamma(\cdot)$ of $X_i^{(n)}$ is given by the following equation [41]:

$$\gamma(k) = \mathbb{E}X_i^{(n+k)}X_i^{(n)} = \frac{1}{2}\left[|k-1|^{2H} - 2|k|^{2H} + |k+1|^{2H}\right] \tag{8}$$

for a discrete-time random process ($k, n \in \mathbb{Z}$). For $H = 0.5$, Equation (8) is reduced to $\gamma(k) = 0$ for $k \neq 0$, where an ordinary Brownian motion is recovered. Note that fractional Brownian motion is not a Markov process based on Equation (6), and there is (in general) no independence in fractional Gaussian noises $X_i^{(n)}$ (as demonstrated in Equation (8)). However, for simplicity, the 3 components of vector $\xi^{(n)}$ are extracted from three independent realizations of fractional Gaussian noises, which means $\mathbb{E}X_i^{(n+k)}X_j^{(n)} = 0$ for $i \neq j$. And in this manuscript, the fractional Gaussian noise $X_i^{(n)}$ was generated by the fBm-0.3.0 package programmed by Christopher Flynn by using the Davies Harte method [42].

The dynamics of "bottle-brush"-type polymers in cylindrical confinement were investigated systematically. The initial configurations of our simulation systems were prepared by using 2×10^6 time steps with long standard Brownian dynamics (BD) equilibration in an overdamped regime, which can be achieved simply by setting $H = 0.5$ for $\xi^{(n)}$ in Equation (5). Then, fractional Brownian dynamics simulations (based on Equation (5)) of 1×10^7 time steps were carried out for various Hurst parameters. Only the last 8×10^6 time steps' simulation results were used for further analysis. Moreover, the time step Δt was set to 0.01 for all our simulations unless otherwise stated.

3. Results and Discussion

3.1. Potential Energy vs. Time for Standard and Fractional Brownian Dynamics Simulations

As described above, we use a simple "bottle-brush" polymer model in combination with fractional Gaussian noises to simulate the anomalous diffusive behaviors of polymer segments. As demonstrated in Figure 2a,b, the potential energy of the simulation system reaches a plateau rapidly for polymer models with both FENE bonds and harmonic bonds as Δt varies from 0.005 to 0.02. The potential energy curves in Figure 2a,b indicate that system equilibrium is obtained by a standard Brownian dynamics simulation since a fractional Brownian motion is reduced to an ordinary Brownian motion by setting the Hurst parameter H to 0.50. The potential energy of the polymer model with FENE bonds is much larger than that of the polymer model with harmonic bonds, which implies that the polymer model is much "stiffer" with FENE bonds.

To test the suitability of the fractional Brownian dynamics simulation, the influence of fractional Gaussian noises on the potential energy of the simulation system is examined by varying the Hurst parameter H. The potential energy curves for both the FENE and the harmonic bond models are plotted against simulation time for different H, as shown in Figure 2c,d. Again, a fractional Brownian dynamics simulation can rapidly achieve the equilibrium state, similar to a standard Brownian dynamics simulation. The potential energy increases for both FENE and harmonic bonds as the Hurst parameter H increases from 0.35 to 0.65. Note that fractional Brownian motion with $H \in (0, 0.5)$ is subdiffusive, while fractional Brownian motion with $H \in (0.5, 1)$ is an accelerated Brownian motion [38] or superdiffusive. As the Hurst parameter approaches 1.0, the fractional Brownian motion becomes a ballistic-like motion. As a result, although the bead positions are updated based on Equation (5), where the inertia of the polymer beads is neglected, the polymer beads act like they have random "momenta", owing to the highly correlated fractional Gaussian noises for the large Hurst parameter $H \in (0.5, 1)$. Since the direction and magnitude of the correlated fGn changes less frequently than white noise for a large $H \in (0.5, 1)$, the correlated fGn works as an external random "force" or "impulse" exerted on the polymer beads, which drives the system out of equilibrium. Such random "impulses" greater than thermal fluctuations cause the increase in potential energy of the simulation system for large Hurst parameters, as seen in Figure 2c,d. As shown by the pink curve in Figure 2d, the time it takes to equilibrate the simulation system is much longer for $H = 0.65$ than that for a smaller H, and the fluctuations in the potential energy are stronger for $H = 0.65$ because the correlated fGn condenses the polymer beads for the large Hurst parameter. Snapshots of the condensed polymer model with $H = 0.65$ are presented as the insets of Figure 2c,d.

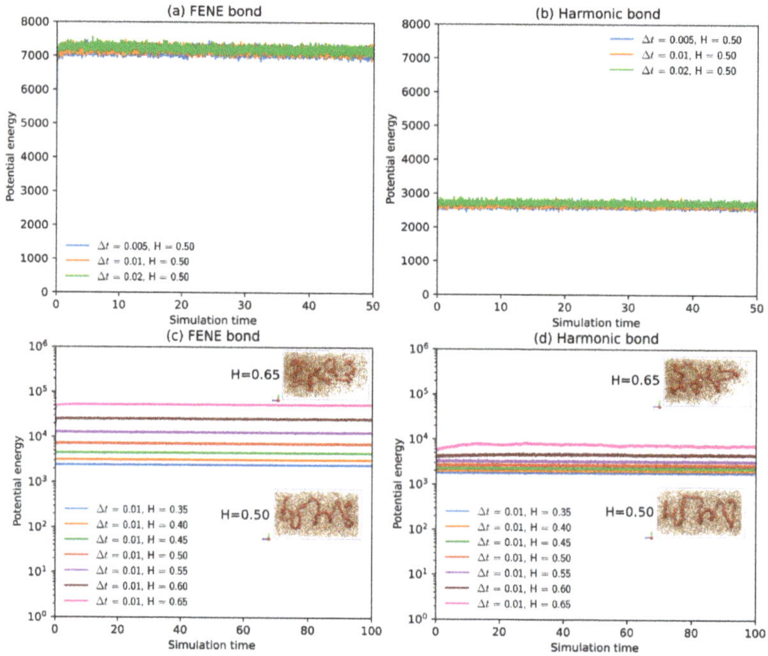

Figure 2. (**a**) The potential energy of a polymer model with FENE bonds as a function of simulation time for different time steps Δt, obtained by standard Brownian dynamics simulation ($H = 0.5$) in an overdamped regime (Equation (5)). (**b**) The potential energy of a polymer model with harmonic bonds as a function of simulation time for different time steps Δt. (**c**) The potential energy of a polymer model with FENE bonds as a function of simulation time for different Hurst parameters, obtained by fractional Brownian dynamics simulation. (Inset) snapshots of polymer models in cylindrical confinement with $H = 0.65$ and $H = 0.50$. (**d**) The potential energy of a polymer model with harmonic bonds as a function of simulation time for different Hurst parameters by fractional Brownian dynamics simulation. (Inset) snapshots of polymer models for $H = 0.65$ and $H = 0.50$.

Bacterial chromosomes are highly dynamic structures [43] whose center of mass is precisely located at the center of the cell volume [21]. To compare this observation [21] to our simulation results, the two-dimensional projected number densities of the polymer beads in the xz-plane for various Hurst parameters are plotted in Figure 3. As can be seen from Figure 3d,k, the projected number densities of the polymer beads, which are obtained by standard Brownian dynamics simulations ($H = 0.5$) for the models with both FENE and harmonic bonds, exhibit shapes with grooves. These simulation results agree with the helical shape of this simple polymer model predicted by Langevin dynamics simulations [22]. As the Hurst parameter H decreases from 0.5 to 0.35, i.e., entering the subdiffusion regime, the projected number density of polymer beads becomes more evenly distributed across the whole cross-sectional area of the cylindrical confinement, which is demonstrated in Figure 3a–d,h–k. On the other hand, as H increases from 0.5 to 0.65, the polymer beads are close to one end of the cylindrical confinement, acting as though they are pushed by external forces, as shown in Figure 3d–g,h–n. So, the center of mass of all the polymer beads is no longer positioned at the center of the cell volume (Figure 3f,g,m,n). The condensation of the polymer beads for a large H increases the potential energy of the simulation system, mainly due to the excluded volume interactions among the polymer beads, as shown in Figure 2c,d. Although switching from harmonic bonds to FENE bonds increases the potential energy significantly, the effects of the bond types on the projected number density of the polymer beads are negligible, as demonstrated in Figure 3f,g,m,n.

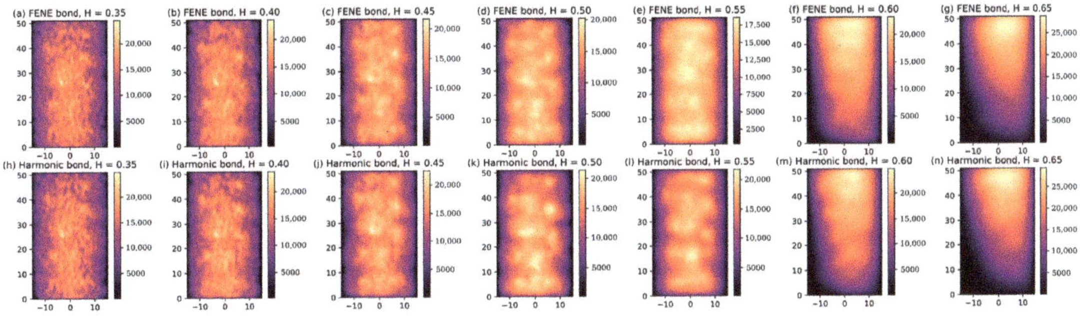

Figure 3. (**a–n**) Projected number density profiles in xz plane of polymer beads on polymer models with FENE and harmonic bonds for different Hurst parameters H.

3.2. Diffusion Coefficient and Exponent α

Since fluorescence microscopy experiments record the two-dimensional diffusive behaviors of bacterial chromosomes in the focusing plane [3–5], to compare our fractional Brownian dynamics simulations directly to previous experimental observations, the diffusion coefficients, as well as the exponent α characterizing the anomalous diffusion, are obtained by fitting the 2D ensemble-averaged mean-squared displacement of the polymer beads against time lag based on Equation (9) as follows:

$$MSD_{2d} = 4D_{xz}t^{\alpha} \tag{9}$$

where MSD_{2d} is the averaged 2D mean-squared displacement of the polymer beads in cylindrical confinement, D_{xz} is the ensemble-averaged 2D diffusion coefficient in the xz plane, t is the time lag, and α is the exponent characterizing the anomalous diffusion. And for fractional Brownian motion, $\alpha = 2H$. The averaged 2D MSD as a function of time lag for different Hurst parameters with both FENE and harmonic bonds are plotted in Figure 4a. Those MSDs show that the polymer beads connected by FENE bonds diffuse slightly faster than those connected by harmonic bonds, but this speeding-up effect is insignificant. One example of the exponential fitting of 2D MSD against time lag according to Equation (9) is presented in the inset of Figure 4b for $H = 0.50$. It turns out that although the diffusions of polymer beads are affected by the neighboring beads through bonding energy as well as the excluded volume potentials, the diffusive behaviors of those polymer beads can still be well characterized by Equation (9). As shown in Figure 4b, the fitted diffusion coefficient of polymer beads increases as the Hurst parameter H increases, which is not surprising based on the fact that the larger the Hurst parameter H, the more correlated the fractional Gaussian noises are. Figure 4c demonstrates that the exponent α also increases as the Hurst parameter H increases for $H \in (0.40, 0.60)$. Previous experimental observations show that

α for chromosomal loci is about 0.38∼0.40 [3–5], which agrees very well with the simulation results for $H \in (0.45, 0.50)$. This agreement indicates that the weak viscoelasticity of cytoplasm is sufficient for the subdiffusion of chromosomal loci in vivo. The fitted diffusion coefficient and exponent α also show that polymer beads connected by FENE bonds diffuse slightly faster than beads with harmonic bonds, especially for large Hurst parameters ($H > 0.50$).

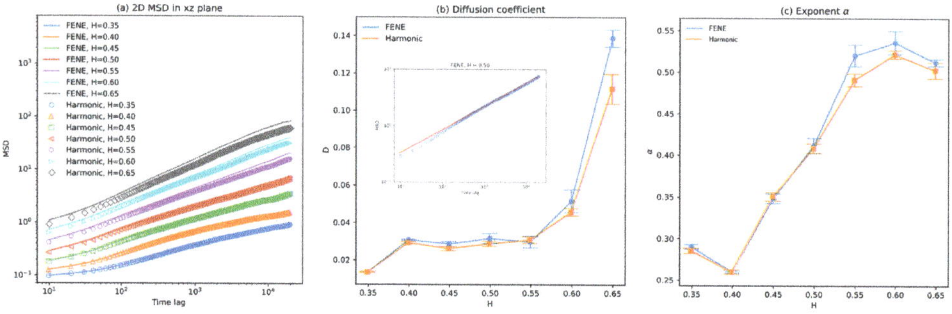

Figure 4. (**a**) Averaged 2D mean-squared displacement in xz plane as a function of time lag for polymer models with both FENE and harmonic bonds for different Hurst parameters. (**b**) Fitted diffusion coefficient as a function of Hurst parameter H for polymer models with FENE and harmonic bonds. (Inset) one example of fitting of averaged 2D MSD against time lag according to Equation (9). (**c**) Fitted exponent α as a function of Hurst parameter H for polymer models with FENE and harmonic bonds (the error bars are standard deviations).

Although various random walks models such as CTRW [28], fBM [30], OD [27], and diffusing diffusivity [6] can reproduce anomalous subdiffusion, only fractional Brownian motion [29,30] seems to be promising concerning the anticorrelated velocity of chromosomal loci and cytoplasmic particles. The effects of the polymer model structures and excluded volume interactions between the polymer segments on the velocity autocorrelation function were examined in this work. The velocity autocorrelation function of a discrete-time process with a time step Δt is given by the equation below [3]:

$$C_v^{(\Delta t)}(t) = \frac{1}{\Delta t^2}\langle (\mathbf{r}(t+\Delta t) - \mathbf{r}(t)) \cdot (\mathbf{r}(\Delta t) - \mathbf{r}(0)) \rangle \quad (10)$$

where $\mathbf{r}(t)$ is the bead position at time t and $C_v^{(\Delta t)}(t)$ is the velocity autocorrelation function with a time step $\Delta t = 0.01$ at simulation time t. One polymer bead in the middle of the backbone chain (shown as the "backbone" in Figure 5) as well as one polymer bead on the side loop attached to the end bead of the backbone chain (shown as the "side-loop" in Figure 5) are chosen for the calculations of velocity autocorrelation functions. The normalized velocity autocorrelation functions $C_v^{(\Delta t)}(\tau)/C_v^{(\Delta t)}(\tau = 0)$ are demonstrated in Figure 5 for polymer models with both FENE bonds and harmonic bonds. As can be seen in Figure 5, the velocity autocorrelation function reduces to 0 for $H = 0.5$ within the first time step $\tau = \Delta t = 0.01$. It remains zero thereafter for ordinary Brownian motion ($H = 0.5$) is an uncorrelated Markovian process. For small Hurst parameters ($H < 0.5$), the velocity is anticorrelated, i.e., the velocity autocorrelation function has a negative part for $\tau > 0$, which agrees with previous experimental measurements of chromosomal loci [3,30]. As H increases from 0.35 to 0.50, this anticorrelation becomes weaker, resulting from a weaker "elasticity" of the cytoplasm. If the Hurst parameter H is above 0.50, the velocity of the polymer bead becomes correlated; that is, the velocity autocorrelation function remains positive for $\tau > 0$. Again, the polymer beads act as if they have "momenta" for $H > 0.50$, although the masses of polymer beads are neglected in our simulations based on Equation (5).

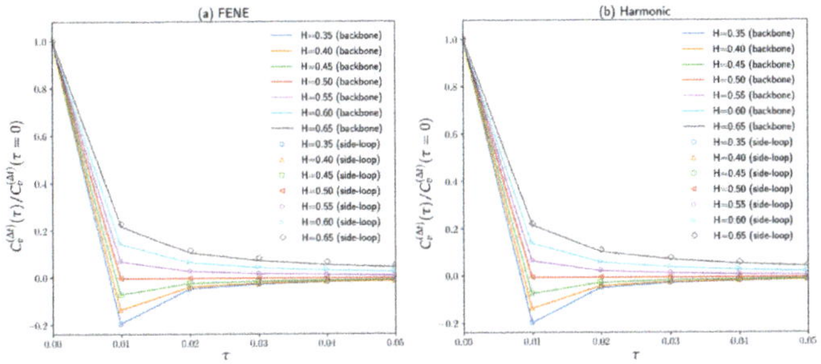

Figure 5. (**a**) Velocity autocorrelation function $C_v^{(\Delta t)}(\tau)$ of both a backbone chain bead and a side loop bead for different Hurst parameters with FENE bonding energy. (**b**) Velocity autocorrelation function $C_v^{(\Delta t)}(\tau)$ of both a backbone chain bead and a side loop bead for different Hurst parameters with harmonic bonds.

The diffusion coefficients, as well as the exponents α of those two polymer beads on the backbone chain and the side loop as described above, are obtained by averaging over three independent runs and are plotted as functions of Hurst parameters in Figures 6 and 7. Similarly, the fitted diffusion coefficient of the polymer beads increases significantly as H increases, just like the fitted diffusion coefficient based on the ensemble-averaged MSD. The side loop polymer bead diffuses faster than the backbone polymer bead since the number density of the polymer beads around the backbone chain is larger than that around the side loop attached to the end of the backbone chain bead, which can be seen from Figure 3. Because the number density of the polymer beads is evenly distributed across the cell volume for small H, the error bars of the fitted diffusion coefficients are also small for $H < 0.5$. The exponent α of the side loop polymer bead is close to 0.5, which agrees with the predictions by the simple Rouse model combined with the Langevin dynamics simulation [29]. However, the exponent α of the backbone polymer bead is 0.30~0.35 for a small $H \leq 0.5$, owing to the interactions between the backbone polymer bead and the condensed polymer segments around the backbone chain.

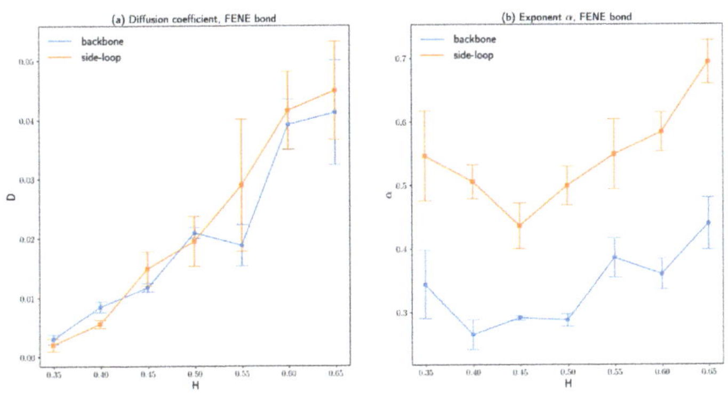

Figure 6. (**a**) The diffusion coefficients of a backbone chain bead and a side loop bead on a polymer model with FENE bonds for various Hurst parameters. (**b**) The exponent α characterizes anomalous diffusion of a backbone chain bead and a side loop bead on a polymer model with FENE bonds for various Hurst parameters (the error bars are standard deviations).

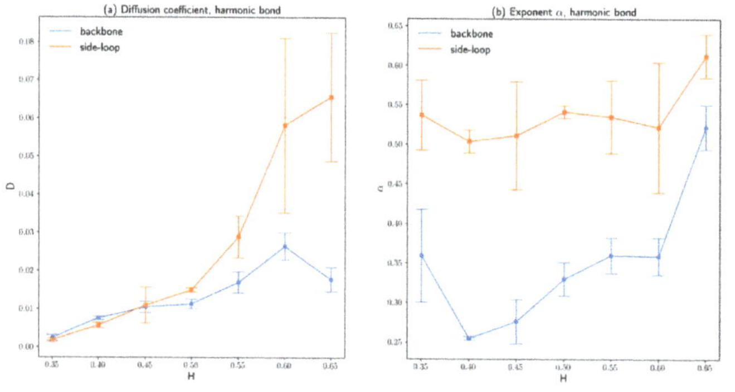

Figure 7. (**a**) The diffusion coefficients of a backbone chain bead and a side loop bead on a polymer model with harmonic bonds for various Hurst parameters. (**b**) The exponent α characterizing anomalous diffusion of a backbone chain bead and a side loop bead on a polymer model with harmonic bonds for various Hurst parameters (the error bars are standard deviations).

By comparing Figures 6 and 7, it can be concluded that the types of bonds connecting consecutive polymer beads have insignificant effects on the diffusive behaviors of polymer beads. Our simulation results show that polymer beads' exponent α can differ depending on their positions. This variation in α agrees with previous experimental results [4,5].

3.3. Displacement Distributions of Polymer Beads with Fractional Brownian Noises

Fractional Brownian motion is a Gaussian process with stationary increments [28,41]. So, the displacement distribution of a single tracer performing fractional Brownian motion is Gaussian. The major difference between the anomalous diffusive behavior of chromosomal loci [4,5] and cytoplasmic particles [6] is that chromosomal loci are affected not only by viscoelastic cytoplasm but also by interactions between DNA segments such as DNA bridging [25]. To investigate how the displacement distributions of polymer beads are affected by the viscoelasticity of cytoplasm as well as the polymer dynamical structure, the one-dimensional displacement distributions in the z-axis direction of both a backbone chain bead and a side loop bead are plotted in Figures 8 and 9. Simulations show that the normalized 1D displacement ($\Delta z/\delta$) distributions agree very well with normal distribution for different Hurst parameters H, as demonstrated in Figures 8 and 9. This agreement indicates that polymer beads exhibit a non-Brownian yet Gaussian subdiffusion (for $H \neq 0.5$). The normalized 1D displacement distributions ($\Delta z/\delta$) over different time lags (τ = 1, 10, 100, 1000) plotted in Figures 8 and 9 show that the displacement distribution of both the backbone chain bead and side-side loop bead can remain Gaussian for a wide range of τ. However, the displacement distribution starts to deviate from the Gaussian distribution for large displacement ($\Delta z > 3\delta$), especially for $\tau = 1000$, as shown in Figure 9d. Such deviations can be attributed to insufficient sampling as well as cylindrical confinement. Moreover, the bonding energy (FENE/harmonic) has negligible effects on displacement distributions, which agrees with our previous analysis on diffusion coefficients and exponent α.

Recent experimental results show that the displacement distributions of cytoplasmic particles [6] as well as chromosomal loci [44] follow a Laplace distribution. Many mathematical combinations of random variables can produce the Laplace random variable. A straightforward example is that the Laplace random variable can be represented by the product of a Gaussian random variable and the square root of an exponential random variable. Moreover, RNA-protein particle diffusivities exhibit an exponential distribution [6], which might contribute to the Laplace distribution of cytoplasmic particle displacement. Chubynsky [33] developed a diffusing diffusivity model that results in an exponential diffusivity distribution for specific conditions. Hence, the diffusing diffusivity model has attracted more attention recently [32] as it has great potential to elucidate the origin of the non-Gaussian behaviors of cytoplasmic particles and chromosomal loci. However, the simple version of the diffusing diffusivity model cannot explain the anticorrelated velocity of tracers in the cytoplasm, as such a simple model does not have a "memory" as the fractional Brownian motion model does. Using fractional Gaussian noise for each polymer bead during our fractional Brownian dynamics simulations, we implicitly assume that the viscoelasticity of cytoplasm remains unchanged across the cell, which is not true as the cytoplasm is a heterogeneous environment [35]. It is possible as well as reasonable to combine the fractional Brownian dynamics as described above with a random diffusivity of the chromosomal locus to model the non-Brownian and non-Gaussian subdiffusion in vivo [6,44]. One possible method to include diffusing or random diffusivity into this fractional Brownian dynamics simulation is assigning a different radius a_{bead} to polymer beads in different areas within the cylindrical confinement to simulate the heterogeneity of the interior environment of the bacterial cell.

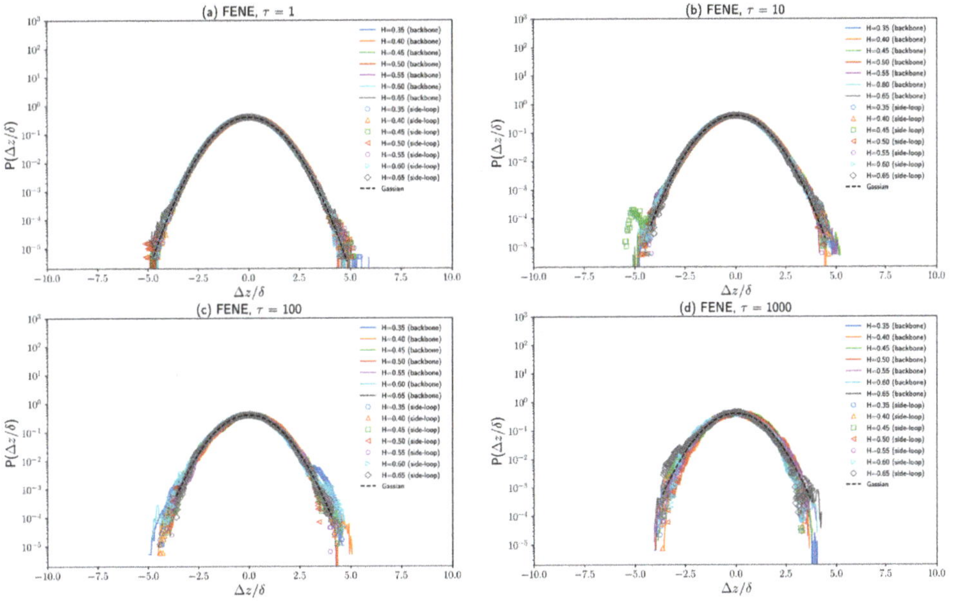

Figure 8. **(a–d)** One-dimensional displacement ($\Delta z/\delta$) distributions of a backbone chain bead and a side loop bead on a polymer with FENE bonds for different Hurst parameters over different time lags (τ = 1∼1000). δ is the standard deviation of displacement in z-axis direction (Δz). The Gaussian distribution is plotted as a black dashed curve.

In this article, in order to simulate the anomalous diffusion of chromosomal loci, we use fractional Gaussian noise instead of white noise (the Brownian force) to simulate the effects of viscoelastic cytoplasm on polymer segments to reduce the number of degrees of freedom of the simulation system. This implicit model of cytoplasm speeds up simulations dramatically by averaging over nonessential degrees of freedom. However, the significant reduction in degrees of freedom obstructs the accurate computation of system entropy. The crowding effect, which is crucial in chromosome organization [45,46], cannot be studied directly in our simple model. Furthermore, since fractional Gaussian noise is highly correlated, generating fractional Gaussian noise is much more complex than generating ordinary Gaussian random numbers. For the simulation results presented in this manuscript, the Hurst parameter H is varied only between 0.35 and 0.65, mainly due to the fact that generating fractional Gaussian noise becomes extremely slow for a small $H < 0.35$

and large $H > 0.65$. The fBm-0.3.0 package we used for fGn generation switches from the fast Davies Harte method [42] to the slow Hosking algorithm [41] for small and large Hurst parameters. In addition, the fGn generation has to be carried out at the beginning of fractional Brownian dynamics simulations, which is computationally expensive. In contrast, the white noise generation in ordinary Brownian dynamics simulations can be performed on the fly. So, it is not easy to apply this fractional Brownian dynamics method to larger systems, for example, a modified version of this "bottle-brush"-type polymer model of bacterial chromosomes developed by Swain et al. [21] which contains about 40,000 polymer beads and extra crowders.

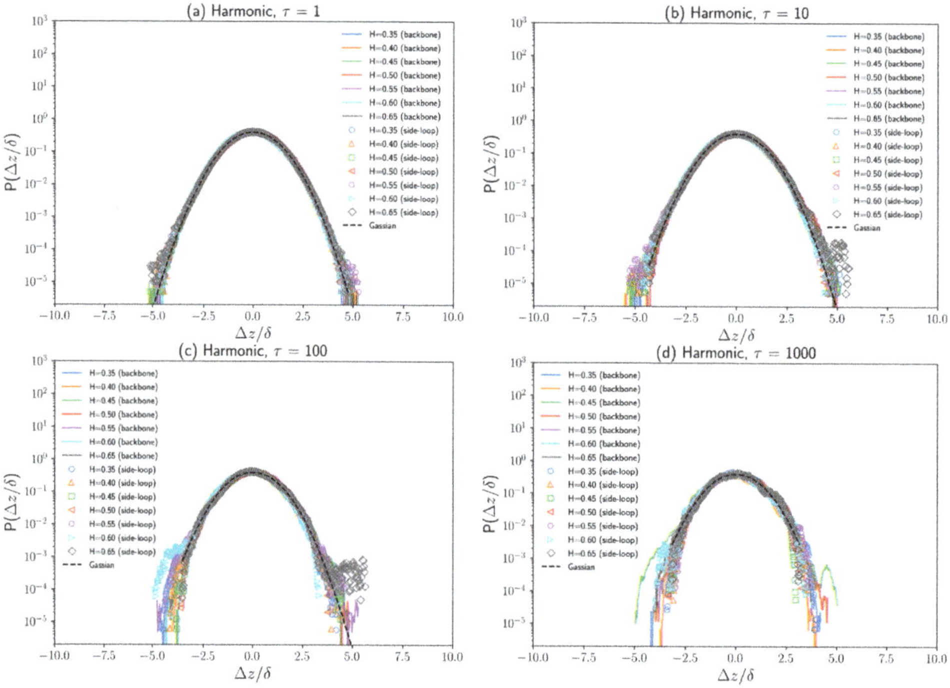

Figure 9. (**a**–**d**) One-dimensional displacement ($\Delta z/\delta$) distributions of a backbone chain bead and a side loop bead on a polymer with harmonic bonds for different Hurst parameters over different time lags (τ = 1~1000). δ is the standard deviation of displacement in z-axis direction (Δz). The Gaussian distribution is plotted as a black dashed curve.

Since chromosomal loci perform subdiffusion in cells, it seems less attractive to carry out fractional Brownian dynamics simulations for a large $H > 0.50$, where fractional Brownian particles perform accelerated Brownian motion. However, as the nonequilibrium metabolic activities play a crucial role in cytoplasmic subdiffusion and can be modeled by a random force [37], the diffusion of polymer beads might be simulated by an overlay of an fGn with $H < 0.5$ and an fGn with $H > 0.50$, where both viscoelastic cytoplasm and metabolic activity are modeled by fractional Gaussian noise. Nevertheless, mapping the random forces exerted on polymer segments to the metabolic activities that occur in cells is highly challenging.

Another problem of this fractional Brownian dynamics simulation is the sufficient sample size. Although ergodicity is not broken within fractional Brownian motion [28], fBm is a self-similar process [41]. As a result, for a large H, fBm is ballistic-motion-like, as shown by the green curve in Figure 1b. To obtain the accurate ensemble-averaged diffusivity, a large sample size is required for fractional Brownian motion with a large H compared to a standard Brownian dynamics simulation since analytical solutions to accelerated Brownian motion are only available for one-dimensional problems [38]. This sample size problem also explains the large standard deviations of the fitted diffusion coefficients for $H > 0.55$, as shown in Figures 6 and 7.

4. Conclusions

Using a simple "bottle-brush"-type polymer model combined with fractional Gaussian noises, the anomalous diffusive behaviors of polymer segments in viscoelastic fluid are investigated systematically by using a homemade fractional Brownian dynamics simulation code. The standard Brownian dynamics simulations are carried out by setting the Hurst parameter H to 0.5. Moreover, the standard Brownian dynamics simulation results are compared to previous experimental and theoretical studies to verify the fractional Brownian dynamics code. By varying the Hurst parameter H, the dynamical structures and the anomalous diffusivity of the polymer model are explored, from a subdiffusive regime to a superdiffusive regime. Our fractional Brownian dynamics simulation demonstrates that both the subdiffusive behaviors of polymer beads ($\alpha = 0.38 \sim 0.40$) and the anticorrelated velocity of chromosomal loci can be attributed to the weak viscoelasticity ($H \approx 0.45$) of cytoplasm. It is crude to assume that fractional Gaussian noises for different polymer beads are independent ($\mathbb{E}X_i^{(n+k)}X_j^{(n)} = 0$ for $i \neq j$) since polymer beads that are close to each other have a similar environment and thus experience correlated fractional Gaussian noises ($\mathbb{E}X_i^{(n+k)}X_j^{(n)} \neq 0$ for small $|\mathbf{r}_i - \mathbf{r}_j|, k \approx 0$). However, our assumption of an independent fGn might be sufficient for the simulations of the subdiffusion of chromosomal loci because the exponent $\alpha = 0.38 \sim 0.40$ characterizing the anomalous subdiffusion of chromosomal loci can be reproduced with a Hurst parameter $H \in (0.45, 0.50)$, where the autocorrelation function of fGn (Equation (8)) decays rapidly. Since our simulations do not include the heterogeneity of cytoplasm and metabolic activity, this simple model does not reproduce the Laplace displacement distribution of cytoplasmic anomalous diffusion. Nevertheless, this fractional Brownian dynamics simulation method can be used to investigate many crucial biological reactions, such as chromosome segregation, facilitated diffusion [39], and the responses of bacterial chromosomes to external mechanical forces [47], as the viscoelasticity of cytoplasm plays a vital role in such processes.

Author Contributions: Conceptualization, S.Y. and R.C.; methodology, S.Y.; software, S.Y.; validation, R.C. and G.W.; formal analysis, S.Y. and X.M.; investigation, G.W. and X.M.; resources, X.M.; data curation, R.C.; writing—original draft preparation, S.Y.; writing—review and editing, R.C.; visualization, S.Y.; supervision, G.W.; project administration, X.M.; funding acquisition, S.Y. All authors have read and agreed to the published version of the manuscript.

Funding: This research was funded by the National Natural Science Foundation of China under grant number 21903087.

Institutional Review Board Statement: Not applicable.

Informed Consent Statement: Not applicable.

Data Availability Statement: Data are contained within the article.

Acknowledgments: We acknowledge support from the NSFC and the Doctoral Program of Innovation and Entrepreneurship in Jiangsu Province.

Conflicts of Interest: The authors declare no conflicts of interest. The funders had no role in the design of the study; in the collection, analyses, or interpretation of data; in the writing of the manuscript; or in the decision to publish the results.

References

1. Soh, S.; Byrska, M.; Kandere-Grzybowska, K.; Grzybowski, B.A. Reaction-diffusion systems in intracellular molecular transport and control. *Angew. Chem. Int. Ed.* **2010**, *49*, 4170–4198. [CrossRef]
2. Schavemaker, P.E.; Boersma, A.J.; Poolman, B. How important is protein diffusion in Prokaryotes? *Front. Mol. Biosci.* **2018**, *5*, 93. [CrossRef]
3. Weber, S.C.; Spakowitz, A.J.; Theriot, J.A. Bacterial chromosomal loci move subdiffusively through a viscoelastic cytoplasm. *Phys. Rev. Lett.* **2010**, *104*, 27–30. [CrossRef]
4. Javer, A.; Long, Z.; Nugent, E.; Grisi, M.; Siriwatwetchakul, K.; Dorfman, K.D.; Cicuta, P.; Cosentino Lagomarsino, M. Short-time movement of *E. coli* chromosomal loci depends on coordinate and subcellular localization. *Nat. Commun.* **2013**, *4*, 3003. [CrossRef] [PubMed]
5. Javer, A.; Kuwada, N.J.; Long, Z.; Benza, V.G.; Dorfman, K.D.; Wiggins, P.A.; Cicuta, P.; Lagomarsino, M.C. Persistent super-diffusive motion of *Escherichia coli* chromosomal loci. *Nat. Commun.* **2014**, *5*, 3854. [CrossRef] [PubMed]
6. Lampo, T.J.; Stylianidou, S.; Backlund, M.P.; Wiggins, P.A.; Spakowitz, A.J. Cytoplasmic RNA-protein particles exhibit non-Gaussian subdiffusive behavior. *Biophys. J.* **2017**, *112*, 532–542. [CrossRef] [PubMed]
7. Parry, B.R.; Surovtsev, I.V.; Cabeen, M.T.; O'Hern, C.S.; Dufresne, E.R.; Jacobs-Wagner, C. The bacterial cytoplasm has glass-like properties and is fluidized by metabolic activity. *Cell* **2014**, *156*, 183–194. [CrossRef] [PubMed]
8. Okumus, B.; Landgraf, D.; Lai, G.C.; Bakhsi, S.; Arias-Castro, J.C.; Yildiz, S.; Huh, D.; Fernandez-Lopez, R.; Peterson, C.N.; Toprak, E.; et al. Mechanical slowing-down of cytoplasmic diffusion allows in vivo counting of proteins in individual cells. *Nat. Commun.* **2016**, *7*, 11641. [CrossRef]
9. Wlodarski, M.; Raciti, B.; Kotar, J.; Lagomarsino, M.C.; Fraser, G.M.; Cicuta, P. Both genome and cytosol dynamics change in *E. coli* challenged with sublethal rifampicin. *Phys. Biol.* **2017**, *14*, 015005. [CrossRef]
10. Wlodarski, M.; Mancini, L.; Raciti, B.; Sclavi, B.; Lagomarsino, M.C.; Cicuta, P. Cytosolic crowding drives the dynamics of both genome and cytosol in *Escherichia coli* challenged with sub-lethal antibiotic treatments. *iScience* **2020**, *23*, 101560. [CrossRef]
11. Yu, I.; Mori, T.; Ando, T.; Harada, R.; Jung, J.; Sugita, Y.; Feig, M. Biomolecular interactions modulate macromolecular structure and dynamics in atomistic model of a bacterial cytoplasm. *eLife* **2016**, *5*, e19274. [CrossRef]
12. Yoo, J.; Winogradoff, D.; Aksimentiev, A. Molecular dynamics simulations of DNA–DNA and DNA–protein interactions. *Curr. Opin. Struct. Biol.* **2020**, *64*, 88–96. [CrossRef] [PubMed]
13. Marrink, S.J.; Corradi, V.; Souza, P.C.T.; Ingólfsson, H.I.; Tieleman, D.P.; Sansom, M.S.P. Computational modeling of realistic cell membranes. *Chem. Rev.* **2019**, *119*, 6184–6226. [CrossRef] [PubMed]
14. Uusitalo, J.J.; Ingólfsson, H.I.; Akhshi, P.; Tieleman, D.P.; Marrink, S.J. Martini coarse-grained force field: Extension to DNA. *J. Chem. Theory Comput.* **2015**, *11*, 3932–3945. [CrossRef]
15. Sengar, A.; Ouldridge, T.E.; Henrich, O.; Rovigatti, L.; Šulc, P. A primer on the oxDNA model of DNA: When to use it, how to simulate it and how to interpret the results. *Front. Mol. Biosci.* **2021**, *8*, 693710. [CrossRef]
16. Shinoda, W.; Discher, D.E.; Klein, M.L.; Loverde, S.M. Probing the structure of PEGylated-lipid assemblies by coarse-grained molecular dynamics. *Soft Matter* **2013**, *9*, 11549–11556. [CrossRef]
17. Frembgen-Kesner, T.; Elcock, A.H. Computer simulations of the bacterial cytoplasm. *Biophys. Rev.* **2013**, *5*, 109–119. [CrossRef]
18. Ando, T.; Skolnick, J. Crowding and hydrodynamic interactions likely dominate in vivo macromolecular motion. *Proc. Natl. Acad. Sci. USA* **2010**, *107*, 18457–18462. [CrossRef] [PubMed]
19. Hasnain, S.; McClendon, C.L.; Hsu, M.T.; Jacobson, M.P.; Bandyopadhyay, P. A new coarse-grained model for *E. coli* cytoplasm: Accurate calculation of the diffusion coefficient of proteins and observation of anomalous diffusion. *PLoS ONE* **2014**, *9*, e106466. [CrossRef]
20. Słyk, E.; Skóra, T.; Kondrat, S. How macromolecules softness affects diffusion under crowding. *Soft Matter* **2022**, *18*, 5366–5370. [CrossRef]
21. Swain, P.; Mulder, B.M.; Chaudhuri, D. Confinement and crowding control the morphology and dynamics of a model bacterial chromosome. *Soft Matter* **2019**, *15*, 2677–2687. [CrossRef] [PubMed]
22. Chaudhuri, D.; Mulder, B.M. Spontaneous helicity of a polymer with side loops confined to a cylinder. *Phys. Rev. Lett.* **2012**, *108*, 268305. [CrossRef] [PubMed]
23. Chaudhuri, D.; Mulder, B.M. Molecular dynamics simulation of a feather-boa model of a bacterial chromosome. *Methods Mol. Biol.* **2018**, *1837*, 403–415. [PubMed]
24. Berlatzky, I.A.; Rouvinski, A.; Ben-Yehuda, S. Spatial organization of a replicating bacterial chromosome. *Proc. Natl. Acad. Sci. USA* **2008**, *105*, 14136–14140. [CrossRef] [PubMed]
25. Subramanian, S.; Murray, S.M. Subdiffusive movement of chromosomal loci in bacteria explained by DNA bridging. *Phys. Rev. Res.* **2023**, *5*, 023034. [CrossRef]
26. Ernst, D.; Hellmann, M.; Köhler, J.; Weiss, M. Fractional Brownian motion in crowded fluids. *Soft Matter* **2012**, *8*, 4886–4889. [CrossRef]
27. Mardoukhi, Y.; Jeon, J.-H.; Metzler, R. Geometry controlled anomalous diffusion in random fractal geometries: Looking beyond the infinite cluster. *Phys. Chem. Chem. Phys.* **2015**, *17*, 30134–30147. [CrossRef]
28. Magdziarz, M.; Weron, A.; Burnecki, K.; Klafter, J. Fractional Brownian motion versus the continuous-time random walk: A simple test for subdiffusive dynamics. *Phys. Rev. Lett.* **2009**, *103*, 180602. [CrossRef]

29. Weber, S.C.; Theriot, J.A.; Spakowitz, A.J. Subdiffusive motion of a polymer composed of subdiffusive monomers. *Phys. Rev. E* **2010**, *82*, 011913. [CrossRef]
30. Weiss, M. Single-particle tracking data reveal anticorrelated fractional Brownian motion in crowded fluids. *Phys. Rev. E* **2013**, *88*, 010101. [CrossRef] [PubMed]
31. Metzler, R.; Klafter, J. The random walk's guide to anomalous diffusion: A fractional dynamics approach. *Phys. Rep.* **2000**, *339*, 1–77. [CrossRef]
32. Chechkin, A.V.; Seno, F.; Metzler, R.; Sokolov, I.M. Brownian yet non-Gaussian diffusion: From superstatistics to subordination of diffusing diffusivities. *Phys. Rev. X* **2017**, *7*, 021002 [CrossRef]
33. Chubynsky, M.V.; Slater, G.W. Diffusing diffusivity: A model for anomalous, yet Brownian, diffusion. *Phys. Rev. Lett.* **2014**, *113*, 098302 [CrossRef]
34. dos Santos, M.A.F.; Junior, L.M. Random diffusivity models for scaled Brownian motion. *Chaos Solit. Fractals* **2021**, *144*, 110634. [CrossRef]
35. Spakowitz, A.J. Transient anomalous diffusion in a heterogeneous environment. *Front. Phys.* **2019**, *7*, 119. [CrossRef]
36. Baldovin, F.; Orlandini, E.; Seno, F. Polymerization induces non-Gaussian diffusion. *Front. Phys.* **2019**, *7*, 124. [CrossRef]
37. Fodor, É.; Guo, M.; Gov, N.S.; Visco, P.; Weitz, D.A.; van Wijland, F. Activity-driven fluctuations in living cells. *EPL* **2015**, *110*, 48005. [CrossRef]
38. Metzler, R.; Klafter, J. Accelerating Brownian motion: A fractional dynamics approach to fast diffusion. *Europhys. Lett.* **2000**, *51*, 492–498. [CrossRef]
39. Florescu, A.-M.; Joyeux, M. Description of nonspecific DNA-protein interaction and facilitated diffusion with a dynamical model. *J. Chem. Phys.* **2009**, *130*, 015103. [CrossRef]
40. Mandelbrot, B.B.; van Ness, J.W. Fractional Brownian motions, fractional noises and applications. *SIAM Rev.* **1968**, *10*, 422–437. [CrossRef]
41. Dieker, A.B.; Mandjes, M. On spectral simulation of fractional Brownian motion. *Probab. Eng. Inf. Sci.* **2003**, *17*, 417–434. [CrossRef]
42. Davies, R.B.; Harte, D.S. Tests for Hurst Effect. *Biometrika* **1987**, *74*, 95–101. [CrossRef]
43. Ha, B.-Y.; Jung, Y. Polymers under confinement: Single polymers, how they interact, and as model chromosomes. *Soft Matter* **2015**, *11*, 2333–2352. [CrossRef]
44. Yu, S.; Dorfman, K.D.; Sheats, J.; Sclavi, B.; Cosentino Lagomarsino, M.; Cicuta, P. Subdiffusion of loci and cytoplasmic particles are different in compressed *Escherichia coli* cells. *Commun. Biol.* **2018**, *1*, 176. [CrossRef]
45. Jeon, C.; Jung, Y.; Ha, B.-Y. A ring-polymer model shows how macromolecular crowding controls chromosome-arm organization in *Escherichia coli*. *Sci. Rep.* **2017**, *7*, 11896. [CrossRef]
46. Chauhan, G.; Simpson, M.L.; Abel, S.M. Crowding-induced interactions of ring polymers. *Soft Matter* **2021**, *17*, 16–23. [CrossRef]
47. Si, F.; Li, B.; Margolin, W.; Sun, S.X. Bacterial growth and form under mechanical compression. *Sci. Rep.* **2015**, *5*, 11367. [CrossRef]

polymers

MDPI

Article

Size and Shape of Primary (Bio)Polyelectrolyte Complexes Chitosan/Gelatin: Study Using Small-Angle X-Ray Scattering from Synchrotron Radiation

Aleksandr Podshivalov *, Mikhail Litvinov, Aleksandr Kashurin and Ksenia Danilova

Center for Chemical Engineering, ITMO University, Kronverkskiy Prospekt, 49, 197101 Saint-Petersburg, Russia; mikhail.litvinov.1996@mail.ru (M.L.); pkashurin@mail.ru (A.K.); k.danilova@itmo.ru (K.D.)
* Correspondence: podshivalov@itmo.ru

Abstract: In this work, using small-angle X-ray scattering from synchrotron radiation, the macromolecular structure of chitosan and gelatin polyelectrolytes and their mixtures at various pH values and ratios was studied to determine the size and shape of primary supramolecular (bio)PEC. Analysis of the scattering profiles of the initial solutions of chitosan and gelatin with the building of the pair distance function showed the formation of single-modal distributions with a maximum molecular size of 46 and 32.2 nm, respectively. Ab initio reconstruction of the macromolecule's shape showed the formation of objects shaped like an oblate spheroid. In mixtures of chitosan and gelatin at a pH below the isoelectric point, it was found that the scattering structures correspond to the initial biopolymers. However, it is observed that values of the aspect ratio at a ratio above 1:10 gradually increase, which indicates a slight elongation of the average particle and indirectly indicates the formation of dissipative structures of (bio)PEC. In mixtures at a pH above the isoelectric point, it was shown that at ratios above 1:5, the formation of primary supramolecular complexes is observed, which is accompanied by an increase in zero-scattering intensity by about three times, maximum molecular size by two to two-and-a-half times relative to the initial polymers, and the formation of elongated structures corresponding to the cylinder (swollen spiral). It may be a consequence of the increased efficiency of the polyelectrolyte associative interaction between chitosan and gelatin.

Keywords: chitosan; gelatin; polyelectrolyte complex; SAXS; synchrotron radiation

1. Introduction

Recently, much attention has been paid by researchers in the field of drug delivery systems and gene carriers to the investigations of self-assembly of interpolyelectrolyte complexes (polyelectrolyte coacervates) [1–4]. These complexes are formed because of the electrostatic interactions of oppositely charged polyelectrolytes. They can spontaneously form various complex structures: spherical micelles, complex aggregates (coacervates), and solid-phase supramolecular structures [5,6]. The analysis of the existing mechanisms of coacervation can be combined into the following stages [7,8]: (1) pairing of oppositely charged ions with the formation of intermediate supramolecular assemblies (soluble PEC); (2) rearrangement of soluble complexes into the coacervate phase due to non-electrostatic interactions; and (3) with a non-stoichiometric charge balance, excess chains of polyanions or polycations can attach to the surface of the complex, causing repulsive interactions, which leads to colloidal stability. With a stoichiometric charge balance, subsequent self-assembly occurs, with the formation of a solid-phase supramolecular structure.

Numerous works [9–13] studying the self-assembly of PEC structures mainly concerned the stages of formation of the coacervate phase and investigated the conditions for obtaining these structures (concentration, ratio, pH, and ionic strength), as well as the effect of these conditions on the size, structure, polymorphism, and stability of coacervates. However, at present, not enough work is devoted to the study of the primary stage of PEC formation, which is mainly due to the dissipativity of the primary supramolecular assemblies formed and the high constant rate of binding of ion pairs. At the same time, existing light scattering methods allow us to estimate only the evolution of the overall size of submicron particles, but at the same time they do not provide any information about the structure of aggregates and provide little information about the details of the structure of nanoparticles. Currently, synchrotron or neutron radiation methods are used to study data on the size, shape, and internal structure of primary complexes at the nanoparticle level [6,14–19]. For example, in the work using small-angle X-ray scattering with time resolution [6], the kinetics of coacervation of sodium polyacrylate and polyallylamine hydrochloride were studied and it was established that at the initial stage, the formation of almost neutral particles of primary coacervates occurs, which subsequently assemble into large coacervates with evolution over time comparable to the rate of Brownian coagulation in the presence of salt. In another work [16] using neutron and X-ray scattering, the coacervation of poly(acrylic acid) and poly(N,N-dimethylaminoethylmethacrylate) was studied, and the formation of an ideal Gaussian conformation of the polymer chain was discovered in complex coacervates forming a strongly intertwined network with a cell size much smaller than the radius of gyration of polymers. These works are related to the mechanisms of formation of PEC based on homopolymer polyelectrolytes. However, PEC formed from two-block copolyelectrolytes with charged and neutral blocks is of great interest. In such systems, the stabilization of coacervates strongly depends on the aggregation of neutral blocks. So, for example, in the works [14,17], the kinetic pathways of the formation of micellar-like PEC formed from anionic and cationic neutral block copolymers were studied and it was found that metastable cylindrical aggregates are formed at the first stage of complexation. Further, they regroup when the charge is balanced into stable micellar-like aggregates. At the same time, the rate of regrouping strongly depends on the MW of the neutral block. However, despite the great scientific foundation, the works described above mainly concerned PECs based on synthetic polyelectrolytes.

Among biopolymers, special attention should be paid to chitosan, which is a natural copolymer consisting of polycation units of glucosamine and neutral units of acetylglucosamine, and its complexes with polyampholytic protein hydrolysates containing both neutral, anionic, and cationic units. Among such systems, chitosan-gelatin (bio)PEC is of interest because of their availability for creating drug delivery systems. The formation of (bio)PEC in the chitosan-gelatin system has been extensively studied in the literature, but most studies have focused exclusively on the macroscopic aspects of coacervation. In numerous works [20–29], the process of coacervation at a pH above the isoelectric point of gelatin has been described in detail. For example, it was shown in work [24] that the formation of a coacervate phase between chitosan and gelatin type B is possible in the pH range of 5.5–6.5 with a minimum polymer ratio of 1:10. Moreover, at a ratio of 1:20 and pH = 6, coacervates with a zeta potential close to 0 and a diameter of ~30 μm are formed. In another work [25], it was shown that the formation of the coacervate phase in a mixture of chitosan and gelatin type B solutions at a ratio of 1:2 is accompanied by a significant increase in the turbidity of the solution, with maximum values at pH = 5.5. It is also shown that the formation of the coacervate phase under these conditions leads to an increase in the particle size of the dispersed phase by more than five times compared with mixtures at pH = 3. In our previous work [29], we also studied the formation and structure of the coacervate phase of (bio)PEC in mixtures of dilute chitosan and gelatin type B solutions. It was shown that the formation of stable dispersions of coacervates with a size ranging from 3 to 4.5 μm is observed in the pH range from 5 to 6.4 and a polymer ratio above 1:5. Additionally, using the dynamic light scattering method, it was shown that the coacervate phase of the (bio)PEC consists of submicron particles of primary complexes and micron-sized coacervates.

An analysis of the literature in the field of chitosan-gelatin (bio)PEC shows that the main works focus on the effect of the pH of the medium and composition on the formation of dispersions of complex particles and their morphology, studied mainly by optical microscopy, turbidimetry, and dynamic light scattering. However, the shape and size of the primary (bio)PEC structures have not been studied in detail. In this work, we solve this fundamental problem by applying synchrotron SAXS research with the analysis of pair distance distribution functions (PDDF) to systematically study the evolution of the sizes of primary structures of (bio)PEC depending on the composition and pH of the medium, the mechanisms of their formation at the molecular level, and to study the shape of primary structures of (bio)PEC. This approach will make it possible for the first time to obtain detailed information about the nanostructure of the forming complexes, which is not available for traditional optical methods.

2. Materials and Methods

2.1. Materials

We used gelatin type B produced by Sigma-Aldrich Corporation (St. Louis, MO, USA) (CAS 9000-70-8) with a viscous average molecular weight of M_v = 63.1 ± 8.7 kDa and an isoelectric point of pI = 4.7 ± 0.3, which was determined by viscometric and turbidimetric methods in previous work [29]; chitosan produced by CJSC "BioProgress" (Losino-Petrovsky, Russia) with M_v = 206 ± 1.5 kDa, protonation constant pK_0 = 6.35, and degree of deacetylation of DD = 83.1% [30]; glacial acetic acid, sodium hydroxide produced by JSC "LenReactiv" (Saint Petersburg, Russia) and deionized water.

2.2. Preparation of Chitosan/Gelatin (Bio)Polyelectrolyte Complex Dispersions

To obtain a dispersion of (bio)PEC between chitosan and gelatin 0.5 M acetic acid solutions of chitosan with C = 1 mg/mL and gelatin, with C = 5 mg/mL were prepared over 24 h at a temperature of 35 °C with constant stirring using a magnetic stirrer. After preparation, the polymer solutions were centrifuged using an Eppendorf MiniSpin Plus centrifuge, Eppendorf AG (Hamburg, Germany) at 10,000 rpm for 10 min. Further, the obtained solutions were mixed in the following ratios of 1:1, 1:5, 1:10, 1:15, and 1:20 by weight over 1 h at a temperature of 35 °C with constant stirring. After that, the obtained solutions were adjusted to pH = 3 by titration with 0.5 M acetic acid and to pH = 5.5 with 1 M and 2 M of NaOH solutions. These pH = 5.5 values were chosen because it was previously established [29] that at a pH above the isoelectric point, effective formation of scattering particles (bio)PEC occurs, which manifests itself in significant turbidity of the solution.

2.3. Small-Angle X-Ray Scattering of Synchrotron Radiation

Data on the X-ray scattering of synchrotron radiation were obtained on the beamline of the BioMUR station of the synchrotron radiation source (National Research Center

"Kurchatov Institute", Moscow, Russia) [31]. The BioMUR experimental station consists of optical and experimental parts. In the optical part, a beam is formed in the geometry of the three-slit collimation, which is focused using a Si (111) monochromator mounted at a Bragg angle of 13.5° and equipped with a stepper motor for bending the crystal and focusing the beam in the horizontal direction. The beam energy after the monochromator crystal is 8.58 keV (λ = 0.1445 nm). SAXS data were recorded using a two-dimensional pixel detector, Pilatus-3 1M (DECTRIS, Baden-Daettwil, Switzerland), with a pixel size of 172 µm and a recording area of 168 × 179 mm. The beam size on the detector was approximately 0.5 × 0.35 mm. To measure the intensity of small-angle X-ray scattering, a sample of the solution was placed in a quartz capillary with a volume of 100–200 µL with a wall thickness of no more than 0.01 mm. We used capillary QGCT 2.0 mm with a diameter of 2 mm and a length of 80 mm, manufactured by Capillary Tube Supplies Ltd., Cornwall, UK. The distance between the sample and the detector was 700 mm, which corresponds to the angular range of the scattering vector q = 0.1–4.0 nm^{-1}, calculated according to Equation (1):

$$q = \frac{4\pi \sin\theta}{\lambda}, \tag{1}$$

where 2θ is the scattering angle, and λ is the wavelength of the X-ray radiation. The sample was exposed to monochromatic X-ray radiation for at least 600 s. The transition to the scattering vector modulus scale was performed by measuring the diffraction standard for AgBh. This transformation and the averaging of the two-dimensional diffraction scattering pattern were performed using the Fit2D software v. 17.006 (Andrew Peter Hammersley, European Synchrotron Radiation Facility, Grenoble, France) [32]. The built-in software module "PRIMUS" of the ATSAS software v. 3.2.1 (European Molecular Biology Laboratory, Hamburg, Germany) [33] was used to subtract the scattering contribution of the buffer solution from the scattering curves of the studied samples.

For the obtained experimental SAXS curves, the fractal dimension of the particle was determined by its scattering profile in a double logarithmic format, analyzing the power mode. In this mode, it was shown that $I(q)$ follows the power form [34,35] in Equation (2):

$$I(q) \propto q^{-\alpha}, \tag{2}$$

where α is the exponent associated with the fractal dimension of the scattering structures. For mass fractals, it can be shown that $\alpha = D_m$ and $1 < \alpha < 3$ in three-dimensional space. In addition, the scattering data were qualitatively evaluated using the Kratky plot ($I(q)q^2$ vs. q).

The obtained experimental SAXS curves were analyzed in the built-in software module "GNOM" of the ATSAS software using the indirect Fourier transform of the distribution function of paired distances $P(r)$ according to Equation (3) [36]:

$$P(r) = \frac{r^2}{2\pi^2}\int_0^\infty q^2 I(q)\frac{\sin(qr)}{qr}dq, \tag{3}$$

where r is the minimum distance between particles (atoms) in the structure of the substance. The values of the root-mean-square radius of gyration of ${R_g}^2$ particles were also determined using PDDF as Equation (4):

$$R_g^2 = \frac{\int_0^{D_{max}} r^2 P(r)dr}{2\int_0^{D_{max}} P(r)dr}, \tag{4}$$

where D_{max} is the maximum particle size in the dispersion. Next, the value of the zero-scattering intensity (direct scattering) I_0 was determined as Equation (5):

$$I_0 = 4\pi \int_0^{D_{max}} P(r)dr, \quad (5)$$

2.4. Determination of the Molecular Weight of Biopolymers Based on SAXS Data

The molecular weight of chitosan and gelatin biopolymers was determined through the correlation volume V_c, which was determined as the ratio of the zero-scattering intensity I_0 to the total scattering intensity in Equation (6) [37]:

$$Vc = \frac{I_0}{\int_0^{\infty} I(q)\, q^2\, dq}, \quad (6)$$

MW was determined by the empirical $V_c{}^2/R_g$ ratio using Equation (7):

$$MW = \left(\frac{V_c^2/R_g}{c}\right)^k, \quad (7)$$

where c and k are empirically determined constants based on the fitting of theoretical scattering profiles: $c = 0.1231$ and $k = 1$. The V_c values for gelatin were 884 Å^2, and the calculated values of MW were 58.9 kDa. The values obtained show good convergence with the results of the determination of the average viscous MW. Despite the use of constants for proteins, these constants have also been used to determine the MW of chitosan, as this method can be used for both flexible and extended macromolecules. As a result, for chitosan, $V_c = 2115$ Å^2 and $MW = 213.8$ kDa. The obtained values also converge with the values of the average viscous MW.

2.5. Ab Initio Modelling

Ab initio reconstruction of the shape of scattering objects was carried out using the annealing simulation algorithm implemented in DAMMIN (Dummy Atom Model Minimization) [38]. In this method, a dummy atom model is generated by filling a sphere with a diameter equal to the maximum particle size, D_{max}. In this case, each dummy atom must be attributed either to a particle or to a solvent. Annealing modeling consists of calculating the scattering intensity from an arbitrary starting configuration of fictitious atoms, and then during the annealing process, the theoretical scattering intensity is calculated from random modifications of this configuration in order to minimize χ^2 as a result, that is, to obtain the minimum discrepancy between the experimental and model scattering curves according to Equation (8):

$$\chi^2 = \frac{1}{N-1}\sum_j \left[\frac{I_{exp}(q_j) - cI_{calc}(q_j)}{\sigma(q_j)}\right]^2, \quad (8)$$

where N is the number of experimental points, c is the scaling factor; $I_{exp}(q_j)$ is the experimental intensity; $I_{calc}(q_j)$ is the calculated intensity in the model; $\sigma(q_j)$ is the experimental intensity error. The stability of the reconstructed models was tested by repeated modeling. Averaging of 10 models was carried out using DAMAVER [39], which overlays all models and defines a common shell containing all models.

To reduce the effective number of free parameters in the model, the method searches for configuration X minimizing (9) [40]:

$$f(X) = \chi^2 + \alpha P(x), \tag{9}$$

where *P(x)* is the looseness penalty, and α is the weight of the penalty. The weight of the penalty is selected to have a significant penalty contribution at the end of the minimization. A final compact and interconnected bead configuration is characterized by $P(x) \approx 0.01$, $\alpha \approx 10$, and $\chi^2 \approx 0.6$–2.5. Visualization of the obtained models was carried out in PyMOL.

2.6. Dynamic Light Scattering

The dynamic light scattering (DLS) method was used to determine the particle size distribution in chitosan and gelatin solutions and their mixtures at pH = 5.5. The study was carried out using the particle size analyzer LB 550, Horiba (Edison, NJ, USA) using a laser beam with a wavelength of 650 nm, a power of 5 mW, and a scattering angle of 177°. Each sample was measured at least 5 times. The measurements were carried out at a temperature of 25 °C. Before the measurement, all solutions were additionally diluted with distilled water by a factor of 10.

3. Results

3.1. Analysis of Conformation and Macromolecular Sizes of Individual Chitosan and Gelatin Solutions Using SAXS

Before studying the structure of PECs and the mechanism of their formation, the sizes and the conformations of gelatin and chitosan macromolecules in a buffer solution were first studied. For this purpose, analyses of the individual diluted solutions of gelatin with a concentration of C = 1 mg/mL and chitosan with C = 5 mg/mL at pH = 3 and 5.5 were carried out using small-angle X-ray scattering (SAXS) with synchrotron radiation. It is important to note that during the measurements of solution samples at pH = 3, low-intensity scattering patterns were observed, and they were considered inadequate; therefore, they are not presented in this work. This result may be associated with significant polyelectrolyte swelling of gelatin protein and chitosan macromolecules because of intense repulsive interaction of similarly charged amino groups in acidic media, leading to the formation of a small number of large-sized scattering centers. Thus, the scattering patterns of biopolymer solution samples at pH = 5.5, presented in Figure 1, were subjected to analysis.

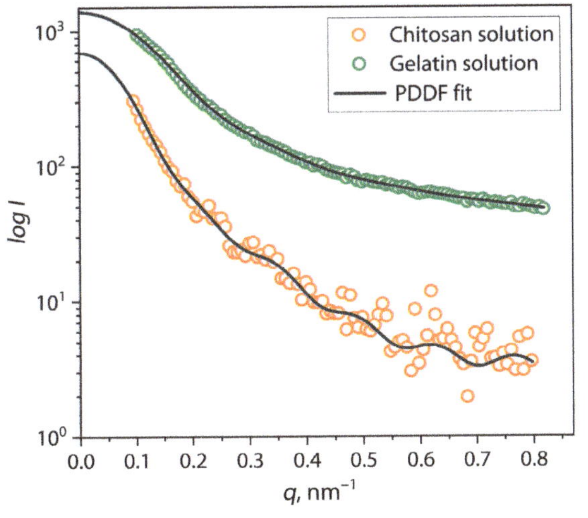

Figure 1. SAXS patterns of individual chitosan and gelatin solutions at pH = 5.5.

The graph shows that the intensity of radiation scattering in a gelatin solution sample significantly exceeds the intensity in a chitosan solution sample. This can be due not only to the difference in polymer concentration but also to their scattering contrast. Gelatin, as a protein polymer, has a higher electron density compared to polysaccharide chitosan, which can lead to higher scattering intensity. In addition, this effect is most likely a consequence of both the low solubility and the high lability of chitosan macromolecules in the buffer solution used at pH = 5.5 due to its polycationic nature, in contrast to proteins in gelatin, which have amphoteric properties. Figure 1 shows that at high values of the scattering vector q, a noisy signal is observed in the scattering pattern (at wide scattering angles) of a chitosan solution sample, which is also associated with a small number of individual chitosan macromolecules compared to associates. Due to the differences in the configuration of the macromolecules of gelatin proteins and chitosan, as well as their different charge balance, under these conditions, gelatin macromolecules are able to effectively swell to form a "true" solution, while chitosan macromolecules can form labile associates, with the formation of a dispersion.

The configuration difference between chitosan and gelatin macromolecules in their dilute solutions is also reflected in a number of works. For example, in work [41], using the DLS and SAXS experiment, it was shown that gelatin solutions are characterized by the so-called collapse-swelling state transition. This transition reflects the fact that gelatin macromolecules outside the isoelectric point swell and transform from a globule into a highly swollen coil [42]. As for chitosan, its key problem is the production of "true" aqueous solutions, which is complicated by its chitinous nature. Chitosan aqueous solutions are typically dispersions containing polymeric macromolecules and labile associates (aggregates) [43–45]. The formation of chitosan aggregates, leading to a decrease in the intensity profile, is mainly caused by the association of *N*-acetyl-*D*-glucosamine units due to hydrophobic interactions, as well as the hydrophilic interaction of *D*-glucosamine units with the formation of hydrogen bonds [44].

For a rough analysis of the fractal dimension and shape of macromolecular coils of biopolymers in solution, using the Porod approach, linear regions of the scattering patterns in the range of $q \leq 0.2$ were analyzed, and the value of the α exponent was determined (Equation (2)). A visual representation of such an analysis is presented in Figure 2.

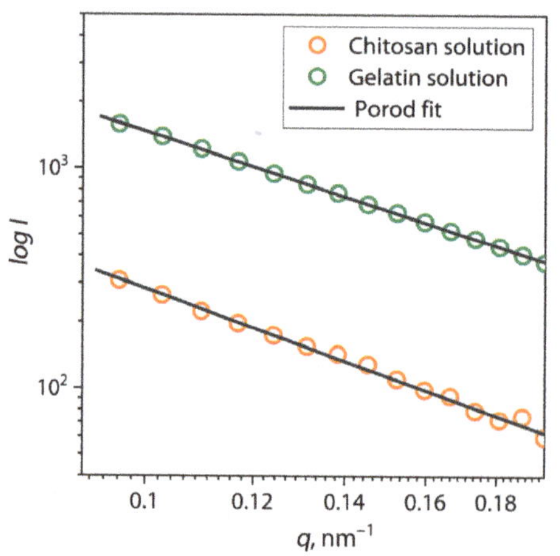

Figure 2. Porod approximation of SAXS patterns of individual chitosan and gelatin solutions at pH = 5.5.

According to this approach, the value of the α exponent (slope coefficient) contains information about the distribution of interatomic distances inside the molecule and allows us to estimate the fractal dimension and surface quality of scattering objects. As a result, the calculated values of the exponent were $\alpha = 2$ for a gelatin solution and $\alpha = 2.26$ for a chitosan solution. Such values correspond to scattering on macromolecules or supramolecular structures such as a random coil in a Θ-solvent [41,46,47]. The results of the analysis of scattering patterns of samples of gelatin and chitosan solutions are also presented in Table 1.

Table 1. Fitting parameters of the pair distance distribution function (PDDF) and the Porod approximation of SAXS patterns of individual chitosan and gelatin solutions at pH = 5.5.

Sample	α	I_0	D_{max}, nm	Estimate (PDDF)	R_g (PDDF), nm	R_g (Viscometric), nm	AR
Gelatin	2.00	1391 ± 50	32.2	0.8332	10.8 ± 0.2	7.5 ± 3.9	1.49
Chitosan	2.26	693 ± 21	46.0	0.6776	17.0 ± 0.1	24.2 ± 4.7	1.35

For example, in the work [47], when analyzing the scattering profiles of mixtures of solutions of the protein rezilin and polypeptide, the values of α equal to 2 are shown. For an aqueous gelatin solution [41], values close to 2 were also observed. In addition, such structures analyzed using synchrotron radiation correspond to theoretical and practical studies of conformational states in chitosan and gelatin, using DLS and TEM. For example, in works [43,44,48,49] using DLS, TEM, and AFM, it was shown that chitosan macromolecules can take conformations from a random coil to a rigid rod, depending on the molecular weight and degree of deacetylation. In work [49], it was shown that the conformation of chitosan aqueous solutions changes from a random coil to a rigid rod, with a decrease in pH from 3.5 to 1.55, which was explained by an increase in the chain charge. In other works [42,50], for gelatin macromolecules at a pH above pI, the main conformation is a swollen random coil. It is worth noting that the conformation of the random coil for chitosan solutions is due to the effect of charge screening in the buffer solution, which reduces its electrostatic repulsion and prevents its elongation into a rod shape. A similar effect, but to a lesser extent (due to the low content of ionogenic groups), is typical for gelatin solutions at a pH below and above the isoelectric point.

For a more detailed study of the equilibrium size and shape of biopolymer macromolecules under selected conditions, small-angle scattering patterns were analyzed using the pair interatomic distance distribution function (PDDF). The PDDF function is a distribution of interatomic distances (r), which was calculated using the inverse Fourier transform of the scattering patterns according to Equation (3), as shown in Figure 2 by dark gray lines. In addition, this approach allows the scattering patterns to be interpolated into the region of the zero value of the scattering vector q (the smallest angles) to determine the value of zero scattering intensity I_0, which is mathematically expressed as Equation (5). The results of the analysis of scattering patterns of samples of gelatin and chitosan solutions using the PDDF function and I_0 values are presented in Table 1. Figure 3 shows the pair distance distribution function for chitosan and gelatin solutions at pH = 5.5.

In general, the PDDF function demonstrates the distribution of pair interatomic distances along their length in the structure of the average scattering object. At maximum convergence of the PDDF function with the experimental scattering pattern in the sample, the limiting interval of such a distribution is the value of D_{max}, corresponding to the maximum diameter of the averaged scattering object. Moreover, the intensity and width of the distribution provide information about the shape of the averaged scattering object.

It can be seen from the figure that the PDDF distribution for a gelatin solution is a wide, single-modal distribution and has a shape close to a normal symmetric distribution, with a maximum intensity value $P(r) = 5.72$ at $r = 10.9$ nm and a maximum diameter $D_{max} = 32.2$ nm for macromolecules of protein in gelatin. The semi-transparent areas on the figure represent the uncertainty range in the PDDF function. At the same time, it is clear that the distribution for the chitosan solution demonstrates a much lower maximum intensity $P(r) = 1.9$ at $r = 19.5$ nm, with a larger width and $D_{max} = 46$ nm. Compared with the curve for a gelatin solution, the graph for a chitosan solution is characterized by a wider and less intense distribution, which, as previously explained, is associated with a different scattering contrast and a lower number of scattering objects in the solution. It is noteworthy that the zero scattering intensity value I_0 in the chitosan solution is reduced by about two times compared to gelatin solution.

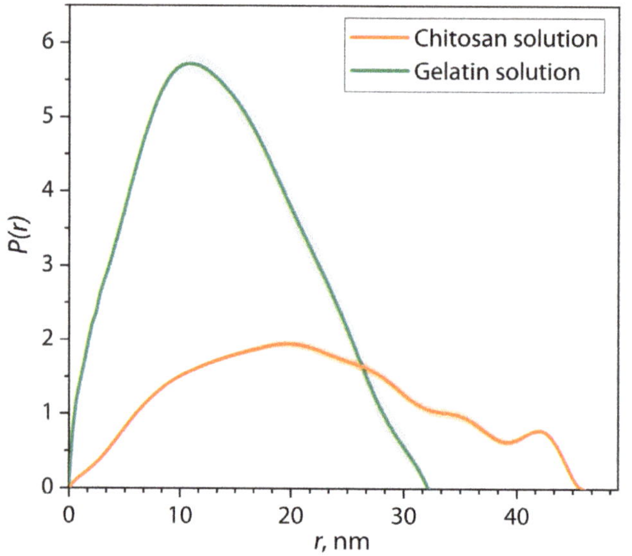

Figure 3. Pair-distance distributions of individual chitosan and gelatin solutions at pH = 5.5.

As part of the PDDF analysis, it is also possible to establish the average value of the hydrodynamic radius R_g for scattering objects using Equation (4). The calculated values were $R_g = 10.8 \pm 0.2$ nm for the gelatin solution and $R_g = 17.0 \pm 0.1$ nm for the chitosan solution. These calculations show large differences in the density of macromolecular coils of biopolymers in solution compared to a comparison of the values of their maximum diameter D_{max}. To numerically estimate the "density" and shape of macromolecular coils of biopolymers in solution, we used the aspect ratio *AR*, calculated using Equation (10) (Table 1). *AR* values show that the macromolecules of gelatin and chitosan, despite the difference in size, have a similar ellipsoidal shape, close to spherical.

According to the literature [38], it is known that the shape of the distribution of the PDDF function also makes it possible to determine the shape of the average scattering object in the sample. PDDF distributions for solutions of gelatin and chitosan in Figure 3 have a dome-shaped form close to a symmetrical distribution, while the distribution for the gelatin solution is more intense and less wide than that for the chitosan solution. This shape corresponds to a close-to-spherical form for gelatin macromolecules and an ellipsoidal flattened form for chitosan coils, which confirms the results of calculations of the α exponent in the Porod analysis and the *AR* ratio.

An additional analysis of the shape of chitosan and gelatin macromolecules was carried out using ab initio reconstruction modeling. Figure 4 shows the obtained models of dummy atoms in front, side, and top projections.

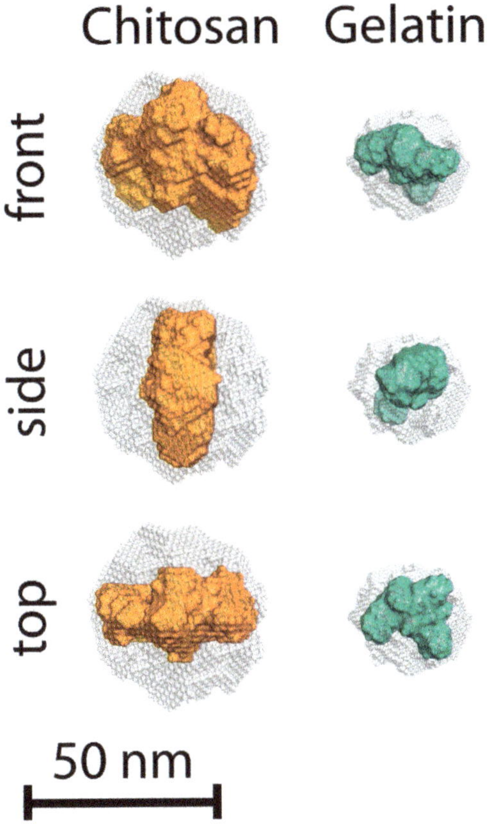

Figure 4. Ab initio models of chitosan and gelatin macromolecules using Equation (8).

The gray semitransparent spheres in the figure framing the obtained models of biopolymer macromolecules correspond to the initial random-phase distribution in the search volume, with a diameter equal to D_{max}. It can be seen from the figure that the scattering structures correspond to their maximum molecular sizes and represent a compressed model of tightly packed dummy atoms (in color). Visually, it is noticeable that chitosan macromolecules are characterized by structures resembling an oblate spheroid, while gelatin is characterized by a more compact and less ordered shape. The modeled macromolecules of gelatin and chitosan, in their shape, correspond well to the structure characteristic of partially unfolded proteins, for which the most advantageous conformation is a random coil. The analysis of scattering profiles and the ab initio reconstruction of the shape of these proteins have been studied in detail in works [51–53].

3.2. Estimation of Self-Assembly of Primary PEC Structures Using SAXS

As described above, in aqueous solutions, the chitosan macromolecules can have a positive charge due to the protonation of open side deacetylated amino groups. Under the same conditions, gelatin polypeptide macromolecules can receive a dual charge: positive during the protonation of open side amino groups of predominantly arginine and lysine in the structure of the α-chains of polypeptides, and negative during the electrolytic dissociation of the carboxyl groups of aspartic and glutamic acids, correspondingly.

Such processes lead to significant swelling of the macromolecules of these polymers due to the strong electrostatic repulsion of similarly charged groups of macroions. Naturally, the strength of the effect of polyelectrolyte swelling of macromolecules depends on the efficiency of protonation of amino groups in the presence of a large number of protons in an acidic medium, and the efficiency of dissociation of acid groups in the presence of hydroxyl ions in an alkaline solution.

In the case when chitosan and gelatin macromolecules are simultaneously present in an acidic medium in a swollen state, the effect of their electrostatic repulsion from each other is observed due to the high positive charge density of equally charged macromolecules or polymer particles. On the contrary, when the macromolecules of chitosan and gelatin are in an alkaline environment, the protonation of amino groups in chitosan and gelatin is almost completely suppressed, but the dissociation of acidic groups in gelatin proteins is significantly enhanced. As a result of these processes, the gelatin proteins swell and repel each other to form a true solution, while the chitosan macromolecules collapse into an uncharged state and strongly associate, forming a dispersion of chitosan particles in the gelatin solution [24,29].

Despite this, there is a "window" in the pH range between the isoelectric point of gelatin proteins and the characteristic protonation constant of chitosan, $pI < pH < pK_0$, in which the protonation of the amino groups of both polymers and the dissociation of the acidic groups of gelatin proteins occur simultaneously. Under such conditions, the chitosan macromolecules are weakly charged positively, and the macromolecules of gelatin proteins partially have a negative charge. These conditions lead to the activation of electrostatic attraction of chitosan and gelatin macromolecules, followed by their ionic bonding at the sites of oppositely charged functional groups and the formation of the primary structures of a (bio)polyelectrolyte complex ((bio)PEC). Such structures, as a rule, consist of several bound macromolecules in an unstable state due to an imbalance of surface charge. Moreover, these structures belong to supramolecular formations with large size and mass relative to the individual macromolecules of biopolymers in solution. The charge imbalance on the surface of the primary (bio)PEC structures leads to their further association with each other due to electrostatic attraction, with the formation of nano- or microdispersion, which is also confirmed by a significant increase in the optical density of the solution due to the intense light scattering of the mixture of chitosan and gelatin when the condition $pI < pH < pK_0$ is met. However, it is clear that methods, such as turbidimetric photometry and dynamic light scattering (DLS), do not allow for accurate assessment of the size and shape of primary (bio)PEC structures.

Establishing the size and shape of primary (bio)PEC structures as a function of the chitosan/gelatin ratio and pH level with higher resolution using small-angle X-ray scattering with high-intensity synchrotron radiation can provide important information for a deeper understanding of the mechanisms of the formation of these structures. For this reason, we studied the scattering patterns of SAXS with synchrotron radiation for samples of (bio)PEC dispersions at pH = 3 and 5.5 (below and above the pI value for gelatin) depending on the chitosan/gelatin mixture ratio. The SAXS scattering patterns for (bio)PEC dispersions at different biopolymer ratios are presented in Figure 5.

Scattering patterns for chitosan/gelatin mixture dispersions at pH = 3, presented in Figure 5a, show that the scattering intensity varies independently of the ratio within a certain range between the values corresponding to the initial biopolymers. A completely different picture is observed for chitosan/gelatin dispersions at pH = 5.5 (Figure 5b). It can be seen that, in general, the intensity of the scattering curves is greater for all dispersions except the 1:1 mixture compared to the curves for dispersions at pH = 3. Moreover, in this case, the effect of the influence of the chitosan/gelatin ratio in the mixture is significantly higher: with an increase in the proportion of gelatin in the mixture, the signal becomes less noisy and more intense.

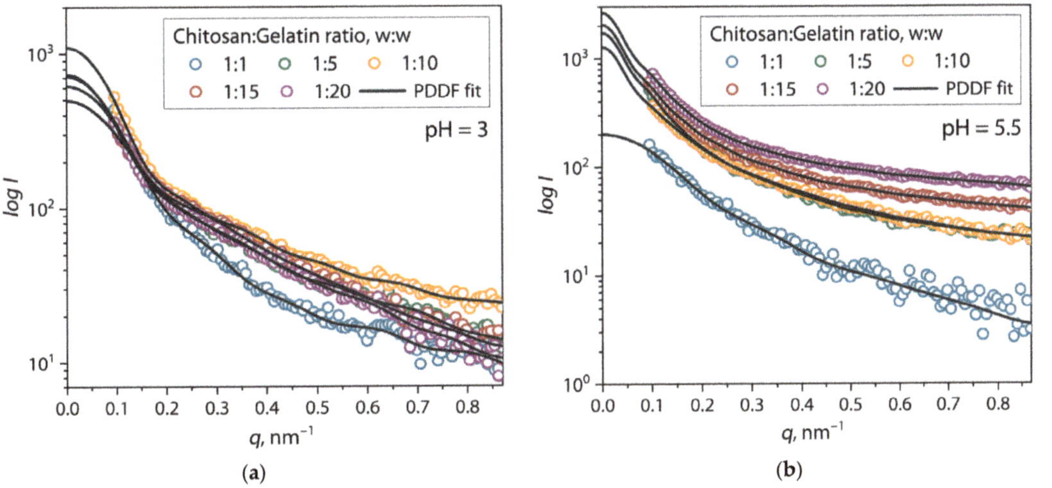

Figure 5. SAXS patterns of chitosan/gelatin mixtures at (**a**) pH = 3 and (**b**) pH = 5.5.

Analyzing and comparing the SAXS results for chitosan/gelatin dispersions at pH = 3 and 5.5, we came to a number of conclusions. First, at pH = 3, the dispersion of the mixtures shows a scattering pattern similar to solutions of individual polymers that did not exhibit significant scattering, as discussed above. In addition, the relatively low scattering intensity for these structures indicates their instability and associative-dissipative nature of existence in an acidic environment with a critically low degree of dissociation of carboxyl groups in the structure of gelatin proteins. Secondly, with an increase to pH = 5.5, the condition $pI < pH < pK_0$ is met, and the formation of primary (bio)PEC structures occurs more efficiently, which leads to an increase in the intensity of X-ray scattering due to the presence of a large number of small-sized scattering centers (particles) in the dispersion. Under these conditions, an increase in the proportion of gelatin in the solution additionally leads to a significant increase in the scattering intensity, which is also associated with an increase in the dispersity of the (bio)PEC phase. Moreover, the observed decrease in the amplitude of scattering intensity fluctuations in the curves indicates the formation of more isotropic and stable dispersions. It is also clear that a mixture with a 1:1 ratio at pH = 5.5 has a lower scattering intensity and a noisier signal compared to other mixture ratios at pH = 5.5. We believe that this feature is associated with a strong imbalance in the density of positive and negative charges in the polymer macromolecules with a predominance of positive charges, which leads to a decrease in the efficiency of formation of (bio)PEC structures. This imbalance is due to a significant difference in the molecular masses of biopolymers (about four times) and a difference in the concentrations of functional groups capable of ionization.

For a rough analysis of the shape of the primary structures of (bio)PECs, the scattering patterns of chitosan/gelatin mixture dispersions were analyzed in the range $q < 0.2$ nm^{-1} using the Porod approach. Porod plots for (bio)PEC dispersions at pH = 3 and 5.5 are presented in Figure 6.

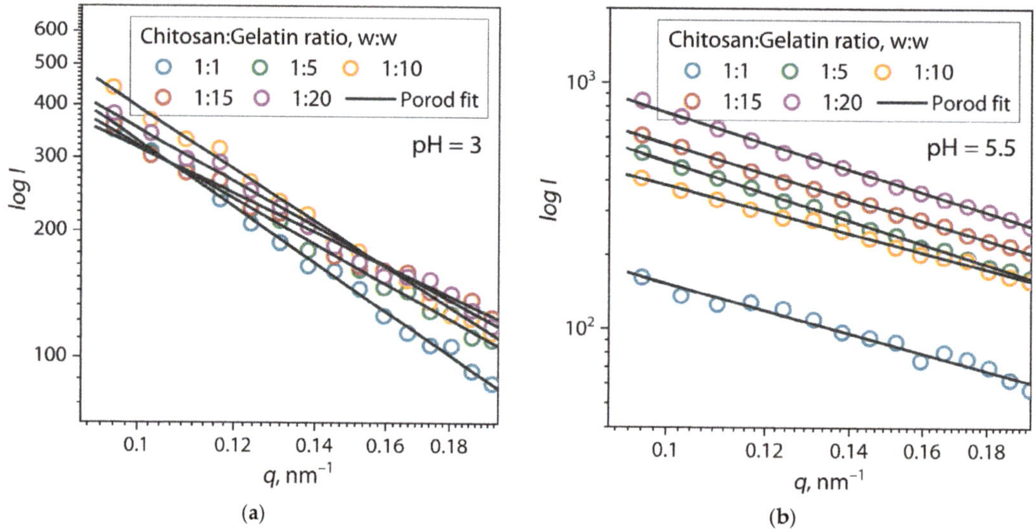

Figure 6. Porod approximation of SAXS patterns of chitosan/gelatin mixtures at (**a**) pH = 3 and (**b**) pH = 5.5.

It can be seen from the figure that these sections of the scattering curves are well described by Equation (2) at $R^2 > 0.99$. The α exponent values for these samples according to Equation (2), as well as the fitting parameters of the PDDF function (Equation (3)) and Equations (4) and (5) for a chitosan/gelatin mixtures, are presented in Table 2.

Table 2. Fitting parameters of the pair distance distribution function (PDDF) and the Porod approximation of SAXS patterns for mixtures of chitosan/gelatin at pH = 3 and 5.5.

Chitosan/Gelatin Ratio, *w/w*	α	I_0	D_{max}, nm	Estimate (PDDF)	R_g (PDDF), nm
pH = 3					
1:1	2.00	615 ± 17	39.3	0.8751	13.5 ± 0.2
1:5	1.63	713 ± 28	42.8	0.7990	15.0 ± 0.3
1:10	1.87	1099 ± 41	44.0	0.8092	16.3 ± 0.2
1:15	1.39	495 ± 15	37.0	0.7202	12.7 ± 0.2
1:20	1.61	736 ± 38	50.0	0.8609	15.9 ± 0.5
pH = 5.5					
1:1	1.38	201 ± 4	31	0.7988	10.7 ± 0.1
1:5	1.33	1731 ± 38	110	0.7391	31.2 ± 0.7
1:10	1.63	1286 ± 38	108	0.7786	30.4 ± 0.7
1:15	1.52	2061 ± 85	110	0.8683	31.8 ± 1.1
1:20	1.58	2657 ± 159	109	0.8078	31.6 ± 1.6

When comparing the α exponent values for (bio)PEC dispersions with the values for individual biopolymer solutions (Table 1), it can be observed that they are generally significantly lower. According to the literature, it is known that at a value of the exponent around $\alpha \approx 2$, dispersed objects on average have a shape close to a random coil (sphere), and at a value of $\alpha \approx 1$, close to a rigid rod (one-dimensional rod). According to these calculations, it is clear that the primary structures of (bio)PEC chitosan/gelatin at pH = 3 have a predominantly spherical shape, while at pH = 5.5 they are rod-shaped. However, the shape of the formed structures is not ideal and is intermediate, regardless of the composition of the dispersion. This behavior may be due to some elongation of high molecular weight chitosan chains due to the attachment of small macromolecules of gelatin proteins on its surface during polyelectrolyte interaction and neutralization of the charge of amino acid residues. As a result, it is likely that gelatin macromolecules behave as large substituents along the chitosan backbone and straighten it due to their volume. This concept seems logical given the fact that the energy of electrostatic attraction and binding between polyelectrolytes is higher than the conformation energy of a random chitosan coil in solution, which leads to a shift in the energy equilibrium.

This behavior is also confirmed by the Kratky plots (Figure 7), which represent the scattering profiles in $I(q)q^2$ vs. q coordinates.

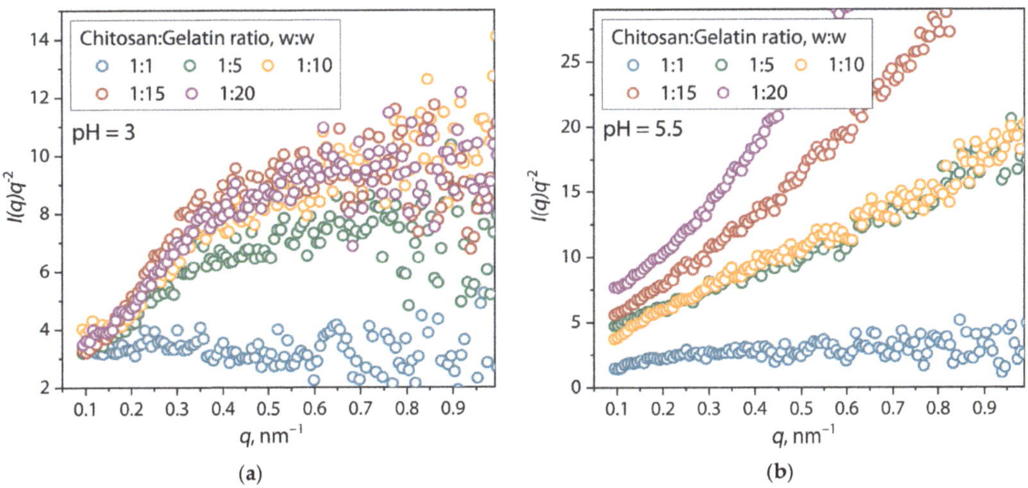

Figure 7. Kratky plots for chitosan/gelatin mixtures at (**a**) pH = 3, (**b**) pH = 5.5.

For mixtures at pH = 3, the Kratky plot initially shows an increase in the $I(q)q^2$ function at low q region, followed by a plateau at high q, which indicates the presence of structures corresponding to partially unfolded conformations of a random coil [41,46,47]. It is also evident that when the proportion of gelatin in the mixture increases above a ratio of 1:1, the intensity of the function increases significantly, but does not depend on the ratio of polymers in the considered range of q. In mixtures at pH = 5.5, the character of the curve changed, showing an increasing slope with increasing fraction of gelatin in the mixture, which indicates the transformation of structural properties from the conformation of a random coil into a fully swollen unfolded spiral [54]. Moreover, the noise of the curve is significantly reduced, indicating the uniformity of the shape of the structural units of the dispersion. For a more accurate analysis of the shape of the structural units of the (bio)PEC dispersion at pH = 3 and 5.5, the analysis of the pair distance distribution function was applied. The corresponding distributions are presented in Figure 8.

The presented distributions for chitosan/gelatin mixtures at pH = 3 (Figure 8a) are monomodal, and their shape is similar to the distributions for individual solutions of biopolymers (Figure 3). Additionally, the D_{max} and R_g values for the presented distributions practically do not change depending on the ratio of chitosan/gelatin and correspond to the values for the initial biopolymers (Table 1). A different pattern is observed for distributions at pH = 5.5 (Figure 8b). In this case, with ratios above 1:5, bimodal distributions of paired distances are formed, which contain two modes: the first mode up to 50 nm, similar to the size of the initial macromolecules of polymers, and the second mode from 50 to 110 nm (depending on the ratio), which is associated with the formation of elongated objects. Thus, such a form of distribution corresponds to dimeric structures consisting of two domains of macromolecules [38,55]. It is also worth noting that the calculated function parameters show an increase in D_{max} and R_g by about two to two-and-a-half times compared to the values for mixtures at pH = 3 at ratios above 1:5. Thus, it is evident that the pH value has a significant influence on the size and shape of the formed primary structures of (bio)PEC chitosan/gelatin.

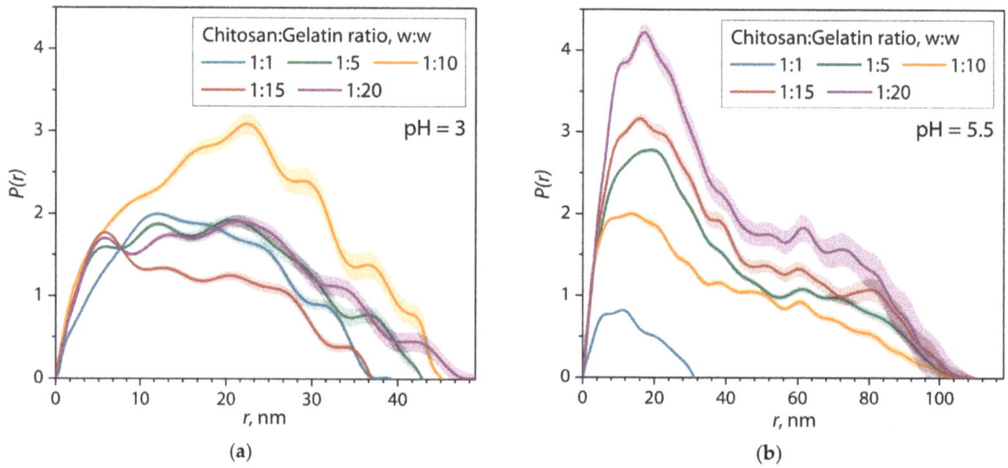

Figure 8. Pair-distance distributions for mixtures of chitosan/gelatin mixtures at (**a**) pH = 3 and (**b**) pH = 5.5.

However, further analyzing the pair distance distribution function, it should be considered that the minimum measured value at $q = 0.1$ nm^{-1} limits the direct determination of particle diameter to about 60 nm. In our case, data analysis using the pair distance distribution function showed that the best fit of the model is observed at D_{max} = 110 nm. Thus, the D_{max} value is derived from extrapolation of data and should be interpreted with caution. Therefore, the dynamic light scattering method (DLS) was used to confirm the results obtained by SAXS data. Figure 9 shows the particle diameter distributions for chitosan and gelatin solutions and their mixtures at pH = 5.5, obtained by the DLS method.

It can be seen from the figure that, in the case of individual polymer solutions, bimodal distributions are formed, which are divided into two ensembles consisting of polymer fractions with lower and higher molecular weights, correspondingly. It is worth noting that the comparison of SAXS and DLS data is coupled with some difficulties related to the peculiarities of sample preparation of polymer solutions and their mixtures for DLS. For the DLS method, polymer solutions are usually strongly diluted in order to eliminate the multiple scattering effect, which can distort the experimental results. In our case, polymer solutions and their mixtures were diluted 10 times. On the other hand, for the SAXS method, multiple X-ray scattering is not typical, since the scattering of X-ray photons on atoms is much weaker than the scattering of irradiation in DLS. In addition, in the case of SAXS, the optimal concentration for obtaining a stable signal is more than 1 mg/mL. Therefore, the analyzed solutions for SAXS measurements were not diluted. Thus, the comparison of DLS and SAXS results is qualitative. Despite all of the above, for chitosan and gelatin solutions, we can observe particles comparable in size to those observed according to SAXS and viscometry data. For mixtures of chitosan and gelatin, we observe a simplification of the character of statistical distributions, as well as an increase in the proportion of particles with an average diameter of about 80 nm, compared with solutions of individual polymers. In addition, the most significant result confirming the effective formation of primary structures of (bio)PEC is the presence of particles with an average diameter exceeding the particle diameters in the initial polymers. For example, for mixtures of chitosan and gelatin at pH = 5.5 and a ratio of 1:10 and 1:20, the formation of particles with an average diameter of more than 100 nm can be observed. It is noteworthy that the strong dilution of solutions for DLS measurements also prevented the subsequent interaction of primary particles of (bio)PEC with the formation of coacervates. Therefore, we do not observe large coacervate particles on the diameter distribution, which is a favorable condition for comparing the results obtained by SAXS and DLS. Thus, the DLS results confirm the results of the SAXS data analysis using the pair distance distribution function shown in Figure 8.

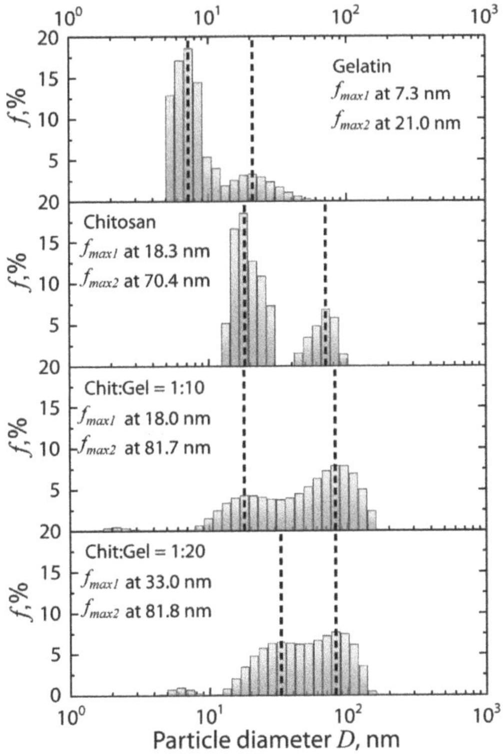

Figure 9. Particle diameter distribution for chitosan and gelatin solutions and their mixtures at pH = 5.5 with different polymer ratios (dashed lines correspond to *f*max).

Further analysis of the shape of structural units in dispersion was carried out using ab initio reconstruction. Figure 10 shows the obtained models of dummy atoms in front projection depending on the ratio of polymers and the pH of the solution.

It can be seen from the figure that the simulation result gives us, as in the case of individual solutions of polymers (Figure 4), a compressed model consisting of tightly packed dummy atoms framed by a sphere (in gray) under initial modeling conditions. The figure shows that the simulation results give particle shapes corresponding to the previously mentioned structures obtained by analyzing with the PDDF function, Kratky plot, and Porod approximation. The particle shape in mixtures of chitosan/gelatin at pH = 3 shows a loose and disordered structure of the oblate spheroid, resembling the shape of the chitosan and gelatin macromolecules. At the same time, the compactness and looseness of the spheroid structure vary regardless of the ratio of polymers. At pH = 5.5, for a ratio of 1:1, a shape similar to that of polymer macromolecules is also observed. However, at higher ratios, there is an increase in longitudinal dimensions in the models. Therefore, elongated structures are formed, the shape of which is close to a cylinder (swollen spiral), which is typical for fully unfolded proteins. Such a form of the structure of primary (bio)PECs is in good agreement with the model proposed by works [14,17], where the formation of metastable elongated aggregates of primary nanoscale PEC based on anionic- and cationic-neutral block copolymers is described in detail.

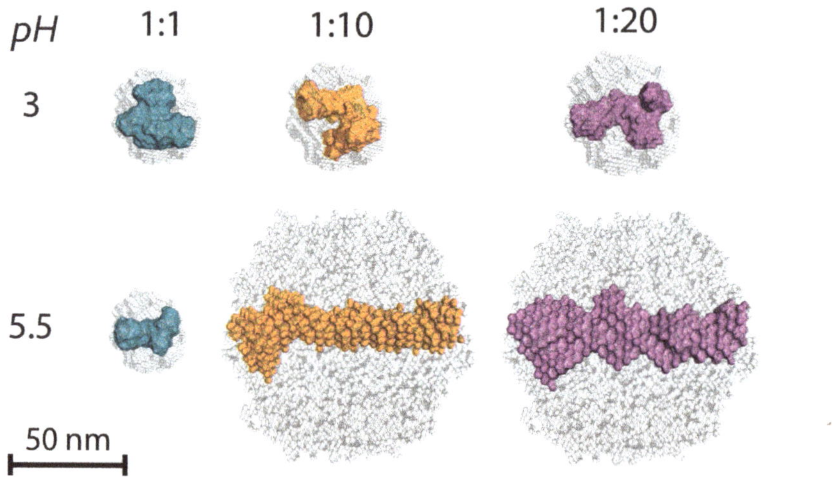

Figure 10. Ab initio models of primary (bio)PEC structures for chitosan/gelatin mixtures with different ratios of polymers at pH = 3 and 5.5.

Additionally, in order to understand how the shape of the analyzed scattering objects changes depending on the pH and composition of the mixture, the aspect ratio *AR* was calculated using Equation (10):

$$AR = \frac{D_{max}}{2R_g} \tag{10}$$

where $AR = 1$ corresponds to the spherical shape of an object. Figure 11 shows the values of AR for the chitosan–gelatin mixture.

Figure 11 shows the dependences of the aspect ratio of the averaged object in dispersion (*AR* (Equation (10)) and the intensity at zero scattering vector (I_0 (Equation (5)) on the ratio of polymers in the mixture at different pH values.

The dependence in Figure 11a shows that at pH = 3, *AR* values fluctuate independently of the ratio of chitosan/gelatin and is close to the values for macromolecules of individual polymers. However, at a ratio above 1:10, *AR* gradually increases, which indicates the slight elongation of the formed average particle. This effect may indirectly confirm the formation of (bio)PEC structures between chitosan and gelatin macromolecules even under acidic conditions, when the dissociation of the acidic groups of gelatin amino acids is suppressed at $pH < pI$. In our previous work [29], this effect was also detected using the dynamic light scattering method on similar dispersions under the same conditions. Here, we believe that under these conditions, structures can be formed consisting of relatively large chitosan macromolecules in the center with one or two gelatin protein macromolecules (based on the values of D_{max} and R_g) attached by ionic bonds along the main chain. With such an interaction, the main chain of the chitosan macromolecule can straighten somewhat, which can be expressed as an increase in the *AR* value of the average object. A qualitative scheme showing the formation of a (bio)polyelectrolyte complex between chitosan and gelatin can be seen in Figure 12. For mixtures at pH = 5.5, there is a clear dependence on an increase in *AR* value from 1.45 to 1.76, with an increase in the chitosan/gelatin ratio from 1:1 to 1:5, after which it remains unchanged. This result, considering the $pH > pI$ conditions of the solution in this case, confirms the proposed structure of (bio)PEC particles and may be a consequence of the increased efficiency of the polyelectrolyte interaction between chitosan and gelatin. Thus, a greater number of gelatin protein macromolecules can be bound to the surface of the chitosan macromolecule at a polymer ratio higher than 1:5 (Figure 12). At the same time, a ratio of 1:1 does not appear due to the insufficient content of negatively charged carboxyl groups for the formation of stable structures with positively charged chitosan amino groups. In this case, as at pH = 3, the process of the formation of (bio)PEC is strongly suppressed.

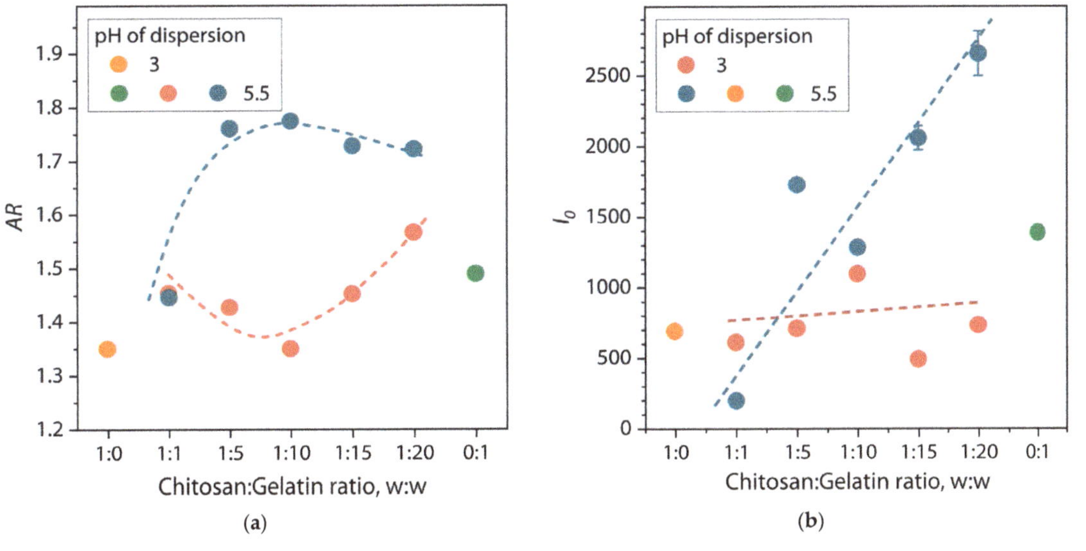

Figure 11. Dependences of (**a**) the aspect ratio *AR* and (**b**) zero scattering intensity I_0 of structural units in dispersions on the chitosan/gelatin ratio with different pH (dashed lines as a guide for the eye).

This conclusion also follows from the dependence in Figure 11b. It is evident that at pH = 3, the I_0 values, which largely depend on the number and size of scattering centers (particles) in the dispersion, change insignificantly for the chitosan/gelatin mixture across the entire range of ratios. However, in the case of pH = 5.5, a sharp and intensive increase in the I_0 values is observed with an increase in the proportion of gelatin in the dispersion. These results also indicate the inefficiency of the polyelectrolyte interaction and the process of formation of (bio)PEC structures at pH = 3, and, conversely, their increased efficiency at pH = 5.5. We believe that at pH = 3, a small number of small (bio)PEC particles are formed, which have low stability under acidic conditions and exist as associative–dissipative structures. As a result, an increase in the proportion of gelatin in the mixture does not significantly affect the number of particles for the scattering intensity. However, at pH = 5.5, the stability of (bio)PEC structures increases significantly, and therefore, with an increase in the proportion of gelatin in the dispersion, their size and quantity can also increase.

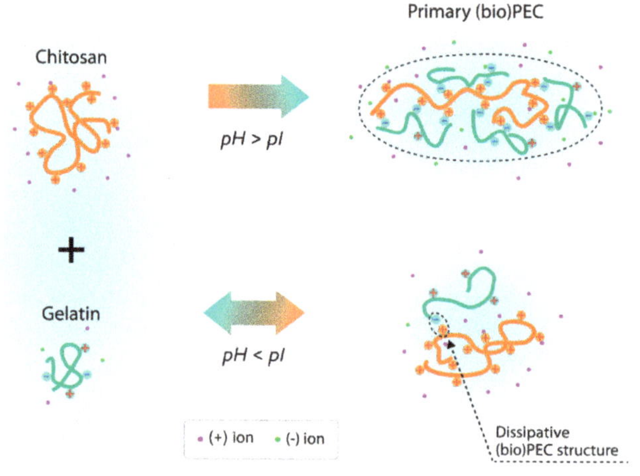

Figure 12. Qualitative scheme of (bio)PEC (borders in dashed circle) formation between chitosan (orange lines) and gelatin (green lines) in solutions with pH below and above pI.

4. Conclusions

In this work, using SAXS synchrotron radiation, we studied the conformational states of chitosan and gelatin polyelectrolyte macromolecules and the features of their behavior in mixtures at different pH levels and ratios to establish the size and shape of primary supramolecular (bio)PECs. Analysis of the scattering profiles of the initial solutions using the Porod approximation showed that the chitosan and gelatin macromolecules correspond to the 2D conformation of a random coil in a Θ-solvent, with the corresponding q-slope values of 2.26 and 2. The building of the pair distance function of the initial chitosan and gelatin showed the formation of single-modal distributions with maximum molecular sizes, D_{max}, of 46 and 32.2 nm, respectively. Ab initio reconstruction of the macromolecules' shape showed the formation of objects shaped like oblate spheroids. In mixtures of chitosan and gelatin at $pH < pI$, it was found that the scattering structures correspond to the initial biopolymers and have a molecular size like them. At the same time, the calculated values of the AR at a ratio above 1:10 gradually increase, which indicates a slight elongation of the average particle. This indirectly indicates the formation of dissipative structures of (bio)PEC even under acidic conditions, when the dissociation of acidic groups of gelatin is suppressed.

However, in mixtures at pH > pI, our results show that at ratios above 1:5, the formation of primary supramolecular complexes is observed, which is accompanied by an increase in I_0 by about three times and D_{max} by two to two-and-a-half times relative to mixtures below pI. The formation of primary structures of (bio)PEC in mixtures at pH > pI was also confirmed using DLS, where we observed the formation of (bio)PEC particles with a diameter of 100–150 nm. Analysis of PDDF showed that the formation of primary PEC is accompanied by the formation of bimodal distributions with a clearly defined peak of the second mode, which corresponds to elongated dimeric structures consisting of two simple macromolecules of the initial biopolymers. Ab initio reconstruction of the PEC shape showed the presence of elongated structures corresponding to a cylinder (swollen spiral). Also, there is a clear dependence on an increase in *AR* value from 1.45 to 1.76, with an increase in the chitosan/gelatin ratio from 1:1 to 1:5, and then it remains unchanged. This result, considering the pH > pI conditions of the solution in this case, confirms the proposed structure of (bio)PEC particles and may be a consequence of the increased efficiency of the polyelectrolyte associative interaction between chitosan and gelatin.

Author Contributions: A.P. was the project manager, obtained experimental SAXS data, and made a significant contribution to the development of the concept and writing of the article. M.L. obtained experimental SAXS data, analyzed SAXS data, and made a significant contribution to the writing of the article. A.K. and K.D. obtained solutions of individual polymers and mixtures and contributed to the writing of the article. All authors have read and agreed to the published version of the manuscript.

Funding: This work was supported by a grant from the Russian Science Foundation (agreement no. 23-73-01277).

Institutional Review Board Statement: Not applicable.

Informed Consent Statement: Not applicable.

Data Availability Statement: The original contributions presented in this study are included in the article. Further inquiries can be directed to the corresponding author.

Conflicts of Interest: The authors declare no conflicts of interest.

References

1. Timilsena, Y.P.; Akanbi, T.O.; Khalid, N.; Adhikari, B.; Barrow, C.J. Complex Coacervation: Principles, Mechanisms and Applications in Microencapsulation. *Int. J. Biol. Macromol.* **2019**, *121*, 1276–1286. [CrossRef] [PubMed]
2. Margossian, K.O.; Brown, M.U.; Emrick, T.; Muthukumar, M. Coacervation in Polyzwitterion-Polyelectrolyte Systems and Their Potential Applications for Gastrointestinal Drug Delivery Platforms. *Nat. Commun.* **2022**, *13*, 2250. [CrossRef] [PubMed]
3. Black, K.A.; Priftis, D.; Perry, S.L.; Yip, J.; Byun, W.Y.; Tirrell, M. Protein Encapsulation via Polypeptide Complex Coacervation. *ACS Macro Lett.* **2014**, *3*, 1088–1091. [CrossRef] [PubMed]
4. Blocher, W.C.; Perry, S.L. Complex Coacervate-Based Materials for Biomedicine. *Wiley Interdiscip. Rev. Nanomed. Nanobiotechnol.* **2017**, *9*, e1442. [CrossRef]
5. Wang, Q.; Schlenoff, J.B. The Polyelectrolyte Complex/Coacervate Continuum. *Macromolecules* **2014**, *47*, 3108–3116. [CrossRef]
6. Takahashi, R.; Narayanan, T.; Sato, T. Growth Kinetics of Polyelectrolyte Complexes Formed from Oppositely-Charged Homopolymers Studied by Time-Resolved Ultra-Small-Angle X-Ray Scattering. *J. Phys. Chem. Lett.* **2017**, *8*, 737–741. [CrossRef]
7. Bakeev, K.N.; Izumrudov, V.A.; Kuchanov, S.I.; Zezin, A.B.; Kabanov, V.A. Kinetics and Mechanism of Interpolyelectrolyte Exchange and Addition Reactions. *Macromolecules* **1992**, *25*, 4249–4254. [CrossRef]
8. Malay, Ö.; Bayraktar, O.; Batigün, A. Complex Coacervation of Silk Fibroin and Hyaluronic Acid. *Int. J. Biol. Macromol.* **2007**, *40*, 387–393. [CrossRef]
9. Elmer, C.; Karaca, A.C.; Low, N.H.; Nickerson, M.T. Complex Coacervation in Pea Protein Isolate-Chitosan Mixtures. *Food Res. Int.* **2011**, *44*, 1441–1446. [CrossRef]
10. Zou, W.; Mourad, F.K.; Zhang, X.; Ahn, D.U.; Cai, Z.; Jin, Y. Phase Separation Behavior and Characterization of Ovalbumin and Propylene Glycol Alginate Complex Coacervates. *Food Hydrocoll.* **2020**, *108*, 105978. [CrossRef]
11. Luo, W.; Huang, H.; Zhang, Y.; Wang, F.; Yu, J.; Liu, Y.; Li, X. Complex Coacervation Behavior and the Mechanism between Rice Glutelin and Gum Arabic at PH 3.0 Studied by Turbidity, Light Scattering, Fluorescence Spectra and Molecular Docking. *LWT* **2021**, *150*, 112084. [CrossRef]

12. Archut, A.; Klost, M.; Drusch, S.; Kastner, H. Complex Coacervation of Pea Protein and Pectin: Contribution of Different Protein Fractions to Turbidity. *Food Hydrocoll.* **2023**, *134*, 108032. [CrossRef]
13. Liu, X.; Haddou, M.; Grillo, I.; Mana, Z.; Chapel, J.P.; Schatz, C. Early Stage Kinetics of Polyelectrolyte Complex Coacervation Monitored through Stopped-Flow Light Scattering. *Soft Matter* **2016**, *12*, 9030–9038. [CrossRef] [PubMed]
14. Amann, M.; Diget, J.S.; Lyngsø, J.; Pedersen, J.S.; Narayanan, T.; Lund, R. Kinetic Pathways for Polyelectrolyte Coacervate Micelle Formation Revealed by Time-Resolved Synchrotron SAXS. *Macromolecules* **2019**, *52*, 8227–8237. [CrossRef]
15. Xu, A.Y.; Melton, L.D.; Ryan, T.M.; Mata, J.P.; Rekas, A.; Williams, M.A.K.; McGillivray, D.J. Effects of Polysaccharide Charge Pattern on the Microstructures of β-Lactoglobulin-Pectin Complex Coacervates, Studied by SAXS and SANS. *Food Hydrocoll.* **2018**, *77*, 952–963. [CrossRef]
16. Spruijt, E.; Leermakers, F.A.M.; Fokkink, R.; Schweins, R.; Van Well, A.A.; Cohen Stuart, M.A.; Van Der Gucht, J. Structure and Dynamics of Polyelectrolyte Complex Coacervates Studied by Scattering of Neutrons, X-Rays, and Light. *Macromolecules* **2013**, *46*, 4596–4605. [CrossRef]
17. Takahashi, R.; Narayanan, T.; Yusa, S.I.; Sato, T. Kinetics of Morphological Transition between Cylindrical and Spherical Micelles in a Mixture of Anionic-Neutral and Cationic-Neutral Block Copolymers Studied by Time-Resolved SAXS and USAXS. *Macromolecules* **2018**, *51*, 3654–3662. [CrossRef]
18. Takahashi, R.; Narayanan, T.; Yusa, S.I.; Sato, T. Formation Kinetics of Polymer Vesicles from Spherical and Cylindrical Micelles Bearing the Polyelectrolyte Complex Core Studied by Time-Resolved USAXS and SAXS. *Macromolecules* **2022**, *55*, 684–695. [CrossRef]
19. Jing, B.; Ferreira, M.; Lin, K.; Li, R.; Yavitt, B.M.; Qiu, J.; Fukuto, M.; Zhu, Y. Ultrastructure of Critical-Gel-like Polyzwitterion-Polyoxometalate Complex Coacervates: Effects of Temperature, Salt Concentration, and Shear. *Macromolecules* **2020**, *53*, 10972–10980. [CrossRef]
20. Remuñán-López, C.; Bodmeier, R. Effect of Formulation and Process Variables on the Formation of Chitosan-Gelatin Coacervates. *Int. J. Pharm.* **1996**, *135*, 63–72. [CrossRef]
21. Basu, S.K.; Kavitha, K.; Rupeshkumar, M. Evaluation of Ketorolac Tromethamine Microspheres by Chitosan/Gelatin B Complex Coacervation. *Sci. Pharm.* **2010**, *78*, 79–92. [CrossRef] [PubMed]
22. Roy, J.C.; Giraud, S.; Ferri, A.; Mossotti, R.; Guan, J.; Salaün, F. Influence of Process Parameters on Microcapsule Formation from Chitosan—Type B Gelatin Complex Coacervates. *Carbohydr. Polym.* **2018**, *198*, 281–293. [CrossRef] [PubMed]
23. Cheng, C.; Tu, Z.; Wang, H. PH-Induced Complex Coacervation of Fish Gelatin and Carboxylated Chitosan: Phase Behavior and Structural Properties. *Food Res. Int.* **2023**, *167*, 112652. [CrossRef] [PubMed]
24. Prata, A.S.; Grosso, C.R.F. Production of Microparticles with Gelatin and Chitosan. *Carbohydr. Polym.* **2015**, *116*, 292–299. [CrossRef]
25. Wang, X.Y.; Wang, C.S.; Heuzey, M.C. Complexation of Chitosan and Gelatin: From Soluble Complexes to Colloidal Gel. *Int. J. Polym. Mater. Polym. Biomater.* **2016**, *65*, 96–104. [CrossRef]
26. Abdul Aziz, F.R.; Jai, J.; Raslan, R.; Subuki, I. Microencapsulation of Citronella Oil by Complex Coacervation Using Chitosan-Gelatin (b) System: Operating Design, Preparation and Characterization. In Proceedings of the MATEC Web of Conferences, Amsterdam, The Netherlands, 23–25 March 2016; Volume 69. [CrossRef]
27. Silva, M.C.; Andrade, C.T. Evaluating Conditions for the Formation of Chitosan/Gelatin Microparticles. *Polimeros* **2009**, *19*, 133–137. [CrossRef]
28. Yin, Y.; Li, Z.; Sun, Y.; Yao, K. A Preliminary Study on Chitosan/Gelatin Polyelectrolyte Complex Formation. *J. Mater. Sci.* **2005**, *40*, 4649–4652. [CrossRef]
29. Litvinov, M.; Kashurin, A.; Podshivalov, A. Study of the Influence of the Composition and PH of the Solution on the Structure and Morphology of Particle Dispersions of a (Bio)Polyelectrolyte Complex between Chitosan and Gelatin. *Colloid. Polym. Sci.* **2025**, *303*, 33–49. [CrossRef]
30. Litvinov, M.; Podshivalov, A.; Kovalev, K. Morphological Study of the Particle-to-Fiber Transition Threshold during Electrohydrodynamic Processing of Chitosan Solution. *J. Macromol. Sci. Part A Pure Appl. Chem.* **2021**, *58*, 804–810. [CrossRef]
31. Peters, G.S.; Gaponov, Y.A.; Konarev, P.V.; Marchenkova, M.A.; Ilina, K.B.; Volkov, V.V.; Pisarevsky, Y.V.; Kovalchuk, M.V. Upgrade of the BioMUR Beamline at the Kurchatov Synchrotron Radiation Source for Serial Small-Angle X-Ray Scattering Experiments in Solutions. *Nucl. Instrum. Methods Phys. Res. Sect. A Accel. Spectrometers Detect. Assoc. Equip.* **2022**, *1025*, 166170. [CrossRef]
32. Hammersley, A.P. FIT2D: A Multi-Purpose Data Reduction, Analysis and Visualization Program. *J. Appl. Crystallogr.* **2016**, *49*, 646–652. [CrossRef]
33. Konarev, P.V.; Petoukhov, M.V.; Volkov, V.V.; Svergun, D.I. ATSAS 2.1, a Program Package for Small-Angle Scattering Data Analysis. *J. Appl. Crystallogr.* **2006**, *39*, 277–286. [CrossRef]

34. Neves, S.; Fonseca, C.P. Determination of Fractal Dimension of Polyaniline Composites by SAXS and Electrochemical Techniques. *Electrochem. Commun.* **2001**, *3*, 36–43. [CrossRef]
35. Schmidt, P.W. Small-Angle Scattering Studies of Disordered, Porous and Fractal Systems. *J. Appl. Crystallogr.* **1991**, *24*, 414–435. [CrossRef]
36. Svergun, D.I. Determination of the Regularization Parameter in Indirect-Transform Methods Using Perceptual Criteria. *J. Appl. Crystallogr.* **1992**, *25*, 495–503. [CrossRef]
37. Rambo, R.P.; Tainer, J.A. Accurate Assessment of Mass, Models and Resolution by Small-Angle Scattering. *Nature* **2013**, *496*, 477–481. [CrossRef]
38. Mertens, H.D.T.; Svergun, D.I. Structural Characterization of Proteins and Complexes Using Small-Angle X-Ray Solution Scattering. *J. Struct. Biol.* **2010**, *172*, 128–141. [CrossRef]
39. Volkov, V.V.; Svergun, D.I. Uniqueness of Ab Initio Shape Determination in Small-Angle Scattering. *J. Appl. Crystallogr.* **2003**, *36*, 860–864. [CrossRef]
40. Svergun, D.I. Restoring Low Resolution Structure of Biological Macromolecules from Solution Scattering Using Simulated Annealing. *Biophys. J.* **1999**, *76*, 2879–2886. [CrossRef]
41. Swain, P.; Ronghe, A.; Bhutani, U.; Majumdar, S. Physicochemical Response of Gelatin in a Coulombic Soup of Monovalent Salt: A Molecular Simulation and Experimental Study. *J. Phys. Chem. B* **2019**, *123*, 1186–1194. [CrossRef]
42. Lin, W.; Yan, L.; Mu, C.; Li, W.; Zhang, M.; Zhu, Q. Effect of PH on Gelatin Self-Association Investigated by Laser Light Scattering and Atomic Force Microscopy. *Polym. Int.* **2002**, *51*, 233–238. [CrossRef]
43. Popa-Nita, S.; Alcouffe, P.; Rochas, C.; David, L.; Domard, A. Continuum of Structural Organization from Chitosan Solutions to Derived Physical Forms. *Biomacromolecules* **2010**, *11*, 6–12. [CrossRef] [PubMed]
44. Korchagina, E.V.; Philippova, O.E. Multichain Aggregates in Dilute Solutions of Associating Polyelectrolyte Keeping a Constant Size at the Increase in the Chain Length of Individual Macromolecules. *Biomacromolecules* **2010**, *11*, 3457–3466. [CrossRef] [PubMed]
45. Blagodatskikh, I.V.; Bezrodnykh, E.A.; Abramchuk, S.S.; Muranov, A.V.; Sinitsyna, O.V.; Khokhlov, A.R.; Tikhonov, V.E. Tikhonov Short Chain Chitosan Solutions: Self-Assembly and Aggregates Disruption Effects. *J. Polym. Res.* **2013**, *20*, 1–10. [CrossRef]
46. Kwok, J.J.; Park, K.S.; Patel, B.B.; Dilmurat, R.; Beljonne, D.; Zuo, X.; Lee, B.; Diao, Y. Understanding Solution State Conformation and Aggregate Structure of Conjugated Polymers via Small Angle X-Ray Scattering. *Macromolecules* **2022**, *55*, 4353–4366. [CrossRef]
47. Whittaker, J.L.; Dutta, N.K.; Knott, R.; McPhee, G.; Voelcker, N.H.; Elvin, C.; Hill, A.; Choudhury, N.R. Tunable Thermoresponsiveness of Resilin via Coassembly with Rigid Biopolymers. *Langmuir* **2015**, *31*, 8882–8891. [CrossRef]
48. Korchagina, E.V.; Philippova, O.E. Effects of Hydrophobic Substituents and Salt on Core-Shell Aggregates of Hydrophobically Modified Chitosan: Light Scattering Study. *Langmuir* **2012**, *28*, 7880–7888. [CrossRef]
49. Pa, J.H.; Yu, T.L. Light Scattering Study of Chitosan in Acetic Acid Aqueous Solutions. *Macromol. Chem. Phys.* **2001**, *202*, 985–991. [CrossRef]
50. Li, Y.I.; Cheng, R. Viscometric Study of Gelatin in Dilute Aqueous Solutions. *J. Polym. Sci. Part B Polym. Phys.* **2006**, *44*, 1804–1812. [CrossRef]
51. Rambo, R.P.; Tainer, J.A. Characterizing Flexible and Intrinsically Unstructured Biological Macromolecules by SAS Using the Porod-Debye Law. *Biopolymers* **2011**, *95*, 559–571. [CrossRef]
52. Dainese, E.; Sabatucci, A.; Cozzani, I. Small Angle X-Ray Scattering: A Powerful Tool to Analyze Protein Conformation in Solution. *Curr. Org. Chem.* **2005**, *9*, 1781–1800. [CrossRef]
53. Prior, C.; Davies, O.R.; Bruce, D.; Pohl, E. Obtaining Tertiary Protein Structures by the Ab Initio Interpretation of Small Angle X-Ray Scattering Data. *J. Chem. Theory Comput.* **2020**, *16*, 1985–2001. [CrossRef] [PubMed]
54. Balu, R.; Mata, J.P.; Knott, R.; Elvin, C.M.; Hill, A.J.; Choudhury, N.R.; Dutta, N.K. Effects of Crowding and Environment on the Evolution of Conformational Ensembles of the Multi-Stimuli-Responsive Intrinsically Disordered Protein, Rec1-Resilin: A Small-Angle Scattering Investigation. *J. Phys. Chem. B* **2016**, *120*, 6490–6503. [CrossRef] [PubMed]
55. Blobel, J.; Bernadó, P.; Svergun, D.I.; Tauler, R.; Pons, M. Low-Resolution Structures of Transient Protein-Protein Complexes Using Small-Angle X-Ray Scattering. *J. Am. Chem. Soc.* **2009**, *131*, 4378–4386. [CrossRef]

polymers

MDPI

Article

The Buffer Capacity of Polyelectrolyte Microcapsules Depends on the Type of Template

Alexey V. Dubrovskii, Aleksandr L. Kim and Sergey A. Tikhonenko *

Institute of Theoretical and Experimental Biophysics Russian Academy of Science, 3, Institutskaya Str., 142290 Puschino, Moscow Region, Russia; dav198@mail.ru (A.V.D.); kimerzent@gmail.com (A.L.K.)
* Correspondence: tikhonenkosa@gmail.com

Abstract: One of the key physicochemical parameters of polyelectrolyte microcapsules (PMCs) is their buffer capacity (BC). The BC of the microcapsules allows for an assessment of the change in protonation state across the entire polyelectrolyte system, which directly impacts the buffer barrier of PMCs, as well as the stability and physical properties of their shell. However, the buffer capacity of PMCs and their behavior under changes in ionic strength and temperature can differ depending on the type of core used to form the microcapsules. As part of this study, we revealed the buffer capacity (BC) of polyelectrolyte microcapsules formed on polystyrene cores (PMC_{Ps}) and studied the influence of ionic strength and environmental temperature on the BC of these capsules. We found that the buffer capacity of PMC_{Ps} differs from the BC of water at a pH above 8; the addition of sodium chloride leads to an increase in buffer capacity in alkaline conditions, and conversely, thermal treatment leads to its decrease at a pH of 9. The results obtained are different from the BC of polyelectrolyte microcapsules formed on $CaCO_3$ cores, which suggests a difference in the physicochemical properties of these types of capsules. The buffer capacity of polyelectrolyte microcapsules depends on the type of template used.

Keywords: polyelectrolyte microcapsules; buffer capacity; polystyrene; ionic strength; temperature

1. Introduction

Polyelectrolyte microcapsules (PMCs) are a type of nano-engineered multifunctional structures of spherical shape [1,2], which are obtained by the method of layer-by-layer deposition of oppositely charged polyelectrolytes on spherical particles ranging in size from 60 nm to several micrometers, followed by their dissolution [3,4]. Thanks to the step-by-step formation of microcapsules, their properties and functionality can be tuned according to the desired characteristics for specific tasks [5–9]. As a result, this type of capsule has become widely used in the medical field [10–12], the food industry [13,14], the development of smart materials [15,16], and other areas.

One of the key physicochemical parameters of polyelectrolyte microcapsules is their buffer capacity. The buffering capacity of a material is defined as its ability to maintain a constant concentration of H^+ ions in solution. The study of the buffer capacity of microcapsules permits the evaluation of changes in the protonation state of the entire polyelectrolyte system, which directly impacts the buffer barrier of PMCs, the stability, and physical properties of their shell [17–19], as well as the interpretation of results of studies with pH-sensitive encapsulated substances [20,21]. We have previously conducted research on the buffer capacity of polyelectrolyte microcapsules, which were formed on $CaCO_3$ cores and consisted of polyelectrolytes polyallylamine and polystyrene sulfonate [22–24]. Of the greatest interest among the results of these studies is the increase in the buffer capacity of polyelectrolyte microcapsules with an increase in NaCl concentration and a decrease in buffer capacity with an increase in temperature, as these environmental conditions significantly influence the physicochemical properties of PMCs.

However, the buffer capacity of polyelectrolyte microcapsules and their behavior under changes in ionic strength and temperature can vary depending on the type of core used to form the PMCs. This assumption is based on the fact that the morphology and other properties of the microcapsule shell depend on the template used [25]. In particular, PMCs formed on polystyrene (PS) cores have a pronounced shell, the thickness of which varies from 10 to 60 nm [26], while PMCs formed on a $CaCO_3$ core are a polyelectrolyte complex of spherical shape with a complex channel-like structure [27]. Also, their morphology varies differently in response to changes in pH and ionic strength: PMCs on a PS template increase in an alkaline medium and decrease at low pH values [26], while PMCs formed on $CaCO_3$ particles do not change their size at different pH values [28,29]; an increase in ionic strength leads to a decrease in the size of PMCs formed on a PS core [30], and microcapsules formed on $CaCO_3$ particles do not change in size [31]. In addition, PMCs formed on PS cores have greater resistance to highly alkaline conditions and can withstand pH 12.5 for 30 min [30], while PMCs based on $CaCO_3$ degrade instantly [30]. Considering the aforementioned differences in the morphology of PMCs formed on PS and $CaCO_3$ cores and their reactions to changes in pH and ionic strength, one cannot extrapolate the behavior of changes in the buffer capacity of PMCs formed on a $CaCO_3$ core to the behavior of changes in the buffer capacity of PMCs formed on a PS core, especially if we consider that this type of capsule can be used in areas such as microsensors or drug delivery systems as smart materials [32–34], where a change in ionic strength and acidity of the medium is implied. Thus, it is necessary to determine the behavior of the buffer capacity of PMCs formed on a PS core (PMC_{PS}).

In connection with the above, the aim of this study is to determine the presence of buffer capacity in PMCs formed on PS cores and to define the behavior of their BC under varying ionic strength and temperature. The results of this study showed that the buffer capacity of PMC_{PS} and its change in response to the ionic strength and temperature of the solution are different compared to the BC of polyelectrolyte microcapsules formed on $CaCO_3$ cores (PMC_{Ca}). Firstly, the buffer capacity of PMC_{PS} is observed only at a pH above 8, while PMC_{Ca} is observed throughout the studied range. Secondly, the BC of PMC_{PS} increases with increasing NaCl concentration, while the BC of PMC_{Ca} has a maximum buffer capacity at a sodium chloride concentration of 1 M. Thirdly, increasing the incubation temperature slightly reduces the BC of PMC_{PS}, while the BC of PMC_{Ca} gradually decreases with increasing temperature. The findings indicate that the buffer capacity of polyelectrolyte microcapsules is significantly influenced by the core material employed in their formation.

2. Materials and Methods

2.1. Materials

Polyelectrolytes sodium polystyrene sulfonate (PSS) and polyallylamine hydrochloride (PAH) with a molecular weight of 70 kDa were purchased from Sigma (St. Louis, MO, USA) (Residual Monomer < 10%). Dimethylformamide (DMFA), sodium chloride, sodium hydroxide, and hydrochloric acid were from "Reakhim". The percentage content of each chemical is a minimum of 99.9%. Polystyrene particles (5 μm) with COOH groups were purchased from Polymer Latex (Saint Petersburg, Russia).

2.2. Preparation of Polyelectrolyte Microcapsules

Polyelectrolyte microcapsules were obtained by layer-by-layer adsorption of negatively or positively charged polyelectrolytes on polystyrene microparticles, followed by their dissolution. Layer-by-layer adsorption of PAH and PSS on the surface of polystyrene microparticles was carried out in polyelectrolyte solutions (concentration 2 mg/mL + 0.5 M NaCl). After each adsorption, polystyrene particles with adsorbed polyelectrolytes were washed three times with 0.5 M NaCl solution, which was necessary to remove unabsorbed polymer molecules. Particles were separated from the supernatant by centrifugation. After applying the necessary number of layers, the polystyrene particles were dissolved in DMFA for 48 h.

After removing the polystyrene core, the polyelectrolyte microcapsules were washed three times in a DMFA–water solution. After each of the 3 washes, the water–DMFA ratio in the solution was changed by increasing the amount of water by 10%. The washes and the procedure of increasing the amount of water were repeated until the DMFA was completely removed from the wash solution. The resulting capsules were washed three times with water to remove the DMFA. The size, number, and ζ-potential of the microcapsules were measured using the dynamic light scattering method on a Zetasizer nano ZS device (Malvern, UK).

2.3. Temperature Treatment

For heating, the suspension of microcapsules was incubated at the corresponding temperature (60 and 90 °C) in a Lauda Ecoline RE 112 (Lauda, Moscow, Russia) thermostat for 60 min. Following the application of the temperature treatment, the polyelectrolyte microcapsules were incubated at room temperature, with the objective of achieving a suspension temperature equal to that of the ambient environment. The subsequent measurement of the buffer capacity of the polyelectrolyte microcapsules was conducted only after the attainment of room temperature.

2.4. Measurement of Buffer Capacity

A suspension of microcapsules $(PSS/PAH)_3$ (about 6.6×10^9 microcapsules in 8 mL of water) at room temperature was titrated with solutions of acids or bases in the pH range from 4 to 9. The titration was carried out manually. Acid or base (with a concentration of 0.001 M or 0.005 M) was added to the solution to change the pH of the solution by 0.02 or more. The buffer capacity was calculated using Equation (1) after assessing the slope of the titration curves at each point by the change in pH between the previous and subsequent introductions [35]:

$$BC = \frac{n_{NaOH}(i+1) - n_{NaOH}(i-1)}{pH(i+1) - pH(i-1)} \quad (1)$$

where n_{NaOH} is the number of moles of NaOH added and $(i+1)$ and $(i-1)$ refer to the previous and subsequent injections, respectively.

2.5. Statistical Processing

Each sample was measured five times. Due to the peculiarity of titration, it is impossible to achieve the same pH value during titration. Therefore, we combined the buffer capacity values obtained from 5 repeated experiments in steps of 0.1 pH values. The graph shows the mean values of the pooled pH points in 0.1 pH steps (X scale) with the mean buffer capacity value for these points (Y scale), with the standard deviation of both the mean pH value (X error bars) and the mean buffer capacity value (Y error bars).

3. Results

The buffer capacity of polyelectrolyte microcapsules allows us to evaluate their ability to maintain a constant hydrogen proton concentration in solution, which is particularly important for maintaining the stability of encapsulated objects such as enzymes [36], fluorescent dyes [20], metal particles [37], etc. We have previously shown the buffer capacity of polyelectrolyte microcapsules formed on a $CaCO_3$ core [24]. However, depending on the type of core used, the buffer capacity of the capsule and its response to external chemical stimuli may differ [25].

In this study, the aim is to determine the presence of buffer capacity (BC) in polyelectrolyte microcapsules formed on a polystyrene (PS) core and to study the behavior of the change in the BC of these capsules under different ionic strengths and temperatures of the solution. For this purpose, polyelectrolyte microcapsules of the composition $(PAH/PSS)_3/PAH$ were prepared using the method of layer-by-layer adsorption of polystyrene sulfonate (PSS) and polyallylamine (PAH) polyelectrolytes onto a PS particle with covalently bound carboxyl groups. In the final stage of PMC creation, the polystyrene core was removed. The basic scheme for obtaining polyelectrolyte microcapsules is presented in Figure 1.

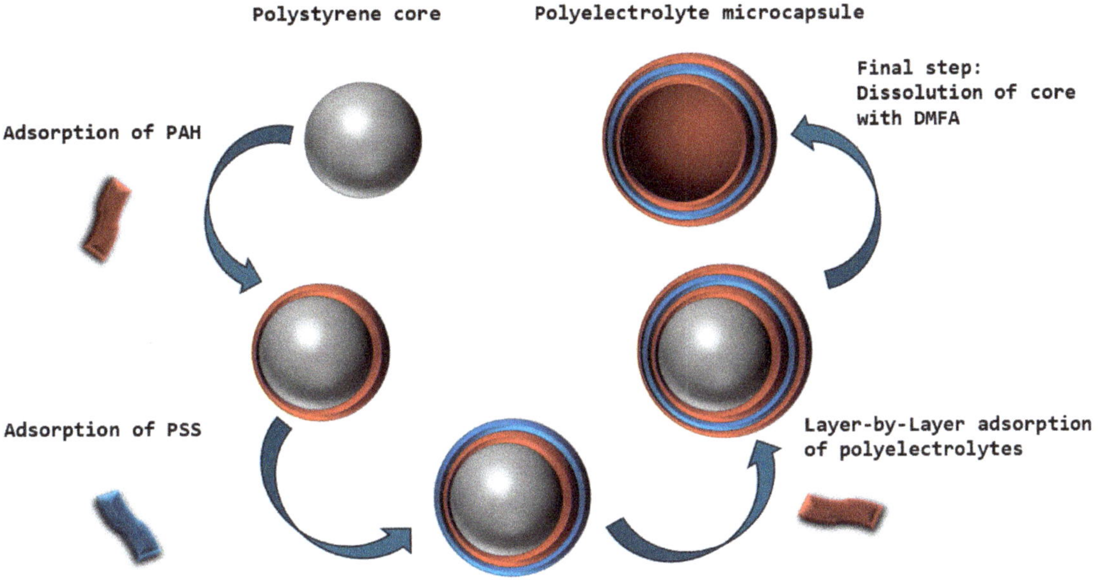

Figure 1. Scheme of preparation of polyelectrolyte microcapsules on a polystyrene core.

The size distribution of the polyelectrolyte microcapsules (PMCs), formed on a polystyrene core (abbreviated as PMC_{PS}), was subsequently determined (Figure 2).

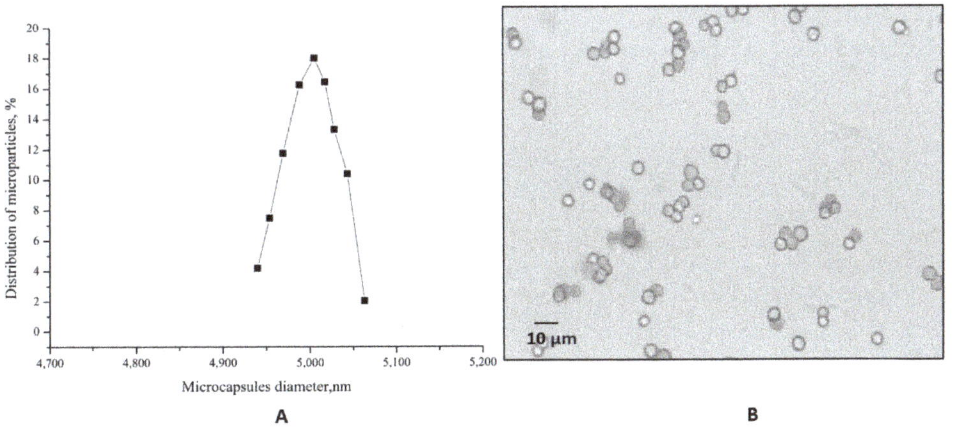

Figure 2. The optical microscopy images of PMCs (**A**). The PMC diameter distribution function (**B**).

The optical microscopy images of PMCs (Figure 2B) demonstrate the morphological homogeneity of the microcapsules. The microcapsules had an average diameter of 5 µm with a 2.7% polydispersity index (Figure 2A) and an ζ-potential of +26 ± 1 mV.

The obtained polyelectrolyte microcapsules, formed on a polystyrene core, were used to determine their buffer capacity. For this, a suspension of PMCs (6.6×10^9 microcapsules in 8 mL of water), consisting of polyallylamine (PAH) and polystyrene sulfonate (PSS), was titrated with alkali or acid in the pH range from 3.5 to 9, and the change in pH of the suspension was determined using a pH meter. The results of determining the buffer capacity for the pH range from 5.5 to 9 are presented in Figure 3.

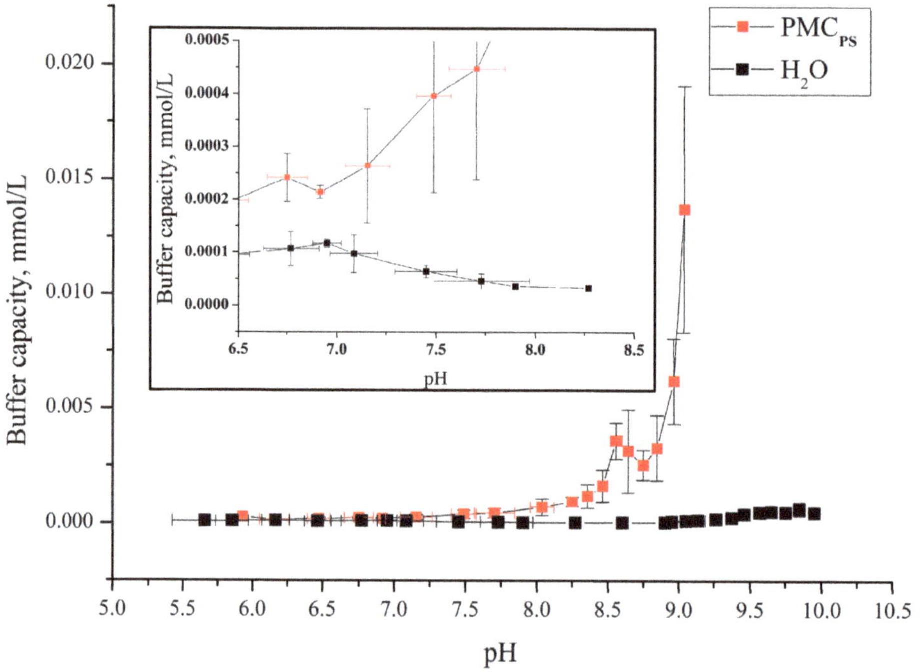

Figure 3. Buffer capacity of PMCs formed on PS with composition $(PAH/PSS)_3/PAH$ in the pH range from 5.5 to 9.

From the figure, it can be seen that the buffer capacity of microcapsules formed on PS particles differs from the BC of water at a pH above 8, which is better seen from the insert in Figure 3. In our previous work, we determined the buffer capacity of polyelectrolyte microcapsules formed on a $CaCO_3$ core (abbreviated as PMC_{Ca}) [24]. Comparing the buffer capacity of these two types of capsules, it was found that the buffer capacity of PMC_{Ca} in the pH range from 5.5 to 9 significantly differs from the presented buffer capacity of PMC_{Ps}. The results of our previous work are presented in Figure 4.

From Figure 4, it can be seen that the buffer capacity of PMC_{Ca} monotonically increases with increasing pH from 6, and at pH 9, the BC increases no more than three times. However, for PMC_{Ps}, the BC starts to rise only after 8, and at pH 9, the buffer capacity increases 10 times. Considering that the unbound PAH regions are responsible for the buffer capacity of polyelectrolyte microcapsules [23,24], the beginning of the increase in the buffer capacity of PMC_{Ca} at a pH lower than that of PMC_{Ps} may be associated with the fact that PMC_{Ps} have a more stable polyelectrolyte complex of polyallylamine with polystyrene sulfonate than PMC_{Ca}. A gradual increase in pH leads to a gradual increase in unbound PAH regions, which leads to an increase in the number of free H^+ donors and an increase in buffer capacity, which is consistent with our previous research [23,24]. However, PMC_{Ps} show a sharp increase in buffer capacity, which is associated with the simultaneous formation of a larger number of unbound PAH regions than in PMC_{Ca}.

Subsequently, we determined the buffer capacity of PMC_{Ps} in the pH range from 4 to 5.5 and found that in these pH values, PMC_{Ps} do not possess buffer capacity. These results are presented in Figure 5.

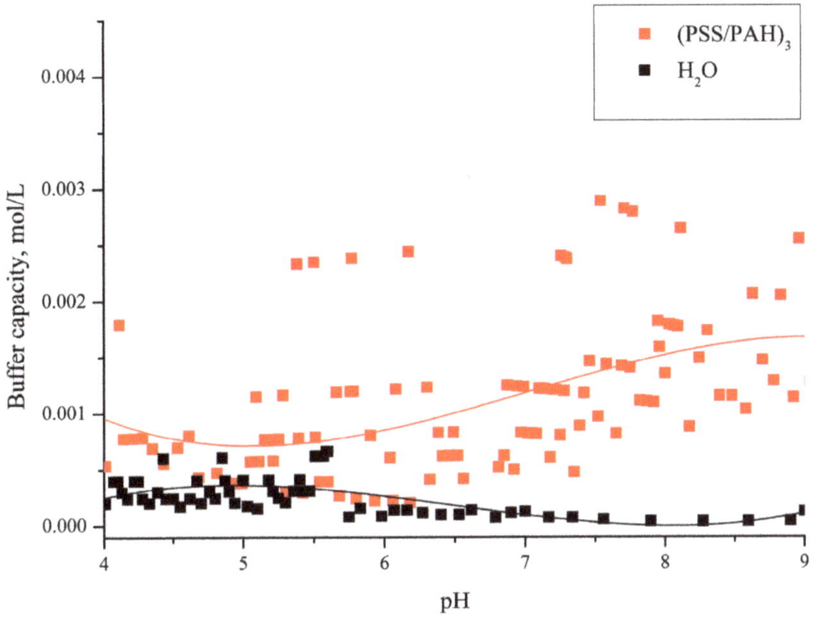

Figure 4. Buffer capacity of microcapsules (PSS/PAH)$_3$ and water in the pH range from 4 to 9. Adapted from [24].

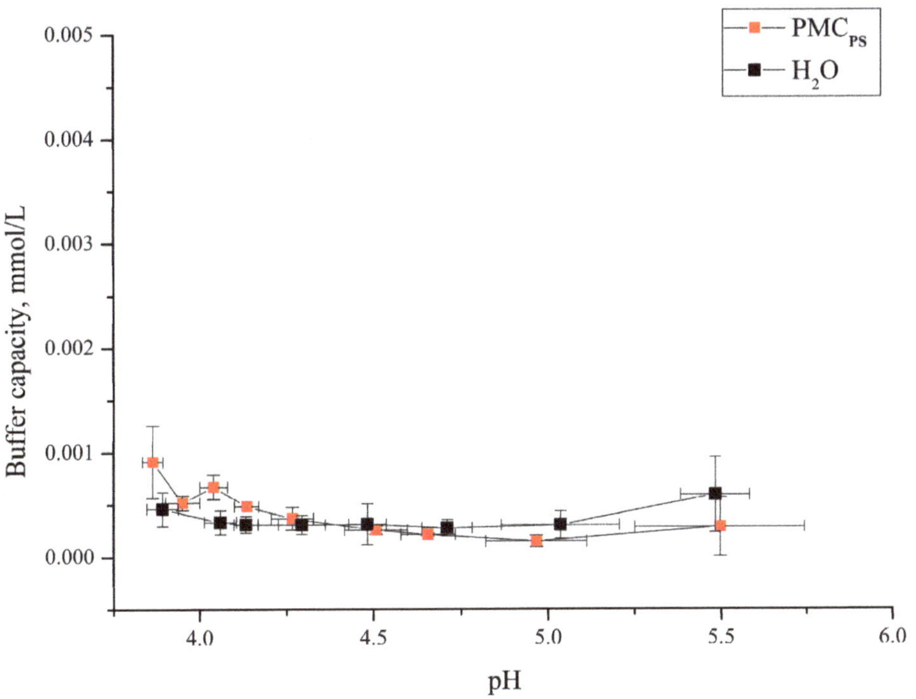

Figure 5. Buffer capacity of PMCs formed on PS with composition (PAH/PSS)$_3$/PAH in the pH range from 3.5 to 5.5.

As can be seen from Figures 4 and 5, the buffer capacity of PMC_{Ps} and PMC_{Ca} in acidic ranges also differ. This effect can also be explained by the fact that the amount of electrostatically bound polyallylamine with polystyrene sulfonate is significantly larger in PMC_{Ps} than in PMC_{Ca}. A larger number of ionic bonds between PSS and PAH chains in PMC_{Ps} is also indirectly confirmed by other works. In particular, several studies have presented the shell thickness for different numbers of layers for PMC_{Ps} and for PMC_{Ca} (without removing cores) [26,27,38]. It was shown that the shell thickness of PMC_{Ca} is 3–5 times larger than in the case of PMC_{Ps}, which is explained by the authors by a denser shell (in PMC_{Ps}) and, as a result, greater interaction between PSS and PAH.

In the next step, the effect of the ionic strength of the solution on the buffer capacity of polyelectrolyte microcapsules formed on PS cores was studied. This step is necessary to confirm the hypothesis put forward in the previous work [23,24] that the BC of PMCs depends on the number of unbound PAH regions in the composition of the microcapsules since, as the ionic strength of the solution increases, the interaction of polyelectrolytes decreases [39–41]. In addition, the obtained results will allow us to take into account the behavior of PMCs and encapsulated substances depending on the level of protonation/deprotonation under different ionic strength conditions. Figure 6 presents the curves of the dependence of the buffer capacity of the PMC_{Ps} composition $(PAH/PSS)_3/PAH$ on pH in the presence of different concentrations of sodium chloride.

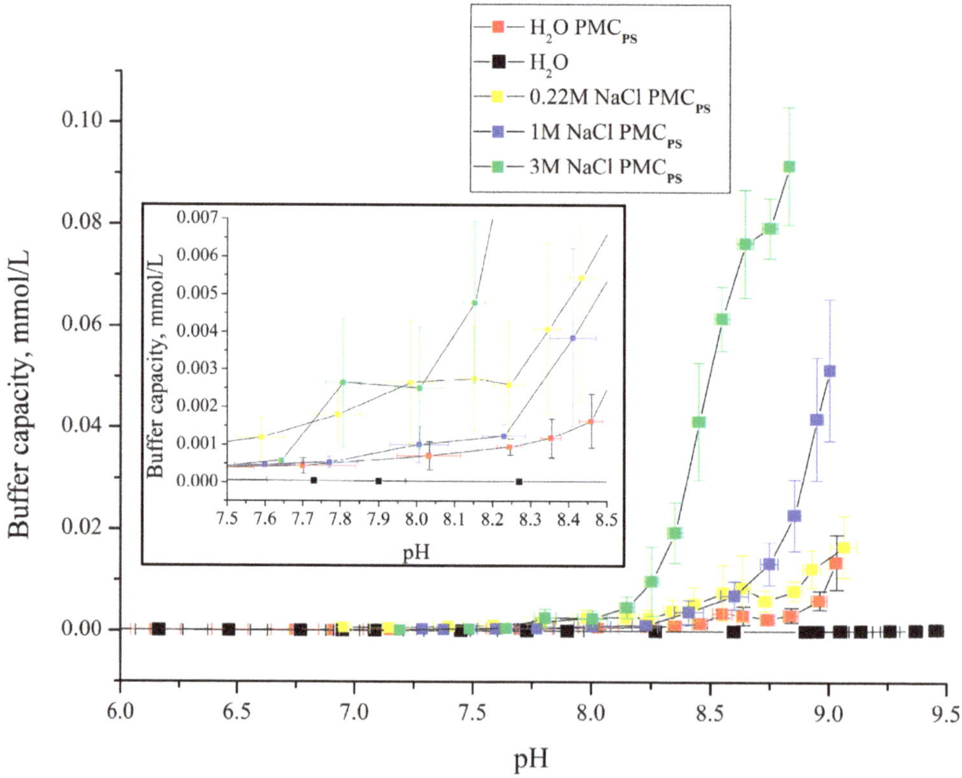

Figure 6. Buffer capacity of PMC_{Ps} composition $(PAH/PSS)_3/PAH$ at different pH in water, 0.22 M, 1 M, and 3 M NaCl solution.

From the figure, it can be seen that the addition of sodium chloride leads to an increase in the buffer capacity of polyelectrolyte microcapsules formed on PS cores, which confirms the hypothesis put forward above. However, the BC of PMC_{Ps} rises with the increase in salt concentration in the range up to 3 M, which does not correspond to the change in the BC of PMC_{Ca} from salt, which we obtained earlier [23,24]. In the previous work, the BC of PMC_{Ca} increased maximally at a salt concentration of 1M, and a further increase in salt concentration did not increase the buffer capacity of this type of capsule. This difference can also be explained by a greater number of PSS-PAH ionic pairs in the PMC_{Ps} shell than in PMC_{Ca}, which prevents the breakage of the ionic bond when the ionic strength increases. In addition, this fact is confirmed by the fact that in all described concentrations of NaCl, an increase in the BC of PMC_{Ps} at pH = 7.9 was observed (see insert in Figure 6), and only the BC values differed, which grew with the increase in the ionic strength of the solution.

Subsequently, the effect of temperature exposure on the buffer capacity of polyelectrolyte microcapsules formed on PS cores was studied. Considering the denser arrangement of polyelectrolytes in the PMC_{Ps} shell [26] than in the PMC_{Ca} [27], it can be assumed that the buffer capacity of these two types of capsules will also differ after temperature treatment. In addition, the size of these types of capsules changed differently after their heating: PMC_{Ps} did not change [32,42], and PMC_{Ca} decreased by 24–28% [43]. Figure 7 presents the curves of the dependence of the buffer capacity of the PMC_{Ps} composition $(PAH/PSS)_3/PAH$ on the pH after temperature treatment at 60 and 90 °C.

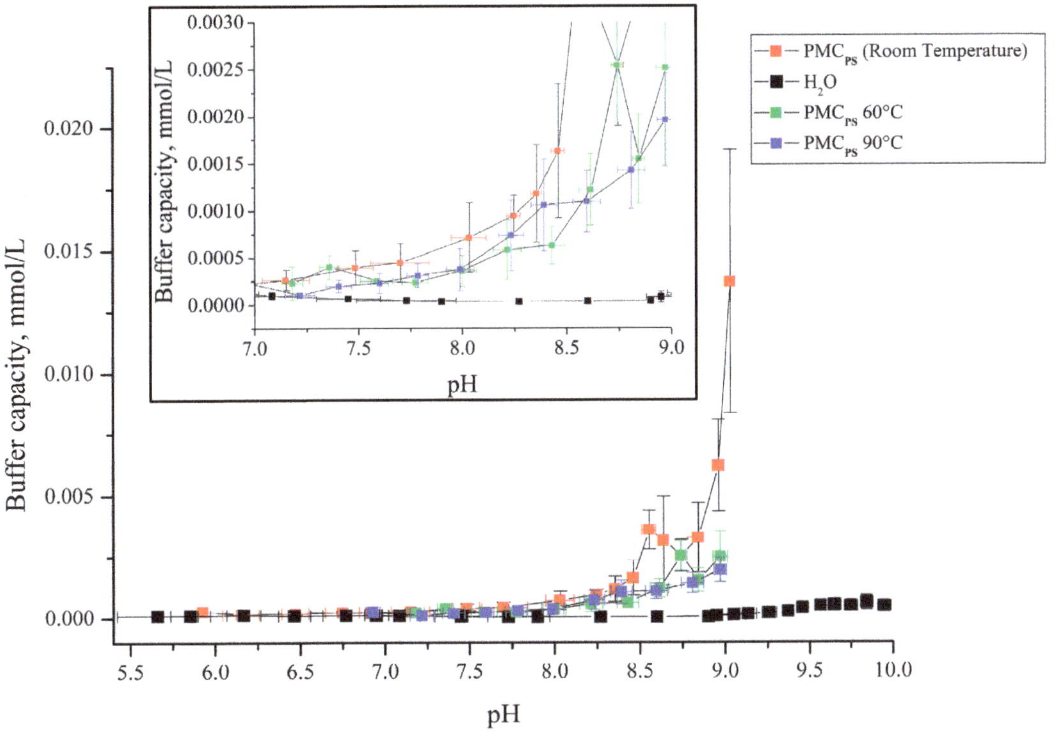

Figure 7. Buffer capacity of PMC_{Ps} composition $(PAH/PSS)_3/PAH$ at different pH after temperature treatment at 60 and 90 °C.

From Figure 7, it can be seen that the buffer capacity of thermally treated polyelectrolyte microcapsules formed on a PS core differs only at a pH above 9 from the buffer capacity of non-heated capsules. According to other authors' works [32,42,43], thermal treatment leads to the breakage and reforming of ionic pairs, which leads to a more compact arrangement of polyelectrolyte molecules. Thus, it is logical to assume that the buffer capacity will change slightly in PMC_{Ps}, the polyelectrolytes of which are compactly arranged [26].

However, at pH 9, thermally treated PMC_{PS} lack a sharp rise in buffer capacity, which suggests an enhancement of the above-described compaction of polyelectrolytes in the shell and an increase in the stability of the polyelectrolyte complex, which allows for reducing the impact of the pH. The previously obtained BC results of PMC_{Ca} [23] after thermal treatment differ from the BC of PMC_{PS}. In the case of PMC_{Ca}, the buffer capacity gradually decreased with increasing temperature, while this gradation was absent in PMC_{PS}. This effect can also be explained by the aforementioned compaction of polyelectrolytes.

Based on the above, it can be unequivocally concluded that the buffer capacity of polyelectrolyte microcapsules formed on $CaCO_3$ and polystyrene cores differ, and these capsules react differently to changes in ionic strength and temperature of the medium, which depends on the compactness of the polyelectrolyte layers.

4. Conclusions

In this study, the buffer capacity (BC) of polyelectrolyte microcapsules formed on polystyrene cores (PMC_{PS}) was determined, and the effect of the ionic strength and medium temperature on the BC of these capsules was studied. It was found that the buffer capacity of PMC_{PS} differs from the BC of water at a pH above 8; the addition of sodium chloride leads to an increase in buffer capacity in alkaline conditions, and temperature treatment, on the contrary, to its decrease at pH 9. By conducting a comparative analysis with the data we obtained earlier [23,24], differences were identified in the buffer capacity of PMC_{PS} and its change in response to the ionic strength and temperature of the solution compared to the BC of polyelectrolyte microcapsules formed on $CaCO_3$ cores (PMC_{Ca}). Firstly, the buffer capacity of PMC_{PS} is observed only at a pH above 8, while PMC_{Ca} is observed throughout the studied range. Secondly, the BC of PMC_{PS} increases with increasing NaCl concentration, while the BC of PMC_{Ca} has a maximum buffer capacity at a sodium chloride concentration of 1M. Thirdly, increasing the incubation temperature slightly reduces the BC of PMC_{PS}, while the BC of PMC_{Ca} gradually decreases with increasing temperature. Such differences are presumably related to the more compact arrangement of polyelectrolytes in the PMC_{PS} shell, which is associated with a greater number of PSS-PAH ionic pairs, complicating the effect of the solution's pH. It can be concluded that the buffer capacity of polyelectrolyte microcapsules depends on the type of template.

The obtained results excellently demonstrate the difference in the physicochemical properties of polyelectrolyte microcapsules formed on $CaCO_3$ and PS cores, which will allow researchers to take into account this difference when choosing a method of encapsulating a substance for the tasks set. In addition, these results will allow consideration of the buffer capacity of PMC_{PS} when interpreting data when including pH-sensitive compounds in various conditions of ionic strength and temperature.

Author Contributions: Conceptualization, S.A.T., A.V.D. and A.L.K.; Data curation, A.V.D., A.L.K. and S.A.T.; Formal analysis, A.V.D. and A.L.K.; Investigation, A.V.D. and S.A.T.; Methodology, A.L.K., S.A.T. and A.V.D.; Project administration, S.A.T.; Supervision, S.A.T.; Visualization, A.V.D. and A.L.K.; Writing—original draft, A.L.K.; Writing—review and editing, A.V.D., A.L.K. and S.A.T. All authors have read and agreed to the published version of the manuscript.

Funding: This research was funded by the State Assignment of the Russian Federation No. 075-00224-24-03.

Institutional Review Board Statement: Not applicable.

Data Availability Statement: Data are contained within the article.

Conflicts of Interest: The authors declare no conflicts of interest.

References

1. Sukhorukov, G.B.; Donath, E.; Davis, S.; Lichtenfeld, H.; Caruso, F.; Popov, V.I.; Möhwald, H. Stepwise polyelectrolyte assembly on particle surfaces: A novel approach to colloid design. *Polym. Adv. Technol.* **1998**, *9*, 759–767. [CrossRef]
2. Sukhorukov, G.B.; Donath, E.; Lichtenfeld, H.; Knippel, E.; Knippel, M.; Budde, A.; Möhwald, H. Layer-by-layer self assembly of polyelectrolytes on colloidal particles. *Colloids Surf. A Physicochem. Eng. Asp.* **1998**, *137*, 253–266. [CrossRef]

3. Köhler, K.; Shchukin, D.G.; Sukhorukov, G.B.; Möhwald, H. Drastic Morphological Modification of Polyelectrolyte Microcapsules Induced by High Temperature. *Macromolecules* **2004**, *37*, 9546–9550. [CrossRef]
4. Gao, C.; Donath, E.; Möhwald, H.; Shen, J. Spontaneous Deposition of Water-Soluble Substances into Microcapsules: Phenomenon, Mechanism, and Application. *Angew. Chem. Int. Ed.* **2002**, *41*, 3789–3793. [CrossRef]
5. Dubreuil, F.; Elsner, N.; Fery, A. Elastic properties of polyelectrolyte capsules studied by atomic-force microscopy and RICM. *Eur. Phys. J. E* **2003**, *12*, 215–221. [CrossRef] [PubMed]
6. Sukhorukov, G.B.; Shchukin, D.G.; Dong, W.; Möhwald, H.; Lulevich, V.V.; Vinogradova, O.I. Comparative Analysis of Hollow and Filled Polyelectrolyte Microcapsules Templated on Melamine Formaldehyde and Carbonate Cores. *Macromol. Chem. Phys.* **2004**, *205*, 530–535. [CrossRef]
7. Köhler, K.; Shchukin, D.G.; Möhwald, H.; Sukhorukov, G.B. Thermal Behavior of Polyelectrolyte Multilayer Microcapsules. 1. The Effect of Odd and Even Layer Number. *J. Phys. Chem. B* **2005**, *109*, 18250–18259. [CrossRef] [PubMed]
8. Skirtach, A.G.; Yashchenok, A.M.; Möhwald, H. Encapsulation, release and applications of LbL polyelectrolyte multilayer capsules. *Chem. Commun.* **2011**, *47*, 12736. [CrossRef]
9. De Geest, B.G.; Skirtach, A.G.; Mamedov, A.A.; Antipov, A.A.; Kotov, N.A.; De Smedt, S.C.; Sukhorukov, G.B. Ultrasound-Triggered Release from Multilayered Capsules. *Small* **2007**, *3*, 804–808. [CrossRef]
10. Verkhovskii, R.; Ermakov, A.; Sindeeva, O.; Prikhozhdenko, E.; Kozlova, A.; Grishin, O.; Makarkin, M.; Gorin, D.; Bratashov, D. Effect of Size on Magnetic Polyelectrolyte Microcapsules Behavior: Biodistribution, Circulation Time, Interactions with Blood Cells and Immune System. *Pharmaceutics* **2021**, *13*, 2147. [CrossRef]
11. Palamarchuk, K.V.; Borodina, T.N.; Kostenko, A.V.; Chesnokov, Y.M.; Kamyshinsky, R.A.; Palamarchuk, N.P.; Yudina, E.B.; Nikolskaya, E.D.; Yabbarov, N.G.; Mollaeva, M.R.; et al. Development of Submicrocapsules Based on Co-Assembled Like-Charged Silica Nanoparticles and Detonation Nanodiamonds and Polyelectrolyte Layers. *Pharmaceutics* **2022**, *14*, 575. [CrossRef] [PubMed]
12. Gileva, A.; Trushina, D.; Yagolovich, A.; Gasparian, M.; Kurbanova, L.; Smirnov, I.; Burov, S.; Markvicheva, E. Doxorubicin-Loaded Polyelectrolyte Multilayer Capsules Modified with Antitumor DR5-Specific TRAIL Variant for Targeted Drug Delivery to Tumor Cells. *Nanomaterials* **2023**, *13*, 902. [CrossRef] [PubMed]
13. Pan, H.M.; Subramanian, A.; Ochs, C.J.; Dewavrin, J.-Y.; Beyer, S.; Trau, D.W. Edible polyelectrolyte microcapsules with water-soluble cargo assembled in organic phase. *RSC Adv.* **2014**, *4*, 35163–35166. [CrossRef]
14. Tan, C.; Selig, M.J.; Lee, M.C.; Abbaspourrad, A. Encapsulation of copigmented anthocyanins within polysaccharide microcapsules built upon removable $CaCO_3$ templates. *Food Hydrocoll.* **2018**, *84*, 200–209. [CrossRef]
15. Khan, A.; Sliem, M.H.; Arif, A.; Salih, M.A.; Shakoor, R.A.; Montemor, M.F.; Kahraman, R.; Mansour, S.; Abdullah, A.M.; Hasan, A. Designing and performance evaluation of polyelectrolyte multilayered composite smart coatings. *Prog. Org. Coat.* **2019**, *137*, 105319. [CrossRef]
16. Popov, A.L.; Popova, N.; Gould, D.J.; Shcherbakov, A.B.; Sukhorukov, G.B.; Ivanov, V.K. Ceria Nanoparticles-Decorated Microcapsules as a Smart Drug Delivery/Protective System: Protection of Encapsulated P. pyralis Luciferase. *ACS Appl. Mater. Interfaces* **2018**, *10*, 14367–14377. [CrossRef] [PubMed]
17. Musin, E.V.; Kim, A.L.; Tikhonenko, S.A. Destruction of polyelectrolyte microcapsules formed on $CaCO_3$ microparticles and the release of a protein included by the adsorption method. *Polymers* **2020**, *12*, 520. [CrossRef] [PubMed]
18. Lulevich, V.V.; Vinogradova, O.I. Effect of pH and Salt on the Stiffness of Polyelectrolyte Multilayer Microcapsules. *Langmuir* **2004**, *20*, 2874–2878. [CrossRef]
19. Kim, B.; Vinogradova, O.I. pH-Controlled Swelling of Polyelectrolyte Multilayer Microcapsules. *J. Phys. Chem. B* **2004**, 8161–8165. [CrossRef]
20. Kazakova, L.I.; Shabarchina, L.I.; Sukhorukov, G.B. Co-encapsulation of enzyme and sensitive dye as a tool for fabrication of microcapsule based sensor for urea measuring. *Phys. Chem. Chem. Phys.* **2011**, *13*, 11110–11117. [CrossRef]
21. Kazakova, L.I.; Shabarchina, L.I.; Anastasova, S.; Pavlov, A.M.; Vadgama, P.; Skirtach, A.G.; Sukhorukov, G.B. Chemosensors and biosensors based on polyelectrolyte microcapsules containing fluorescent dyes and enzymes. *Anal. Bioanal. Chem.* **2013**, *405*, 1559–1568. [CrossRef]
22. Musin, E.V.; Dubrovskii, A.V.; Kim, A.L.; Tikhonenko, S.A. A Study of the Buffer Capacity of Polyelectrolyte Microcapsules Depending on Their Concentration and the Number of Layers of the Polyelectrolyte Shell. *Int. J. Mol. Sci.* **2022**, *23*, 9917. [CrossRef]
23. Dubrovskii, A.V.; Kim, A.L.; Musin, E.V.; Tikhonenko, S.A. A Study of the Buffer Capacity of Polyelectrolyte Microcapsules Depending on Their Ionic Environment and Incubation Temperature. *Int. J. Mol. Sci.* **2022**, *23*, 6608. [CrossRef] [PubMed]
24. Dubrovskii, A.V.; Kim, A.L.; Musin, E.V.; Ramazanov, B.R.; Tikhonenko, S.A. The Discovery of the Buffer Capacity of Various Types of Polyelectrolyte Microcapsules. *Polymers* **2021**, *13*, 4026. [CrossRef] [PubMed]
25. Kim, A.L.; Musin, E.V.; Chebykin, Y.S.; Tikhonenko, S.A. Characterization of Polyallylamine/Polystyrene Sulfonate Polyelectrolyte Microcapsules Formed on Solid Cores: Morphology. *Polymers* **2024**, *16*, 1521. [CrossRef]
26. Déjugnat, C.; Sukhorukov, G.B. pH-Responsive Properties of Hollow Polyelectrolyte Microcapsules Templated on Various Cores. *Langmuir* **2004**, *20*, 7265–7269. [CrossRef]
27. Volodkin, D.V.; Petrov, A.I.; Prevot, M.; Sukhorukov, G.B. Matrix Polyelectrolyte Microcapsules: New System for Macromolecule Encapsulation. *Langmuir* **2004**, *20*, 3398–3406. [CrossRef] [PubMed]

28. Tong, W.; Dong, W.; Gao, C.; Möhwald, H. Charge-Controlled Permeability of Polyelectrolyte Microcapsules. *J. Phys. Chem. B* **2005**, *109*, 13159–13165. [CrossRef]
29. Haložan, D.; Riebentanz, U.; Brumen, M.; Donath, E. Polyelectrolyte microcapsules and coated $CaCO_3$ particles as fluorescence activated sensors in flowmetry. *Colloids Surf. A Physicochem. Eng. Asp.* **2009**, *342*, 115–121. [CrossRef]
30. Heuvingh, J.; Zappa, M.; Fery, A. Salt Softening of Polyelectrolyte Multilayer Capsules. *Langmuir* **2005**, *21*, 3165–3171. [CrossRef]
31. Pechenkin, M.A.; Möhwald, H.; Volodkin, D.V. pH- and salt-mediated response of layer-by-layer assembled PSS/PAH microcapsules: Fusion and polymer exchange. *Soft Matter* **2012**, *8*, 8659. [CrossRef]
32. Prevot, M.; Déjugnat, C.; Möhwald, H.; Sukhorukov, G.B. Behavior of Temperature-Sensitive PNIPAM Confined in Polyelectrolyte Capsules. *ChemPhysChem* **2006**, *7*, 2497–2502. [CrossRef] [PubMed]
33. Haložan, D.; Déjugnat, C.; Brumen, M.; Sukhorukov, G.B. Entrapment of a Weak Polyanion and H^+/Na^+ Exchange in Confined Polyelectrolyte Microcapsules. *J. Chem. Inf. Model.* **2005**, *45*, 1589–1592. [CrossRef]
34. Radtchenko, I.L.; Sukhorukov, G.B.; Leporatti, S.; Khomutov, G.B.; Donath, E.; Möhwald, H. Assembly of Alternated Multivalent Ion/Polyelectrolyte Layers on Colloidal Particles. Stability of the Multilayers and Encapsulation of Macromolecules into Polyelectrolyte Capsules. *J. Colloid Interface Sci.* **2000**, *230*, 272–280. [CrossRef] [PubMed]
35. Richard, I.; Thibault, M.; De Crescenzo, G.; Buschmann, M.D.; Lavertu, M. Ionization Behavior of Chitosan and Chitosan–DNA Polyplexes Indicate That Chitosan Has a Similar Capability to Induce a Proton-Sponge Effect as PEI. *Biomacromolecules* **2013**, *14*, 1732–1740. [CrossRef]
36. Kim, A.L.; Musin, E.V.; Dubrovskii, A.V.; Tikhonenko, S.A. Determination of urea concentration using urease-containing polyelectrolyte microcapsules. *Anal. Methods* **2019**, *11*, 1585–1590. [CrossRef]
37. Pavlov, A.M.; Saez, V.; Cobley, A.; Graves, J.; Sukhorukov, G.B.; Mason, T.J. Controlled protein release from microcapsules with composite shells using high frequency ultrasound—Potential for in vivo medical use. *Soft Matter* **2011**, *7*, 4341–4347. [CrossRef]
38. Sukhorukov, G.B.; Volodkin, D.V.; Günther, A.M.; Petrov, A.I.; Shenoy, D.B.; Möhwald, H. Porous calcium carbonate microparticles as templates for encapsulation of bioactive compounds. *J. Mater. Chem.* **2004**, *14*, 2073–2081. [CrossRef]
39. Scheepers, D.; Chatillon, B.; Borneman, Z.; Nijmeijer, K. Influence of charge density and ionic strength on diallyldimethylammonium chloride (DADMAC)-based polyelectrolyte multilayer membrane formation. *J. Memb. Sci.* **2021**, *617*, 118619. [CrossRef]
40. Ferrand-Drake del Castillo, G.; Hailes, R.L.N.; Dahlin, A. Large Changes in Protonation of Weak Polyelectrolyte Brushes with Salt Concentration—Implications for Protein Immobilization. *J. Phys. Chem. Lett.* **2020**, *11*, 5212–5218. [CrossRef]
41. Sagou, J.-P.S.; Ahualli, S.; Thomas, F. Influence of ionic strength and polyelectrolyte concentration on the electrical conductivity of suspensions of soft colloidal polysaccharides. *J. Colloid Interface Sci.* **2015**, *459*, 212–217. [CrossRef] [PubMed]
42. Park, M.-K.; Deng, S.; Advincula, R.C. Sustained Release Control via Photo-Cross-Linking of Polyelectrolyte Layer-by-Layer Hollow Capsules. *Langmuir* **2005**, *21*, 5272–5277. [CrossRef] [PubMed]
43. Dubrovskii, A.V.; Shabarchina, L.I.; Kim, Y.A.; Sukhorukov, B.I. Influence of the temperature on polyelectrolyte microcapsules: Light scattering and confocal microscopy data. *Russ. J. Phys. Chem. A* **2006**, *80*, 1703–1707. [CrossRef]

polymers

MDPI

Article

Humic Polyelectrolytes Facilitate Rapid Microwave Synthesis of Silver Nanoparticles Suitable for Wound-Healing Applications

Yu Zhang [1], Konstantin S. Larionov [1], Simeng Zhang [1], Nikita A. Sobolev [1], Andrey I. Konstantinov [1], Dmitry S. Volkov [1], Evgeniya V. Suslova [1], Vladimir E. Chernov [2], Anton I. Poloskov [2], Ruslan I. Glushakov [2] and Irina V. Perminova [1,*]

[1] Department of Chemistry, Lomonosov Moscow State University, Leninskie Gory 1-3, 119991 Moscow, Russia; zhangyu13051837552@gmail.com (Y.Z.); konstantin.larionov@chemistry.msu.ru (K.S.L.); zhangsimeng@mail.ru (S.Z.); n.a.sobolev@outlook.com (N.A.S.); konstant@org.chem.msu.ru (A.I.K.); dmsvolkov@gmail.com (D.S.V.); suslova@kge.msu.ru (E.V.S.)

[2] Kirov Military Medical Academy, Academician Lebedev Street 6, 194044 Saint Petersburg, Russia; vechernov@mail.ru (V.E.C.); a.i.poloskov@gmail.com (A.I.P.); glushakoffruslan@yandex.ru (R.I.G.)

* Correspondence: iperm@med.chem.msu.ru

Abstract: This article describes the one-pot microwave synthesis of silver nanoparticles (AgNPs) assisted with natural polyelectrolytes—humic substances (HS). The humic polyelectrolytes served both as chemical reductants for silver ions and as end-capping agents for AgNPs. Three commercially available sodium humates extracted from lignites and leonardite and one sodium fulvate isolated from natural brown water seeped through peat deposits were used in this study. The dynamics of the growth rate of AgNPs was characterised by UV–VIS spectroscopy by measuring the intensity of surface plasmon resonance at 420 nm. Transmission electron microscopy was used to characterise the size and morphology of AgNPs. Dynamic light scattering was used to determine size distributions of the synthesised AgNPs in the solutions. It was established that both conventional and microwave syntheses assisted with the coal humates produced small-size AgNPs in the range from 4 to 14 nm, with the maximum share of particles with sizes of (6 ± 2) nm by TEM estimates. The peat fulvate yielded much larger NPs with sizes from 10 to 50 nm by TEM estimates. DLS measurements revealed multimodal distributions of AgNPs stabilised with HS, which included both single NPs with the sizes from 5 to 15 nm, as well as their dominating aggregates with sizes from 20 to 200 nm and a smaller portion of extra-large aggregates up to 1000 nm. The given aggregates were loosely bound by humic polyelectrolyte, which prevented the coalescence of AgNPs into larger particles, as can be seen in the TEM images. The significant acceleration in the reaction time—a factor of 60 to 70—was achieved with the use of MW irradiation: from 240 min down to 210–240 s. The coal humate stabilised AgNPs showed antimicrobial properties in relation to *S. aureus.* A conclusion was made regarding the substantial advantages of microwave synthesis in the context of time and scaling up for the large-scale production of AgNP-HS preparations with antimicrobial properties suitable for external wound-healing applications.

Keywords: microwave heating; natural polyelectrolytes; humic substances; silver nanoparticles; surface plasmon resonance; TEM

1. Introduction

Humic substances (HSs) are natural polyelectrolytes with many unique physicochemical and biological properties [1–3]. They find applications in nanoscience as polyfunctional macroligands for the formation of nanoparticles and nanocomposites [4–6]. However, the reported applications, except for the systematically studied synthesis of iron oxohydroxides [7–9], are mostly sporadic and scarce. This is because of the extreme complexity of

HS mixtures produced by oxidation from the plant biomacromolecules during stochastically driven humification processes. As a result, HSs are composed of both low- and high-molecular-weight components highly substituted with multiple carboxylic and phenolic groups. This causes their acidic character and polyelectrolytic properties. Due to the above-described complexity of HS compositions, they are operationally defined according to solubility in alkalis and acids [10]. HSs are alkali extracts of organic matter from fossil fuels (e.g., coal, peat), as well as from soils and sediments, which can make up to 90% of OM in the highly oxidised low-rank lignites (e.g., leonardite). HSs, in turn, could be segregated into two further classes: humic acids (HAs), which are not soluble at acidic conditions (pH < 2), and fulvic acids (FAs), which are soluble in the full pH range. In practice, they are isolated by the acidification of HS extracts until 2. The properties of HSs vary largely and are dependent both on the source of their origin and fractional composition. Coal HSs are the most hydrophobic among all other sources due to them having the highest content of aromatic fragments in the coal structure. They are mostly composed of aromatic acids and phenols. Coal HA fraction is enriched with phenols, whereas FA fraction is more acidic and dominated by carboxyl-substituted components. On contrary to the rather homogeneous aromatic structure of coal, the structure of peat is extremely heterogeneous due to the presence of both aromatic lignins and aliphatic polysaccharides. As a result, peat HSs are much more hydrophilic as compared to coal HSs, and peat FAs are the most hydrophilic among all other sources of HSs. Hence, if the application calls for material, which should be small, acidic, hydrophilic, highly charged, peat FAs should be selected; if the application calls for material with more complex, multidentate, phenolic, hydrophobic properties, then coal HAs should be considered.

Despite these substantial differences in the acid–base, redox, as well as the molecular weight properties of HS, which are completely defined by the source and fractionation procedure, they are often used interchangeably without a clear understanding of their specific properties. This undermines the quality of the obtained results and brings about the "bad flair" of HS applications: among the most frequent accusations are low transferability and reproducibility of the results, which kills all high-tech applications of HSs. We understand that reproducibility for technology is above benevolent properties and the lack of toxicity of HSs in the whole range of their concentrations in nature [3]. This is why the task of this research was to demonstrate the boundary conditions that make HS applications predictable and transferrable. For such a demonstration, we used the application of HSs for the synthesis of silver nanoparticles (AgNPs).

AgNPs are promising agents for suppressing resistant strains of pathogenic microorganisms, which makes them of particular importance for the healing of infected wounds [11–13]. Various studies have demonstrated that AgNPs can release Ag^+ and generate reactive oxygen species for bacterial cell membrane destruction. They can also cause enzyme function disruption and DNA damage in bacteria, which ultimately leads to bacterial death [13–17]. There is a number of publications that show that HSs can be successfully used to reduce and stabilise silver NPs [18–23]. All reported studies used different sources and fractions of HSs. Alexandrova et al. [21,22] used coal HAs, mud Has, and shale HAs; Sal'nikov et al. [20] used peat FAs; and Litvin et al. [18,19] used artificial FAs. Our recent publication also presented screening results for a set of humic materials from different sources and fractional composition [23]. However, all reported studies so far used conventional heating technology [18–23]. This kind of synthesis is time- and labour-consuming. In addition, the shape and size of nanoparticles are difficult to control.

Microwave (MW)-assisted heating is of particular importance in this context. It has been widely used as an alternative to conventional heating for the synthesis of various polymers, inorganic compounds, nanoparticles, and biological nanomaterials [24–26]. Conventional heating tends to use a convective gradient to transfer heat from the surface to the reactor interior; moreover, the temperature is higher at the surface, which results in uneven heat distribution and reduces reaction yield [27]. High-energy MW irradiation can accelerate reactions, while modulation of its frequency may increase the selectivity of the formation of the desired reaction product. The use of MW heating offers quite a few advantages over conventional heating, such as instantaneous and rapid heating (deep inside heating), high temperature homogeneity, and selective heating [28]. MW-assisted heating has been also successfully used for the synthesis of AgNP [29–32]. However, there are only a few reports available on the synthesis of MW-assisted humics-based nanomaterials [33,34], but none of them are devoted to the synthesis of AgNPs.

Here, we present rapid a one-pot MW-assisted synthesis of AgNPs that was transferred from the conventional prototype. We used HSs from four different commercial sources to assess the reproducibility of the synthesis with regard to the quality of the final product—the size and antibacterial properties of AgNPs. Three similar humic products—sodium humates from coal extracted with alkali—were used for the purpose of this synthesis. We had to drop two more humates from the initial experimental set—they were manufactured with the use of sodium carbonate extraction instead of NaOH, which formed insoluble silver carbonate and prevented the formation of AgNPs. One more commercial product was used—sodium fulvate from peat—to add a structural gradient to the experimental sample set. The same conditions were applied with regard to the concentration of humic and silver precursors, as well as heating regime and isolation. A use of concentrated HS solutions (12 g/L) enabled the preparation of highly concentrated AgNPs of 40 mM. The sizes of AgNPs synthesised in the presence of all three sodium humates were smaller than 10 nm (6–7 nm), whereas much larger AgNPs were obtained in tne case of sodium fulvate at 20 nm. We provide data on the antimicrobial properties of the HS-stabilised AgNPs and a rationalisation of the prospects of their wound-healing applications.

2. Materials and Methods

2.1. Materials

Three sodium humates were used in this study: sodium humate extracted from lignite, which is marketed under the trade name "Relict" and produced by the Genesis LLC (Novosibirsk, Russia); sodium humate extracted from leonardite, which is marketed under the trademark Powhumus (Humintech GmbH, Grevenbroich, Germany); and sodium humate extracted from lignite, marketed under the trademark Life Force (Life Force Group LLC, Saratov, Russia). The humates were designated as CHG, CHP, and CHL, respectively. The sodium fulvate was extracted by ion exchange from water leachate of underground peat. It is marketed under the trademark Fulvagra (Humintech GmbH, Grevenbroich, Germany). Silver nitrate (Molychem (Mumbai, India), 98%) was used as a silver precursor, and sodium hydroxide (Kemphasol, Moscow, Russia, 95%) was purchased and used as it was.

The contents of elements (CHNS) in the humic materials used in this study were determined using a 2400 Series II CHNS/O elemental analyser (PerkinElmer, Waltham, MA, USA).

The structural group compositions were determined using ^{13}C NMR spectroscopy. ^{13}C NMR spectra were recorded in 0.3 M $NaOD/D_2O$ (99+% isotopic purity, Sigma Aldrich, St. Louis, MI, USA). The spectra were recorded on an Avance-400 NMR spectrometer (Bruker, Berlin, Germany) with a carrier frequency for ^{13}C nuclei of 100 MHz. The INVGATE pulse sequence was used to exclude the nuclear Overhauser effect. A relaxation delay of 8 s was used for the complete relaxation of quaternary carbon atoms. To calculate the structure–group composition, the spectra were divided into nine intervals corresponding to the main structural components of HSs, and the intervals were integrated. The obtained integrals normalised to the whole spectrum represent the quantitative data of the structure–group composition of the studied HSs [35].

The content of silver in the final products was determined using an ICP-AES 720-ES spectrometer (Agilent Technologies, Santa Clara, CA, USA).

UV-visible absorption spectra were recorded using a UV/Vis Cary 50 Probe spectrometer (Varian, Palo Alto, CA, USA).

Transmission electron microscopy (TEM) images were obtained using a JEOL JEM-2100F microscope (JEOL, Akishima, Japan). For TEM investigation, a drop of the AgNPs-HS solution was placed onto a copper grid covered with a carbon film. The drop was brought to dryness (2 h under ambient air conditions), and the TEM image was detected at an accelerating voltage of 200 kV, a point resolution of 0.24 nm, and a line resolution of 0.18 nm. The processing of the image files was performed on more than 100 particles using the image analysis software "ImageJ 1.54d".

Dynamic light scattering (DLS) experiments were carried out using Malvern Zetasizer Nano ZS (Worcestershire, UK). The measurements were performed on the AgNPs-HS samples obtained in this study after diluting them by a factor of 100 with a use of 0.2 M $NaNO_3$ as a background electrolyte. The concentration of HSs was 120 mg/L, the concentration of Ag was 0.4 mM, and ionic strength was 0.2 M. The data treatment was performed using Malvern Software 8.01.

All reactions were carried out in a modified household microwave oven (Galanz MOG-2001M, Foshan, China).

2.2. Synthesis of Silver Nanoparticles

A weight of sodium humate (1 g) was dissolved in 35 mL of distilled water and centrifuged at 10,000 rpm for 10 min to remove any insoluble admixtures. The obtained supernatant was transferred into a 150 mL round-bottom flask, and 35 mL of deionised water was added. The pH of the obtained solution was adjusted to 11 using 1 M NaOH. Then, 15 mL of $AgNO_3$ solution was added at a concentration of 38.7 g/L. The pH value of the obtained solution was adjusted to 10 using 1 M NaOH under constant stirring. The resulting reaction mixture had concentrations of $AgNO_3$ and HSs of 6.8 and 11.7 g/L, respectively, which corresponded to 40 mM on the basis of metallic Ag in all the systems.

The microwave synthesis was conducted in a modified household microwave oven operating at 800 W using a sequence of pulses of 15 s and a pause of 10 s (10 s on/off) to prevent the solution from boiling and splashing due to overheating. The microwave exposure time was up to 240 s. After the completion of the synthesis, the samples were frozen and dried under vacuum. All syntheses were performed in the dark. The general scheme of microwave synthesis is shown in Figure 1.

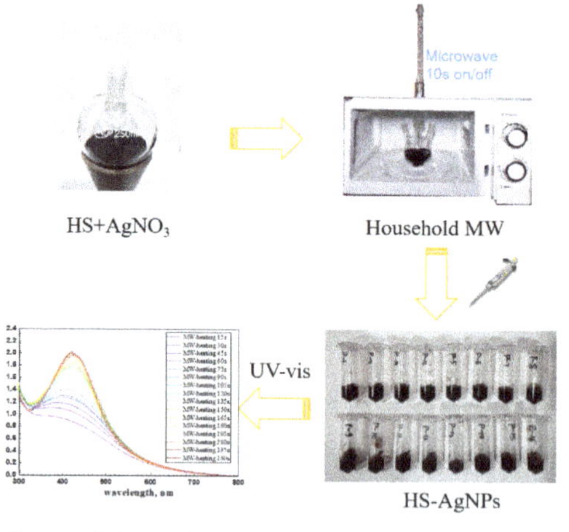

Figure 1. Scheme of microwave synthesis of HS-capped silver nanoparticles.

For comparison, the synthesis of silver NPs by conventional heating was conducted with identical stoichiometries of the reagents in a foil-wrapped, three-necked, round-bottom glass bulb, which was placed into a water bath and heated under continuous stirring at 85 °C for 4 h.

2.3. Screening of Antimicrobial Activity of the Synthesised AgNP-HS Compositions

The antimicrobial tests of the clinical strain of Methicillin-resistant *Staphylococcus aureus* (MRSA) were carried out in 96-well plates using a non-agarised Lysogeny broth (LB) nutrient medium. For the tests, the required volume of 0.9% NaCl solution and 0.142 mL of LB medium were added to each well, and the corresponding amounts of the tested compositions of AgNP-HS with the initial concentrations of HSs of 11,700 μg/mL and 40 mM of AgNPs were added. The range of the resulting concentrations was from 0 to 1500 μg/mL for the concentration of HSs and from 0.7 to 5 mM (from 73 to 551 μg/mL) of AgNPs. MRSA colonies were pre-incubated at 35 °C in an unagarised Lysogeny broth (LB) for 12 h. A total of 0.025 mL of bacterial suspension was added to each well, which was brought to 2 IU according to McFarland and analysed on a Perkin Elmer Victor X5 multimode microplate reader (Perkin Elmer, Waltham, MA, USA) at a wavelength of 600 nm for 24 h.

3. Results and Discussion

3.1. Characteristics of Humic Materials Used in this Study

The humic materials used in this study were characterised with elemental analysis and ^{13}C NMR spectroscopy. Elemental compositions are shown in Table 1, and structural group compositions are shown in Table 2.

Table 1. Elemental compositions of the humic materials used in this study.

	Content of Elements, % (Mass Including Ash)				Atomic Ratios	
Sample	C	H	N	S	H/C	C/N
CHL	39.2 ± 0.2	3.35 ±0.05	0.97 ±0.03	0.48 ± 0.04	1.0	47
CHP	39.6 ±0.5	3.62 ±0.08	0.91 ± 0.03	0.51±0.01	1.1	51
CHG	38.5 ± 0.3	3.78 ± 0.04	0.76 ± 0.01	0.47 ± 0.03	1.2	59
FA	40.2 ± 0.1	4.29 ± 0.03	0.70 ± 0.01	0.67 ± 0.01	1.3	67

Table 2. Carbon distribution in the humic materials used in this study (%) (^{13}C NMR data).

Samples	Chemical Shift, ppm						
	0–56 C_{Alk}	56–89 C_{AlkO}	89–145 C_{Ar}	145–165 C_{ArO}	165–185 COO	185–220 C=O	$\Sigma C_{Ar}/\Sigma C_{Alk}$
CHL	26	10	38	9	15	2	1.3
CHP	9	11	49	11	17	3	3.0
CHG	21	8	47	8	14	3	1.9
FA	28	17	22	7	21	5	0.63

The contents of elements were determined in the samples containing ash (about 20%). This is why the most informative data are the values of atomic ratios (H/C and C/N), which do not depend on the content of ash. The obtained values of the H/C ratio are indicative of much higher aromaticity of all three coal humates under study as compared to the fulvate from peat. This was expected, as peat contains polysaccharides and other aliphatic components.

The ^{13}C NMR spectra are shown in Figure 2. The quantitative distribution of carbon among the main structural groups of the humic materials used in this study is provided in Table 1 in accordance with the carbon type assignments made in the obtained ^{13}C NMR spectra.

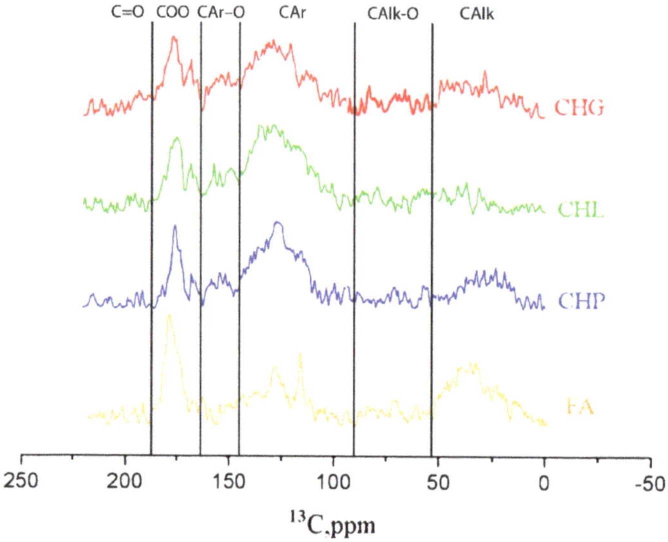

Figure 2. ^{13}C NMR spectra of the humic materials used in this study: sodium humates from leonaridte (CHP) and two lignites (CHG and CHL), and sodium fulvate isolated from peat DOM (FA).

The ^{13}C NMR data are indicative of the dominating contribution of aromatic carbon in all three humates from coal, which is consistent with the highly aromatic character of their source material. The CHG sample was characterised with the highest contribution of aliphatic structures (36%) and the least content of aromatic carbon (47%), while the CHL sample demonstrated the highest content of aromatic carbon (60%) and the least content of aliphatics (20%). Both samples were extracted from lignite. The CHP sample had 55% and 30% of aromatic and aliphatic carbon, respectively. This sample was extracted from leonardite. The least content of aromatic carbon and the highest contents of "carbohydrate" (17%) and aliphatic (28%) carbons were observed for peat FA, which was consistent with the presence of polysaccharides in the source material of this HS. Given the found content of C in the humic materials used in this study (Table 1) and its distribution among the functional groups (Table 2), the concentrations of ArOH in the 11.8 g/L solutions of CHG, CHL, and CHP samples accounted for 34, 42, and 31 mM, respectively, and the FA solution contained 27 mM. The contents of the carboxylic groups in the same samples were 57, 65, 54, and 81 mM, respectively.

The data on elemental and structural-group composition of the humic materials used in this study demonstrated a drastic difference between the sodium fulvate from peat and three humates from coal. In addition, substantial variations in the content of aromatic carbon were observed among the three humates from coal.

3.2. Humic Polyanions—Assisted Synthesis of AgNPs under MW and Conventional Heating Conditions Followed by UV–VIS Spectroscopy

The reported values of the redox potential of HSs vary between 450 and 650 mV, which provides us its reducing properties with regard to Ag^+ (E_0 = 799 mV) [36]. The major reducing units are phenolic and hydroquinonic moieties. Given that humic polyanioins have both reducing (ArOH) and capping groups (COO-) in their structure, the used reagent amounts for all three Ag-HS systems yielded 1:1 to 1:2 ratios for Ag/ArOH stoichiometries depending on the use of one or two electron transitions by phenolic groups, as well as around 2:3 for Ag/COOH stoichiometries with regard to the amount of carboxylic groups. The excessive amounts of both reducing and complexing groups used in this study were to achieve a better conversion rate of ionic to metallic silver and the stabilisation of AgNPs. Formation of AgNPs in the HS solutions can be schematically described by the reaction (1):

$$AgNO_3 + (ArOH)_{HS} + (COOH)_{HS} + 2OH^- = [Ag^0(COO)_{HS}] + ArO\cdot + NO_3^- + 2H_2O \quad (1)$$

Two-electron reduction of Ag^+ ions by quinonoid-like moieties in humic polyanions can be accounted by the following reaction:

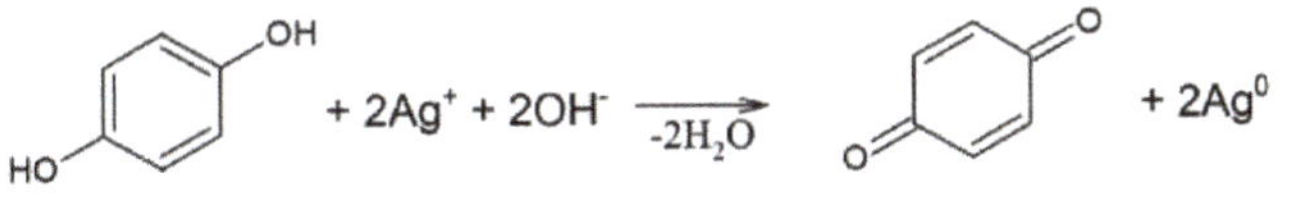

(2)

The typical dynamics of UV–VIS spectra in the Ag-HS system under MW irradiation used in this study are shown in Figure 3a with the example of CHG-humate. Figure 3b shows the same system under conditions of conventional heating. Figure 3a shows a rapid growth in intensity of the surface plasmon resonance (SPR) peak of AgNPs located at 414 nm, along with an increase in the heating time. In the case of the MW-assisted heating, the SPR peak appeared already after the first 15 s of heating. The maximum value of the SPR intensity was reached after 4 min of MW irradiation: the longer treatment did not lead to the further increase in SPR intensity. This is indicative of the complete conversion of ionic silver species into metallic Ag^0 over 4 min of MW heating.

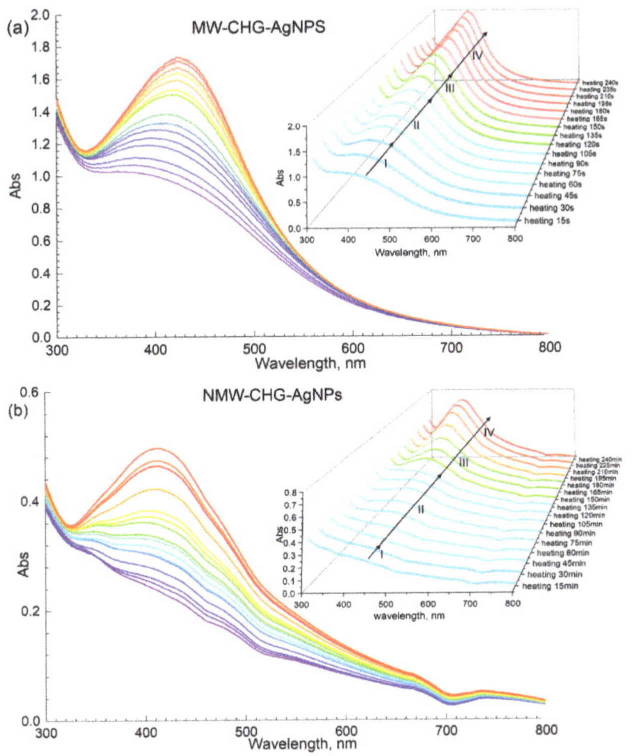

Figure 3. Synthesis of AgNPs mediated with HSs on the example of CHG sample: (**a**) microwave-assisted heating; (**b**) conventional water bath heating (85 °C); concentration of AgNPs at 40 mM, concentration of CHG at 11.8 g/L, pH 9. The Roman numbers from I to IV refer to the four stages of synthesis of AgNPs as described in the text.

Figure 3b demonstrates the dynamics of UV–VIS spectra of the reaction mixture [$AgNO_3$+HS] under conditions of conventional heating (85 °C). The formation of AgNPs occurred for several hours as compared to several minutes for the MW-assisted synthesis. The plots shown in Figure 3a, b demonstrate a much faster nucleation and growth of AgNP under microwave irradiation as compared to conventional heating. Evolution curves of the major SPR parameters, namely, the maximum wavelength and peak area for the four humic materials used in this study, are shown in Figure 4.

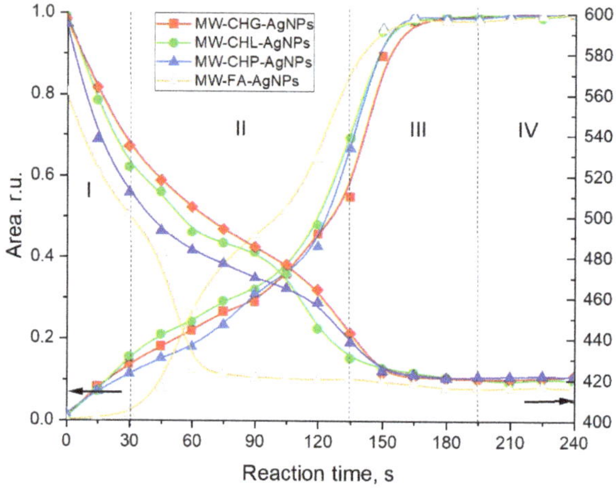

(**a**)

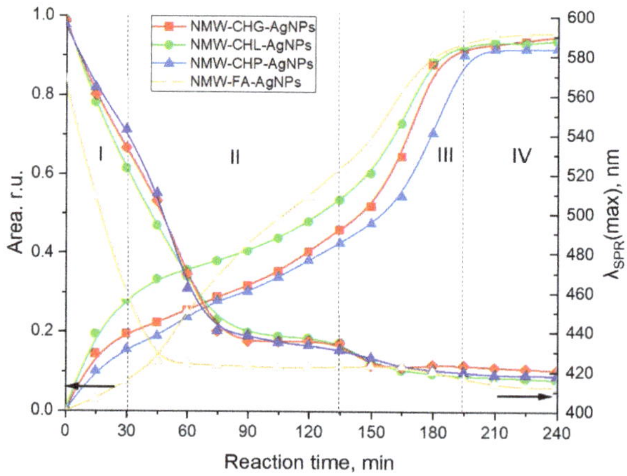

(**b**)

Figure 4. Evolution of the SPR peak of AgNPs during synthesis assisted with the four humic materials: CHG, CHL, CHP, and FAs under different heating conditions: (**a**) the MW-assisted heating; (**b**) the conventional heating at 85 °C. Concentrations: Ag—40 mM, sodium humate—11.8 g/L, pH 9; I (ionic strength) = 0.2 M. The Roman numbers from I to IV refer to the four stages of synthesis of AgNPs as described in the text. The left-headed black arrows point out to the dependences of Area of SPR on reaction time, the right-headed black arrows point out to the dependences of the SPR peak mosition (λ_{SPR}(max)) on reaction time.

AgNP formation in the presence of concentrated solutions of different humic polyanions—humates and fulvates—differed substantially. At the same time, the SPR evolution curves for all three coal humates looked relatively similar. They were characterised by the steady initial growth of the AgNP population without any induction period, which was followed by a moderate slope and, finally, the steep slope of the curve. This type of growth kinetic curve was observed for both MW-assisted and conventional heating processes. At the same time, the evolution curve for the peat fulvate had an initial induction period, which was followed by a rapid growth of the SPR peak area, and then by a short intermediary stabilisation, followed by another very rapid growth period and a final stabilisation.

The initial stage of Ag^+ interaction with the phenol-rich coal humates can be described as an ultrafast stage of reduction of Ag^+ to Ag^0 and nucleation into silver nanoclusters followed by a much slower process of AgNP aggregation (stage II) and, finally, by a fast process of coalescence to a final size of 5–7 nm (stage III) and the stable phase of constant AgNP population density in the solution (stage IV). The same stages for phenol-depleted but sugar-polyol-rich peat fulvate can be described as a slow initial process of reduction of Ag^+ to Ag^0 and nucleation into nanoclusters followed by very fast aggregation, much faster than in the case of the coal humates, with short stabilisation and a very fast coalescence up to the final particle size.

The character of the growth kinetic curves observed for the coal-humate-assisted synthesis of AgNPs was very similar to those reported by Polte et al. [37] and Van Hyning et al. [38]: they also lacked the induction stage. This type of growth curve does not obey the sigmoidal two-stage kinetic model for nucleation and the growth of noble metal nanoparticles, which was developed by the pioneering work of Watzky and Finke [39] that has been widely applied in the literature since [31,32,40,41]. At the same time, the latter model is more applicable to the peat FA.

3.3. AgNP Size Distribution Assessment Using TEM and DLS

For obtaining a better understanding of the sizes of the AgNPs formed and their distributions in the solutions containing high concentrations of HSs and salts, direct size measurements were undertaken using TEM, and the hydrodynamic radii of the particles in the solution were assessed using DLS. The TEM images of the AgNPs at the stable stage IV for all syntheses are shown in Figure 5.

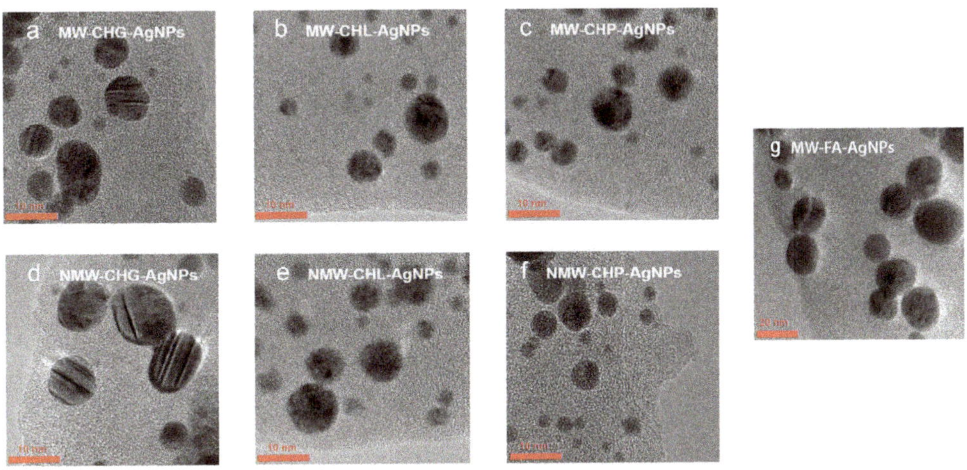

Figure 5. The particle size and distribution of AgNPs synthesised in the presence of three samples of coal humates used in this study under conditions of MW-assisted heating: (**a**) CHG, (**b**) CHL, (**c**) CHP, and under conditions of conventional heating: (**d**) CHG, (**e**) CHL, and (**f**) CHP. The AgNPs synthesised with MW irradiation in the presence of FAs are shown in (**g**).

The six Ag-HS systems prepared with the use of the higher-molecular-weight, phenolic-rich coal humates at the final synthesis stage yielded AgNPs under 10 nm with sizes ranging predominantly from 4 to 8 nm (Figure 5a–f), whereas AgNPs synthesised with a use of low-molecular-weight more aliphatic peat fulvate were much larger in size, ranging between 10 and 30 nm, with a predominant size of 20 nm (Figure 5g). The statistics of the obtained TEM data are shown in Figure 6.

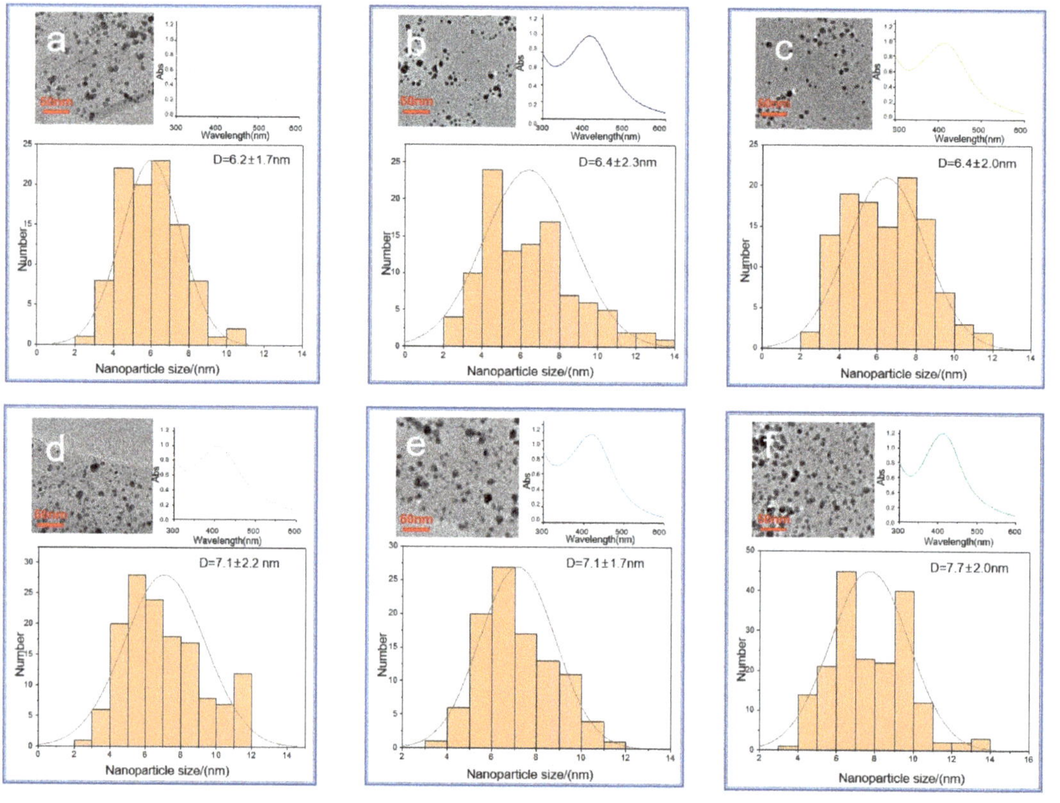

Figure 6. TEM images of the six Ag-HS systems used in this study obtained under MW irradiation (**a–c**) and conventional heating (**d–f**): (**a**) MW-CHG-AgNPs (6.2 ± 1.7) nm; (**b**) MW-CHL-AgNPs (6.4 ± 2.3) nm; (**c**) MW-CHP-AgNPs (6.4 ± 2.0) nm; (**d**) NMW-CHG-AgNPs (7.1 ± 2.2) nm; (**e**) NMW-CHL-AgNPs (7.1 ± 1.7) nm; (**f**) NMW-CHP-AgNPs (7.7 ± 2.0) nm. The histograms show the particle size distribution. The insets in the right upper corner show UV–VIS spectra at the final stage of synthesis.

The comparison of the size distributions of AgNPs obtained with a use of three coal humates under conditions of conventional heating and MW irradiation shows a lack of substantial differences among them: all coal humates showed high performance both in the reduction of ionic silver and in capping the formed AgNPs. Of particular importance for this study is the small size of the obtained Ag NPs, which varied from 2 to 12 nm, with distribution maxima located in the range from 4 to 8 nm. We did not observe a substantial reduction in size and polydispersity between AgNPs obtained with a use of conventional and MW heating, despite a much shorter synthesis time—4 min instead of 4 h.

Given that TEM images do not provide realistic data on sizes of NPs in the solutions, we performed DLS measurements on all AgNP-HS systems used in this study. The systems were diluted down to a factor of 100 to account for the light absorption of HS. We kept the ionic strength of the solutions at the same value of 0.2 M by using sodium nitrate of this concentration as a dilutor. The obtained results are provided in Figure 7.

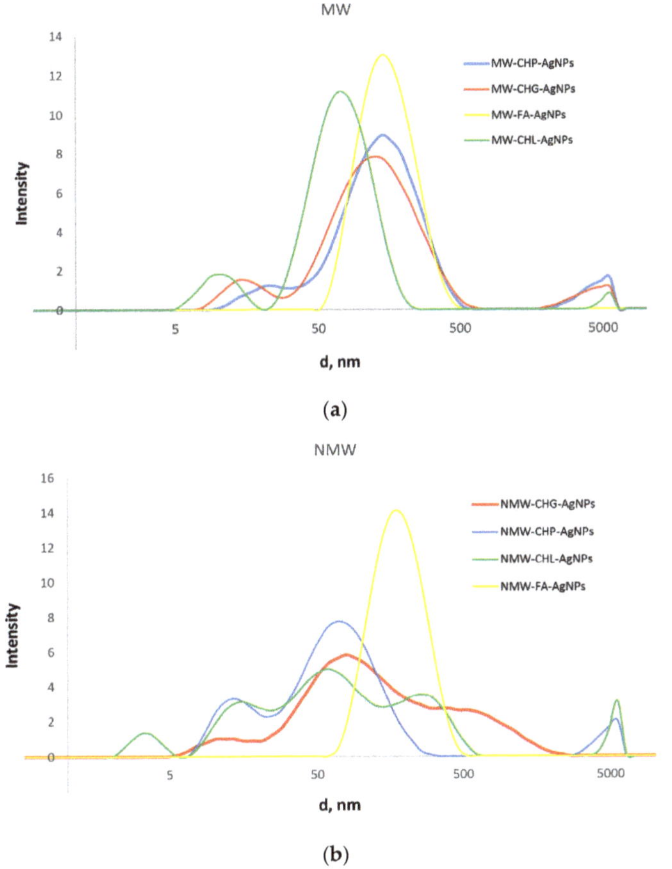

Figure 7. The distribution of hydrodynamic diameters of the particles present in the obtained AgNP-HS systems at the ionic strength of 0.2 M, as measured by DLS. The results are shown for the systems obtained with a use of MW irradiation (**a**) and conventional heating (**b**).

The data shown in Figure 7 reveal substantial differences in the size distribution of the three coal-humate-stabilised AgNPs from the peat-FA-stabilised AgNPs. While the former were characterised by multimodal broad distributions with sizes from 5 to 500–1000 nm, the latter had a much narrower monomodal distribution located between 50 and 500 nm with a maximum at around 200 nm. It should be noted that the both FA-stabilised AgNP systems had identical distribution, regardless of the type of heating—conventional or microwave. At the same time, for the coal humates, the modal distributions of the MW-assisted systems were much narrower as compared to the conventional heating. The three coal humates were characterised by the presence of narrow small-size components from 5 to 15 nm and broad medium-size components from 15 to 200 nm: the medium component had three modes for the conventional heating and monomodal structure for the MW heating. The CHG- and CHL-stabilised AgNPs had very similar broad-band distributions under conventional heating, whereas CHL-stabilised AgNPs in the case of MW synthesis had a single-band medium component with the smallest size located at around 50 nm. Quantitative characteristics of the obtained distributions (average sizes of the components) are summarised in Table 3.

Table 3. The average values of sizes (d) and zeta-potential (Z) for eight AgNP-HS systems used in this study (ionic strength—0.2 M; concentration of HS—118 mg/L; concentration of Ag—0.4 mM).

Sample	d-Avg *, nm	d-Pk1-Avg *, nm	d-Pk2-Avg *, nm	d-Pk3-Avg *, nm	Z-Potential, mV
NMW-CHG-AgNPs	53	141	3	2918	$-(44 \pm 12)$
NMW-CHP-AgNPs	87	177	2861	1	$-(27 \pm 19)$
NMW-CHL-AgNPs	20	34	1912	1	$-(30 \pm 14)$
NMW-FA-AgNPs	128	263	17	4612	$-(43 \pm 17)$
MW-CHG-AgNPs	80	146	3352	6	$-(35 \pm 19)$
MW-CHP-AgNPs	27	59	3	3500	$-(28 \pm 24)$
MW-CHL-AgNPs	58	124	12	1	$-(34 \pm 29)$
MW-FA-AgNPs	137	233	29	3645	$-(36 \pm 18)$

* d-Avg refers to the average particle size of the whole distribution; d-Pk1, d-Pk2, and d-Pk3 refer to the average particle size of the main, second, and third peaks of the distributions shown in Figure 7, respectively.

The obtained DLS data are indicative of the presence in the solutions of AgNPs stabilised with the coal humates of a small portion of the single AgNPs with sizes between 5 and 15 nm, which corroborate well with the sizes seen on TEM images (Figures 5 and 6). At the same time, a much larger portion of the AgNPs in solution was present in the form of loose aggregates bound together with humic macroligands. These aggregates had sizes from 20 to 200 nm, which were displayed as very intense broad peaks with shoulders reaching out up to 1000 nm in size. These shoulders were particularly distinct for CHG- and CHL-stabilised AgNPs. The values of zeta-potential measured for all systems showed good colloidal stability of all the systems under study (the values of potential were near or below 30 mV), with the highest values (below 40) found for CHG and FAs (Table 3).

The important issue is if the AgNPs stabilised with humic polyanions remain re-dispersible within aggregates observed in DLS curves. For answering this question, we conducted theoretical calculations of colloidal stability of the humics-stabilised AgNPs with a use of the model developed by our group for interactions of gold NPs stabilised with HSs [5]. As in the case of AuNPs, we considered the interaction of two identically charged spherical AgNPs with a radius a. We used the following expression for the total energy of interaction of colloidal NPs (V_{tot}) expressed in kBT units (kB and T are the Boltzmann constant and absolute temperature, respectively):

$$V_{tot} = V_{vdw} + V_{el-st} + V_{solv} \quad (3)$$

The first term refers to attraction Van der Waals forces; the second term refers to electrostatic interactions—both terms are accounted for by the DLVO theory of colloidal stability of particles in the solution [42]. The third term refers to solvation forces, which are not accounted by the DLVO theory. The solvation forces account for the contribution of steric factors such as a mechanical barrier created by the sorbed polymer on the surface of a particle. For the sake of calculations, let r be the distance between centres of the interacting particles, and $h = r - 2a$ is the shortest distance between their surfaces. The attraction potential (V_{vdW}) is given by the Hamaker expression for spherical particles [43]—we used the reported value of the silver/water/silver Hamaker constant $A = 4 \times 10^{-20}$ J [38]. The Ohshima expression [44–46] was used for the calculation of V_{el-st}, since the condition $a/r_D < 5$ was fulfilled. In this study the ratio a/r_D was 4.8, while the particle radius was 3 nm (estimated from the TEM results in Figure 5), and r_D (the Debye radius) was equal to 0.63 nm for the ionic strength of 0.2 M used in our studies. The Bjerrum length l_b was set to 0.6 nm because the syntheses were performed at T = 350 K. Parameters for the calculation of solvation force potential were: l = 0.5 nm and $B_s = 2.2 \times 10^{-2}$ J·m^{-2}. We paid particular attention to solvation forces while at the distances under few nanometres they may substantially contribute to interparticle interaction and provide stability of the dispersion. In the published model [5], the steric repulsion of soft adsorbed layers was approximated by a smooth repulsive potential V_{solv} instead of the infinite vertical potential wall, which was used by Chou and Zukoski [47].

Thus, the solvation repulsive forces potential can be written in the framework of Derjaguin approximation (h ≪ a) [42] as follows:

$$V_{solv} = 2\pi \frac{B_s a l}{k_B T} exp\left(-\frac{h}{l}\right) \quad (4)$$

with the decay length $l = 0.5$ nm and parameter $B_s = 22 \times 10^{-3}$ J m^{-2}. Full mathematical expressions are given in the source publication by Polyakov et al. (2017) [5] in the ESI. The model was applied for the qualitative explanation of how particle surface potential, which increases along with a sorption of humic polyanions at the surface of NP, affects the dispersion stability. Figure 8a shows a set of pairwise interaction potentials that corresponded to different ψ_0 values (in *V*) for the case when only Van der Waals attractive forces and electrostatic repulsion was accounted for (two terms of Equation (3)), and Figure 8b shows the same set of interaction potentials calculated with the use of three terms, which took care of repulsive solvation forces.

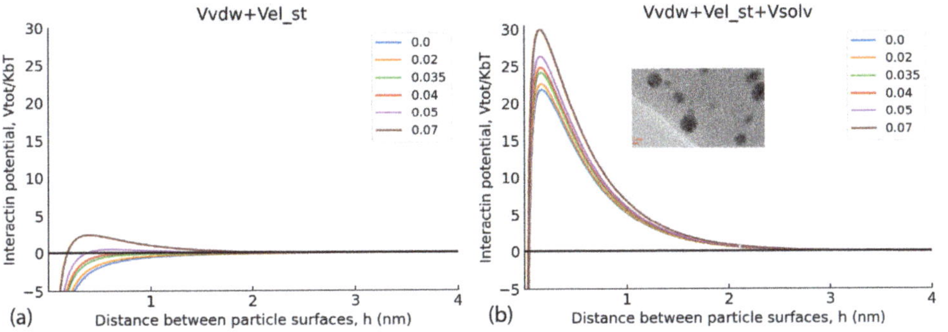

Figure 8. Pairwise AgNP interaction potentials at different values of surface potential ψ_0. (**a**) the model calculations without consideration of repulsive solvation forces potential (V_{solv}), (**b**) the interaction potentials calculated with the use of three terms including repulsive solvation forces.

Figure 8a shows that under the used experimental conditions (high ionic strength 0.2 M and small Debye radius), a use of the two-term interaction model prognosticated irreversible flocculation of nanoparticles—all curves were below 0, which contradicted experimental data. At the same time, incorporation of the third term responsible for steric repulsion yielded very different results. The curves of interaction potential in Figure 8b show that at all values of interaction potential, AgNPs would coalesce only when approaching each other by less than 0.4 nm. Otherwise, the potential barrier was too large to allow the AgNPs to irreversibly aggregate. The sorbed layer of humic polyanions prevents such a close proximity of the particles. This corroborates the TEM data (Figure 6)—the particles existed in the dispersed state due to thermal motion of particles. So, the conceptual results of the modelling indicate the reversible character of the aggregates of AgNPs stabilised with HS, which was able to be observed in the DLS curves. They can redisperse under conditions of smaller ionic strength.

In total, the obtained results demonstrate that the use of MW irradiation allowed us to reduce the synthesis time from 240 min down to 240 s, or more precisely, to 210 s; there was no change observed in the system during the last 30 s of heating. So, the MW-assisted synthesis proceeded by a factor of 60 to 70 faster as compared to the conventional heating one. It should also be noted that the use of coal humates turned out to be much more preferential for achieving smaller sizes of AgNPs as compared to peat fulvate. This could be connected both to different reducing groups present in the structure of fulvates and humates, as well as to the higher molecular weights of humates versus fulvates. Given both the smaller sizes of AgNPs, stabilised with the coal humates, and the much higher availability of coal humates on the market, antibacterial activity was tested only for six coal-humate-stabilised AgNPs.

3.4. Assessment of the Antibacterial Activity of the AgNPs Stabilised with Coal Humates Synthesised in This Study

The antibacterial effects of the compositions of AgNPs synthesised in the presence of coal humates were studied in relation to the methicillin-resistant strain of *Staphylococcus aureus* (MRSA). The suppression of the growth of this clinical strain was tested with the six AgNP-HS compositions in the concentration range of HSs from 0 to 200 mg/L and of AgNPs from 0 to 73.5 mg/L. The results on minimum inhibitory concentration (MIC) determination for the AgNPs under study are shown in Figure 9.

NMW-CHP-Ag

CHP, mg/L	0	20	40	60	80	100	200
Ag, µg ml^{-1}	0	7.4	14.7	22.0	29.4	36.7	73.5

MW-CHP-Ag

CHP, mg/L	0	20	40	60	80	100	200
Ag, µg ml^{-1}	0	7.4	14.7	22.0	29.4	36.7	73.5

NMW-CHL-Ag

CHL, mg/L	0	20	40	60	80	100	200
Ag, µg ml^{-1}	0	7.4	14.7	22.0	29.4	36.7	73.5

MW-CHL-Ag

CHL, mg/L	0	20	40	60	80	100	200
Ag, µg ml^{-1}	0	7.4	14.7	22.0	29.4	36.7	73.5

NMW-CHG-Ag

CHG, mg/L	0	20	40	60	80	100	200
Ag, µg ml^{-1}	0	7.4	14.7	22.0	29.4	36.7	73.5

MW-CHG-Ag

CHG, mg/L	0	20	40	60	80	100	200
Ag, µg ml^{-1}	0	7.4	14.7	22.0	29.4	36.7	73.5

Figure 9. The change in optical density in 24 h when measuring the minimum inhibitory concentration of the synthesised compositions of AgNP-HS in the LB medium with the concentrations of HSs from 0 to 200 mg/L shown in the upper raw of the tables and concentration of AgNPs from 0 to 73.5 mg/L in relation to the clinical strain of MRSA. Green indicates complete suppression of bacterial growth, and red indicates the absence of inhibition.

It can be seen that the silver concentration which suppressed bacterial growth (shown in green in Figure 9) upon the addition of all six compositions accounted for 73.5 mg/L. The highest antibacterial activity—the lowest MIC—was observed for MW-CHL-Ag composition: it accounted for 22 mg/L of Ag. It should be noted that this composition was characterised with the smaller size AgNP aggregates as compared to CHG and CHP (Figure 7b). The obtained MIC values lay within those reported for other AgNP-based compositions against MRSA strains [14]. In total, the obtained data provide a good promise for wound applications of AgNP-HS compositions. Further studies are planned to reduce the size of AgNPs stabilised with HSs and to optimise MICs of the AgNP-HS compositions.

4. Conclusions

This contribution reports for the first time the results of feasibility studies on the MW-assisted synthesis of concentrated solutions (40 mM) of small-size AgNPs in the presence of natural humic polyelectrolytes. The specific goal of this study was to develop rapid, large-scale, high-yield synthesis of AgNPs of high antibacterial activity suitable for wound-healing applications. For achieving this goal, we used MW irradiation as a source of energy to accelerate both the reduction rate of ionic silver to metallic particles and its growth rate until a final size of the particles in the range from 4 to 10 nm. This brought about a factor of 60 to 70 in in synthesis time: we were able to shorten it down to 3–4 min instead of

240 min under conditions of conventional heating with a use of a water bath. We used three commercially available sodium humates extracted from coal and one peat fulvate as the solo reducing and capping agents for producing AgNPs. The use of humic materials has several advantages: First of all, by their beneficial biological profile: they are natural polyfunctional polyelectrolytes which do not pose toxic effects to living organisms in the whole range of their natural concentrations. In addition, they possess pronounced antioxidant activity due to enrichment with phenolic groups. We have recently reported that combinations of HSs with AgNPs [23] bring about enhanced antioxidant effects and reduced cytotoxicity. The commercial humates are reasonably priced and provide a rich source of widely available natural polyelectrolytes suitable for biomedical applications. A direct comparison of the coal humates with peat fulvates has demonstrated an influence of the source of humic material and a technique of its isolation on the molecular weight and structural group composition of the isolated humic materials. The obtained results are very promising for the use of coal-humate-based AgNP compositions in wound-healing applications, which are very demanding both with regard to the concentration of the antibacterial agent—metallic Ag—in the medication and regarding the amount of medication needed for the efficient clinical treatment of infected wounds because of the large size and slow healing process of infected wounds.

5. Patents

The patent application "Preparation of antibacterial compositions of AgNPs capped with HS" is underway in its submission to the Russian Patent Office. It results from the work reported in this manuscript.

Author Contributions: Conceptualisation, I.V.P. and Y.Z.; methodology, Y.Z. and N.A.S.; software, K.S.L.; validation, I.V.P., D.S.V. and N.A.S.; investigation, Y.Z., S.Z., A.I.K., A.I.P., V.E.C., E.V.S. and D.S.V.; resources, D.S.V. and E.V.S.; writing—original draft preparation, Y.Z., K.S.L. and I.V.P.; writing—review and editing, I.V.P. and R.I.G.; visualisation, K.S.L.; supervision, I.V.P. and R.I.G.; project administration, N.A.S.; funding acquisition, I.V.P. All authors have read and agreed to the published version of the manuscript.

Funding: This research was funded by the Russian Science Foundation, grant number 20-63-47070.

Institutional Review Board Statement: Not applicable.

Data Availability Statement: All data resulting from this study are provided in this manuscript.

Acknowledgments: The authors would like to thank Maria G. Chernysheva (Dept. of Chemistry, Lomonosov MSU) for invaluable assistance with DLS measurements. We also thank the anonymous reviewers whose precise professional comments allowed us to significantly improve the quality of the manuscript. The research was carried out using the equipment of the MSU Shared Research Equipment Centre "Technologies for obtaining new nanostructured materials and their complex study", "Nanochemistry and Nanomaterials" purchased by MSU in the framework of the Equipment Renovation Program (National Project "Science and Universities"), and the MSU Program of Development.

Conflicts of Interest: The authors declare no conflicts of interest.

References

1. Olk, D.C.; Bloom, P.R.; Perdue, E.M.; McKnight, D.M.; Chen, Y.; Farenhorst, A.; Senesi, N.; Chin, Y.-P.; Schmitt-Kopplin, P.; Hertkorn, N.; et al. Environmental and Agricultural Relevance of Humic Fractions Extracted by Alkali from Soils and Natural Waters. *J. Environ. Qual.* **2019**, *48*, 217–232. [CrossRef] [PubMed]
2. Havelcová, M.; Mizera, J.; Sýkorová, I.; Pekar, M. Sorption of metal ions on lignite and the derived humic substances. *J. Hazard. Mater.* **2009**, *161*, 559–564. [CrossRef] [PubMed]
3. Perminova, I.V. From green chemistry and nature-like technologies towards ecoadaptive chemistry and technology. *Pure Appl. Chem. PAC* **2019**, *91*, 851–864. [CrossRef]
4. Baigorri, R.; García-Mina, J.M.; Aroca, R.F.; Alvarez-Puebla, A.A. Optical Enhancing Properties of Anisotropic Gold Nano-plates Prepared with Different Fractions of a Natural Humic Substance. *Chem. Mater.* **2008**, *20*, 1516–1521. [CrossRef]

5. Polyakov, A.Y.; Lebedev, V.A.; Shirshin, E.A.; Rumyantsev, A.M.; Volikov, A.B.; Garshev, A.V.; Goodilin, E.A.; Perminova, I.V. Non-classical growth of water-redispersible spheroidal gold nanoparticles assisted by leonardite humate. *CrystEngComm* **2017**, *19*, 876–886. [CrossRef]
6. Venezia, V.; Verrillo, M.; Gallucci, N.; Di Girolamo, R.; Luciani, G.; D'Errico, G.; Paduano, L.; Piccolo, A.; Vitiello, G. Ex-ploiting bioderived humic acids: A molecular combination with ZnO nanoparticles leads to nanostructured hybrid interfaces with enhanced pro-oxidant and antibacterial activity. *J. Environ. Chem. Eng.* **2023**, *11*, 108973. [CrossRef]
7. Pomogailo, A.D.; Kydralieva, K.A.; Zaripova, A.A.; Muratov, V.S.; Dzhardimalieva, G.I.; Pomogailo, S.I.; Golubeva, N.D.; Jorobekova, S.J. Magnetoactive Humic-Based Nanocomposites. In *Macromolecular Symposia*; WILEY-VCH Verlag: Weinheim, Germany, 2011; Volume 304, pp. 18–23. [CrossRef]
8. Illés, E.; Tombácz, E. The effect of humic acid adsorption on pH-dependent surface charging and aggregation of magnetite nanoparticles. *J. Colloid Interface Sci.* **2006**, *295*, 115–123. [CrossRef] [PubMed]
9. Kulikova, N.A.; Polyakov, A.Y.; Lebedev, V.A.; Abroskin, D.P.; Volkov, D.S.; Pankratov, D.A.; Klein, O.I.; Senik, S.V.; Sorkina, T.A.; Garshev, A.V.; et al. Key roles of size and crystallinity of nanosized iron hydr(oxides) stabilized by humic substances in iron bioavailability to plants. *J. Agric. Food Chem.* **2017**, *65*, 11157–11169. [CrossRef]
10. Stevenson, F.J. *Humus Chemistry*, 2nd ed.; John Wiley and Sons: New York, NY, USA, 1994; 512p.
11. Anees, A.; Sachi Das, S.; Khatoon, A.; Tahir Ansari, M. Bactericidal Activity of Silver Nanoparticles: A Mechanistic Review. *Mater. Sci. Energy Technol.* **2020**, *3*, 756–769. [CrossRef]
12. El-Aassar, M.R.; Ibrahim, O.M.; Fouda, M.M.G.; El-Beheri, N.G.; Agwa, M.M. Wound Healing of Nanofiber Comprising Polygalacturonic/Hyaluronic Acid Embedded Silver Nanoparticles: In-Vitro and In-Vivo Studies. *Carbohydr. Polym.* **2020**, *238*, 116175. [CrossRef]
13. Paladini, F.; Pollini, M. Antimicrobial Silver Nanoparticles for Wound Healing Application: Progress and Future Trends. *Materials* **2019**, *12*, 2540. [CrossRef] [PubMed]
14. Swolana, D.; Wojtyczka, R.D. Activity of Silver Nanoparticles against *Staphylococcus* spp. *Int. J. Mol. Sci.* **2022**, *23*, 4298. [CrossRef] [PubMed]
15. Slavin, Y.N.; Asnis, J.; Hńfeli, U.O.; Bach, H. Metal nanoparticles: Understanding the mechanisms behind antibacterial activity. *J. Nanobiotechnol.* **2017**, *15*, 65. [CrossRef] [PubMed]
16. Deng, H.; McShan, D.; Zhang, Y.; Sinha, S.S.; Arslan, Z.; Ray, P.C.; Yu, H. Mechanistic study of the synergistic antibacterial activ-ity of combined silver nanoparticles and common antibiotics. *Environ. Sci. Technol.* **2016**, *50*, 8840–8848. [CrossRef] [PubMed]
17. Wang, E.; Huang, Y.; Du, Q.; Sun, Y. Silver Nanoparticle Induced Toxicity to Human Sperm by Increasing ROS (Reactive Oxygen Species) Production and DNA Damage. *Environ. Toxicol. Pharmacol.* **2017**, *52*, 193–199. [CrossRef] [PubMed]
18. Litvin, V.A.; Galagan, R.L.; Minaev, B.F. Kinetic and Mechanism Formation of Silver Nanoparticles Coated by Synthetic Humic Substances. *Colloids Surf. A Physicochem. Eng. Asp.* **2012**, *414*, 234–243. [CrossRef]
19. Litvin, V.A.; Minaev, B.F. Spectroscopy study of silver nanoparticles fabrication using synthetic humic substances and their antimicrobial activity. *Spectrochim. Acta Part A Mol. Biomol. Spectrosc.* **2013**, *108*, 115–122. [CrossRef]
20. Sal'nikov, D.S.; Pogorelova, A.S.; Makarov, S.V.; Vashurina, I.Y. Silver ion reduction with peat fulvic acids. *Russ. J. Appl. Chem.* **2009**, *82*, 545–548. [CrossRef]
21. Aleksandrova, G.P.; Lesnichaya, M.V.; Dolmaa, G.; Klimenkov, I.V.; Sukhov BGRegdel, D.; Trofimov, B.A. Silver-containing nanocomposites with antioxidant activity based on humic substances of different origin. *Russ. Chem. Bull. Int. Ed.* **2017**, *66*, 143–149. [CrossRef]
22. Alexandrova, G.P.; Dolmaa, G.; Enkhbadral, U.; Grishenko, G.L.; Tserenpi, S.; Sukhov, B.G.; Regdel, D.; Trofimov, B.A. A new humic acid remedy with addition of silver nanoparticles. *Mong. J. Chem.* **2012**, *13*, 7–11. [CrossRef]
23. Zykova, M.V.; Volikov, A.B.; Buyko, E.E.; Bratishko, K.A.; Ivanov, V.V.; Konstantinov, A.I.; Logvinova, L.A.; Mihalyov, D.A.; Sobolev, N.A.; Zhirkova, A.M.; et al. Enhanced Antioxidant Activity and Reduced Cytotoxicity of Silver Nanoparticles Stabilized by Different Humic Materials. *Polymers* **2023**, *15*, 3386. [CrossRef] [PubMed]
24. Bartoli, M.; Frediani, M.; Briens, C.; Berruti, F.; Rosi, L. An Overview of Temperature Issues in Microwave-Assisted Pyrolysis. *Processes* **2019**, *7*, 658. [CrossRef]
25. Gawande, M.B.; Shelke, S.N.; Zboril, R.; Varma, R.S. Microwave-Assisted Chemistry: Synthetic Applications for Rapid Assembly of Nanomaterials and Organics. *Acc. Chem. Res.* **2014**, *47*, 1338–1348. [CrossRef] [PubMed]
26. Sun, J.; Wang, W.; Yue, Q. Review on Microwave-Matter Interaction: Fundamentals and Efficient Microwave-Associated Heat-ing Strategies. *Materials* **2016**, *9*, 231. [CrossRef] [PubMed]
27. Liu, Q.; Gao, M.R.; Liu, Y.; Okasinski, J.S.; Ren, Y.; Sun, Y. Quantifying the Nucleation and Growth Kinetics of Microwave Nanochemistry Enabled by in Situ High-Energy X-Ray Scattering. *Nano Lett.* **2016**, *16*, 715–720. [CrossRef] [PubMed]
28. Praneeth, N.V.S.; Paria, S. Microwave-Assisted One-Pot Synthesis of Anisotropic Gold Nanoparticles with Active High-Energy Facets for Enhanced Catalytic and Metal Enhanced Fluorescence Activities. *CrystEngComm* **2018**, *20*, 4297–4304. [CrossRef]
29. Nishioka, M.; Miyakawa, M.; Kataoka, H.; Koda, H.; Sato, K.; Suzuki, T.M. Continuous Synthesis of Monodispersed Silver Nanoparticles Using a Homogeneous Heating Microwave Reactor System. *Nanoscale* **2011**, *3*, 2621–2626. [CrossRef]
30. Helmlinger, J.; Heise, M.; Heggen, M.; Ruck, M.; Epple, M. A Rapid, High-Yield and Large-Scale Synthesis of Uniform Spherical Silver Nanoparticles by a Microwave-Assisted Polyol Process. *RSC Adv.* **2015**, *5*, 92144–92150. [CrossRef]

31. Thanh, N.T.K.; Maclean, N.; Mahiddine, S. Mechanisms of Nucleation and Growth of Nanoparticles in Solution. *Chem. Rev.* **2014**, *114*, 7610–7630. [CrossRef]
32. Özkar, S.; Finke, R.G. Silver Nanoparticles Synthesized by Microwave Heating: A Kinetic and Mechanistic Re-Analysis and Re-Interpretation. *J. Phys. Chem. C* **2017**, *121*, 27643–27654. [CrossRef]
33. Kostyukhin, E.M.; Nissenbaum, V.D.; Abkhalimov, E.V.; Kustov, A.L.; Ershov, B.G.; Kustov, L.M. Microwave-Assisted Synthesis of Water-Dispersible Humate-Coated Magnetite Nanoparticles: Relation of Coating Process Parameters to the Properties of Nanoparticles. *Nanomaterials* **2020**, *10*, 1558. [CrossRef]
34. da Silva, L.C.; Abate, G.; Oliveira, N.A.; Fantini, M.C.D.A.; Masini, J.C.; Mercuri, L.P.; Olkhovyk, O.; Jaroniec, M.; Matos, J.R. Microwave synthesis of FDU-1 silica with incorporated humic acid and its application for adsorption of Cd^{2+} from aqueous solutions. *Stud. Surf. Sci. Catal.* **2005**, *156*, 941–950. [CrossRef]
35. Hertkorn, N.; Permin, A.; Perminova, I.; Kovalevskii, D.; Yudov, M.; Petrosyan, V.; Kettrup, A. Comparative Analysis of Partial Structures of a Peat Humic and Fulvic Acid Using One- and Two-Dimensional Nuclear Magnetic Resonance Spectroscopy. *J. Environ. Qual.* **2002**, *31*, 375–387. [CrossRef] [PubMed]
36. Cao, X.; Ma, C.; Zhao, J.; Guo, H.; Dai, Y.; Wang, Z.; Xing, B. Graphene oxide mediated reduction of silver ions to silver nanoparticles under environmentally relevant conditions: Kinetics and mechanisms. *Sci. Total Environ.* **2019**, *679*, 270–278. [CrossRef] [PubMed]
37. Polte, J.; Xenia Tuaev, X.; Wuithschick, M.; Fischer, A.; Thuenemann, A.F.; Rademann, K.; Kraehnert, R.; Emmerling, F. For-mation Mechanism of Colloidal Silver Nanoparticles: Analogies and Differences to the Growth of Gold Nanoparticles. *ACS Nano* **2012**, *6*, 5791–5802. [CrossRef] [PubMed]
38. Van Hyning, D.L.; Klemperer, W.G.; Zukoski, C.F. Silver Nanoparticle Formation: Predictions and Verification of the Ag-gregative Growth Model. *Langmuir* **2001**, *17*, 3128–3135. [CrossRef]
39. Watzky, M.A.; Finke, R.G. Transition Metal Nanocluster Formation Kinetic and Mechanistic Studies. A New Mechanism When Hydrogen Is the Reductant: Slow, Continuous Nucleation and Fast Autocatalytic Surface Growth. *J. Am. Chem. Soc.* **1997**, *119*, 10382–10400. [CrossRef]
40. Finney, E.E.; Finke, R.G. The Four-Step, Double-Autocatalytic Mechanism for Transition-Metal Nanocluster Nucleation, Growth, and Then Agglomeration: Metal, Ligand, Concentration, Temperature, and Solvent Dependency Studies. *Chem. Mater.* **2008**, *20*, 1956–1970. [CrossRef]
41. Xiong, Y.; Siekkinen, A.R.; Wang, J.; Yin, Y.; Kim, M.J.; Xia, Y. Synthesis of Silver Nanoplates at High Yields by Slowing Down the Polyol Reduction of Silver Nitrate with Polyacrylamide. *J. Mater. Chem.* **2007**, *17*, 2600–2602. [CrossRef]
42. Israelachvili, J.N. *Intermolecular and Surface Forces*, 2nd ed.; Academic Press: New York, NY, USA, 1992.
43. Hamaker, H.C. The London—Van der Waals attraction between spherical particles. *Physica* **1937**, *4*, 1058–1072. [CrossRef]
44. Ohshima, H. Effective surface potential and double-layer interaction of colloidal particles. *J. Colloid Interface Sci.* **1995**, *174*, 45–52. [CrossRef]
45. Lee, K.; Sathyagal, A.N.; McCormickLee, A.V. A closer look at an aggregation model of the Stöber process. *Colloids Surf. A* **1998**, *144*, 115–125. [CrossRef]
46. Kim, T.; Lee, K.; Gong, M.-S.; Joo, S.-W. Control of gold nanoparticle aggregates by manipulation of interparticle interaction. *Langmuir* **2005**, *21*, 9524–9528. [CrossRef]
47. Chow, M.K.; Zukoski, C.F. Gold sol formation mechanisms: Role of colloidal stability. *J. Colloid Interface Sci.* **1994**, *165*, 97–109. [CrossRef]

polymers

MDPI

Article

Ultrasensitive Detection of PSA Using Antibodies in Crowding Polyelectrolyte Multilayers on a Silicon Nanowire Field-Effect Transistor

Galina V. Presnova [1], Denis E. Presnov [2,3,*], Mariya M. Ulyashova [1], Ilia I. Tsiniaikin [2], Artem S. Trifonov [2], Ekaterina V. Skorb [4], Vladimir A. Krupenin [2], Oleg V. Snigirev [2] and Maya Yu. Rubtsova [1,*]

[1] Department of Chemistry, Lomonosov Moscow State University, 119991 Moscow, Russia; gkovba@enzyme.chem.msu.ru (G.V.P.); mmu@enzyme.chem.msu.ru (M.M.U.)
[2] Faculty of Physics, Lomonosov Moscow State University, 119991 Moscow, Russia; ii.tcinyaykin@physics.msu.ru (I.I.T.); artem.trifonov@physics.msu.ru (A.S.T.); krupenin@phys.msu.ru (V.A.K.); snigirev.oleg@physics.msu.ru (O.V.S.)
[3] D.V. Skobeltsyn Institute of Nuclear Physics, Lomonosov Moscow State University, 119991 Moscow, Russia
[4] Infochemistry Scientific Center of ITMO University, 191002 Saint Petersburg, Russia; skorb@itmo.ru
* Correspondence: denis.presnov@physics.msu.ru (D.E.P.); myr@enz.chem.msu.ru (M.Y.R.)

Abstract: Immunosensors based on field-effect transistors with nanowire channels (NWFETs) provide fast and real-time detection of a variety of biomarkers without the need for additional labels. The key feature of the developed immunosensor is the coating of silicon NWs with multilayers of polyelectrolytes (polyethylenimine (PEI) and polystyrene sulfonate (PSS)). By causing a macromolecular crowding effect, it ensures the "soft fixation" of the antibodies into the 3-D matrix of the oppositely charged layers. We investigated the interaction of prostate-specific antigen (PSA), a biomarker of prostate cancer, and antibodies adsorbed in the PEI and PSS matrix. In order to visualize the formation of immune complexes between polyelectrolyte layers using SEM and AFM techniques, we employed a second clone of antibodies labeled with gold nanoparticles. PSA was able to penetrate the matrix and concentrate close to the surface layer, which is crucial for its detection on the nanowires. Additionally, this provides the optimal orientation of the antibodies' active centers for interacting with the antigen and improves their mobility. NWFETs were fabricated from SOI material using high-resolution e-beam lithography, thin film vacuum deposition, and reactive-ion etching processes. The immunosensor was characterized by a high sensitivity to pH (71 mV/pH) and an ultra-low limit of detection (LOD) of 0.04 fg/mL for PSA. The response of the immunosensor takes less than a minute, and the measurement is carried out in real time. This approach seems promising for further investigation of its applicability for early screening of prostate cancer and POC systems.

Keywords: polyelecrolytes; macromolecular crowding; antibodies; silicon nanowire field-effect transistor; prostate-specific antigen (PSA); immunosensor

1. Introduction

The development of personalized medicine and modern healthcare necessitates the creation of novel techniques for real-time detection of low concentrations of particular biomarkers (proteins, nucleic acids, etc.) in clinical biochemistry to facilitate early diagnosis. The development of biosensors based on field-effect transistors with a nanowire channel (NWFETs) has advanced significantly in the last few years. The key feature is the shape of the channel—a very thin semiconductor nanowire; which allows the detection of charged biomolecules on or near its surface by changing the nanowire conductivity. The thinner the nanowire, the larger part of its volume the field penetrates, significantly changing its conductivity, which leads to an increase in the transistor sensitivity. Distinct electronic characteristics, nanoscale size, and real-time measurement without biomarker labeling make NWFETs an appealing diagnostic platform [1]. Numerous studies have demonstrated silicon nanowire advantages over other sensor types since the first reported experiments [2]. They are unique in that they have high charge/field sensitivity [3], the ability to measure in a wide temperature range up to room temperature, which is crucial for studying biological objects, and spatial resolution limited only by the diameter of the nanowire [4–6]. Biosensor platforms are continually improving as a result of ongoing advancements in nanotechnology techniques. The primary benefits of silicon-on-insulator (SOI) nanowires are their reproducibility and reliability in the manufacturing process, as well as their compatibility with industrial silicon technology, which enables the production of measuring electronics in close proximity to the sensor. Numerous FET-based devices have been created recently for use in biomedical diagnostics, with the aim of identifying different biomarkers of cancer, acute myocardial infarction, bacterial and viral infections, and others [7–11].

In order to identify biological molecules, NWFETs frequently use antibodies to capture particular biomarkers, and the formation of immune complexes changes the nanowires' conductivity. The sensitivity of immunosensors based on NWFETs is very high; they can detect individual molecules at a level ranging from micromoles to attomoles or even lower up to detecting single molecules [12–15].

Optimizing the response of these biosensors, increasing the signal-to-noise ratio, and overcoming the theoretical Nernst limit receive particular attention [16–19]. However, there is a significant restriction concerning the Debye screening length (λ_D), which is the average distance at which the electric field enters the buffer. The λ_D characterizes the thickness of an electrical double layer close to the gate dielectric/buffer interface in an electrolyte buffer [20,21]. The ionic strength of the buffer solution has a significant impact on this parameter [22]. For instance, in high-ionic strength solutions such as human serum (analog $1 \times$ PBS), λ_D is approximately 0.7 nm, which prevents the efficient use of complete immunoglobulin molecules, which have a thickness of approximately 2 nm [23]. This results in the necessity of using low-ionic-strength buffers (λ_D in 0.01 PBS is approximately 7 nm). However, under these circumstances, the effectiveness of antibody interactions with the analyte may be reduced.

By covering the planar nanowires with additional layers with gaps and nanopores, it is possible to reduce the shielding effect of the charged molecules being analyzed [24]. Applying polymers, like polyethylene glycol, to cover the surface of nanowires is an additional tactic that has been demonstrated to extend the effective Debye screening length [25]. There might be more benefits to coating biosensor surfaces with polymers. Antibodies that are immobilized or adsorb directly onto the solid surface of the immunosensor are known to exhibit decreased affinity for the antigen as a result of steric effects and decreased active center mobility, particularly when interacting with large biomarker molecules. In order to improve antibody binding capacity, the surface of the sensor can be modified with charged polyelectrolytes adsorbing proteins through electrostatic interactions [26,27]. Antibodies can be "soft fixed" in a 3D matrix on the surface of sensors formed by multilayers of oppositely charged polyelectrolytes, preserving some degree of "internal disorder" and the polypeptide chains' mobility [28]. Because of the phenomenon known as "molecular crowding", the high specificity and activity of antibodies placed in layers of charged polyelectrolytes are ensured [29]. Due to the small volume of the hydrogel nanolayer as well as the "tightness" effect, the interaction between molecules in such an environment occurs more quickly than in the volume of the solution. In the process of developing an electrochemical biosensor for the detection of the encephalitis virus, the benefits of immobilizing specific antibodies into the matrix of successive layers of positively charged polyethylenimine (PEI) and negatively charged polystyrene sulfonate (PSS) were recently demonstrated [30].

This work investigates the coating of silicon with layers of PEI and PSS polyelectrolytes to create NWFETs for detecting prostate-specific antigen (PSA). PSA represents a molecular biomarker for the clinical diagnosis of prostate cancer. PSA serum concentrations higher than the upper limit of 4.0 ng/mL, particularly if they exceed 10 ng/mL, suggest a possible cancer risk and need to be confirmed further. Furthermore, a postoperative serum PSA level exceeding 0.1 ng/mL would indicate a recurrence risk for cancer, making it especially crucial to reduce the detection limit for PSA to nanogram levels or below.

2. Materials and Methods

2.1. Materials

All chemicals and organic solvents were of analytical grade. Poly(sodium 4-styrenesulfonate) (PSS, Mw 500 kDa), polyethyleneimine (PEI, Mw 70 kDa, 50 wt% water solution), tetrachloroauric acid, bovine serum albumin (BSA), and Tween-20 were purchased from Sigma-Aldrich (St. Louis, MO, USA). Sodium citrate dihydrate was purchased from MP Biomedicals (Eschwege, Germany). PSA and two clones of mouse monoclonal antibodies to PSA and monoclonal antibodies to TSH (nonspecific to PSA) were provided by JSC "NVO Immunotek" (Moscow, Russia). The specificity of the antibodies towards PSA was confirmed by the manufacturer. All water used in experiments was purified with a Milli-Q system (Millipore, Billerica, MA, USA).

Gold nanoparticles (AuNPs) with a diameter of 20–25 nm were prepared by the Frens method [31], based on the reduction of chloroauric acid by sodium citrate. They were characterized spectrophotometrically (λ_{max} = 521 nm and λ_{max} = 519 nm) and by using scanning electron microscopy (SEM).

Conjugates of AuNPs with antibody half-fragments containing thiol groups were prepared as described in [8].

2.2. Silicon Surface Modification by Polyelectrolytes and Adsorption of Antibodies in the Matrix of Polymers

The concentrations of the reagents were used in accordance with the method developed earlier for the formation of a polyelectrolyte matrix of PEI and PSS on the surface of screen-printed electrodes [30]. A total of 5 µL of PEI solution (1 mg/mL in water) was spotted onto the silicon surface, dried, and rinsed three times with water to remove excess PEI. Then, 5 µL of antibodies against PSA (100 µg/mL) in PBS pH 7.4 was dropped on the modified surface of silicon, dried, and washed three times with PBS containing 0.1% Tween20 (PBST). In the next step, 5 µL of PSS solution (2 mg/mL in water) was added, dried, and washed similarly to PEI solution.

2.3. Characterization of a Silicon Surface Covered by Multilayers of Polyelectrolytes and Antibodies

The characterization of silicon samples coated with layers of polyelectrolytes with adsorbed antibodies was performed using successive stages of sandwich immunoassay with the use of AuNPs as labels by SEM and AFM methods. A piece of silicon covered with layers of polyelectrolytes and antibodies adsorbed between them was incubated in a solution of PSA (20 ng/mL in PBST) at 37 °C for 1 h. After washing, it was incubated in a solution of a conjugate of second antibodies with AuNPs at 37 °C for 1 h. Antibodies nonspecific to PSA and their conjugate with AuNPS were used as controls.

SEM images of the silicon surface were obtained using a field emission scanning electron microscopy (FESEM) Supra40 (Carl Zeiss, Germany) with an InLens secondary electron detector integrated in the microscope column.

The tapping mode of SmartSPM-1000 (AIST-NT, Novato, CA, USA) was used for obtaining the topography profiles. The diamond single crystal cantilevers AFM Probe ART™ D300 used in this study have a typical spring constant of 10 N/m and a rounded tip of 5–10 nm radius. The spring constant of each cantilever was determined using a technique based on measuring the change in resonant frequency of the fundamental mode of vibration [32]. The scratching was studied by performing several (40–80) scratching runs at a load of 1.0 µN. The scratch depth was estimated as the maximum level decreasing in the relatively unscratched area of the film.

2.4. Fabrication of NW FETs and Electrical Measurements

For the fabrication of our nanowire field-effect transistor, we used a well-known top-down approach [33,34] that involved electron-beam lithography to create all components of the structure, reactive ion etching to transfer the pattern into the silicon layer, and thin film deposition. The process is fully CMOS-compatible; see the Supplementary Materials for more details. Brief description: commercially available SoiTec Unibond® silicon-on-insulator (SOI) wafers had a 110 nm-thick top Si layer with a resistivity of 10 Ω cm. It was isolated from the 725 μm thick Si handle wafer with the same resistivity by a 200 nm thick buried oxide layer of SiO_2. This wafer was used as a bottom gate to adjust the working point of the transistor. The transistor structures were formed in the top active monocrystalline silicon layer by electron-beam lithography using FESEM Supra40 (Carl Zeiss) equipped with the ELPHY Quantum (RAITH) pattern generator. A nanowire pattern 80–90 nm wide and 3–5 μm long was drawn in a thin layer of positive electron beam resist A4 PMMA 950K. After development, a 15 nm Al film was deposited in open windows, and the rest was removed by a lift-off process. This metal layer was used as a mask during the following anisotropic reactive ion etching of the active Si layer in fluorine-containing plasma and later dissolved in a weak alkaline solution. The top views of the nanowire transistor structure were obtained using FESEM. Particular attention was paid to insulating the titanium conductive leads from the gate handle substrate and the reaction liquid medium. To do this, additional layers of silicon oxide were successively deposited onto the sample (excluding its central part). For measurements, the sample was fixed on a special holder, and the contacts were connected by ultrasonic bonding. The aluminum wires were then completely coated with sealant.

Modification of NWs with polyelectrolytes and antibodies was performed as described in Section 2.2. Measurements of the transistor transport characteristics, as well as measurements of the NWFET responses based on them, were carried out using a home-made measuring system consisting of a preamplifier, a voltage setting unit, and an interface unit providing communication with the control computer. Measurements were performed at positive voltages on the main (substrate) gate, which corresponds to the inverse (electronic) conductivity channel of the transistor. The AgCl reference electrode served as an additional gate for the transistor; it was immersed in the solution drop to adjust its operating point.

The operating point on the signal characteristic of the transistor was determined by the setting voltages (V_{sd}, V_g, and V_{ref}), and the current changes caused by external influence (changes in pH, formation of immune complexes) were recorded by a current preamplifier. The preamplifier's own low noise level (30 fA/$\sqrt{\text{Hz}}$) made it possible to register ultra-low current changes up to picoamps. Details are presented in the Supplementary Materials.

2.5. Determination of PSA by the NWFETs

Measurements on the FETs with nanowire channels modified by polyelectrolytes were performed in static mode without fluid flow, which provides stability and reproducible real-time response. First, a baseline was obtained from the flowing buffer solution (0.01 × PBS buffer pH 7.4). After it was stabilized, various concentrations of PSA in the same buffer were applied, and the change in the current was registered. BSA (1 ng/mL) was used for control experiments. The conductivity of the NW was modulated by the voltages at the main gate silicon substrate of the SOI wafer and the AgCl reference electrode. The limit of PSA detection (LOD) was calculated as the mean conductivity for the PSA-free solution (0 ng/mL PSA) plus 3 SD (n = 10).

3. Results and Discussion

3.1. Principle of Immunosensor Based on Field-Effect Transistors with Nanowire Channel Coated by Layers of Oppositely Charged Polyelectrolytes

The basis for determining the analyte on NW FETs is its direct capture by immobilized antibodies on a nanowire surface. Total charge on the NW surface varies as a result of immune complexes formation, which leads to a change in NW conductivity and variations in transport current. Measurements on nanowires were carried out in solutions with low ionic strength because the Debye length was strongly dependent on it. We propose to coat the surface of nanowires with charged polyelectrolytes to facilitate the functioning of antibodies. Antibody molecules can be "fixed softly" in a pseudo-homogeneous medium consisting of layers of oppositely charged polyelectrolytes, which preserves their mobility more than immobilization via covalent bonds.

The principle of determining PSA on immunosensors based on FETs with NW channels modified by layers of charged polyelectrolytes is shown in Figure 1. Modification of silicon with polyelectrolytes was carried out using a well-known layer-by-layer technique. Cationic polyelectrolyte PEI was suggested for the formation of the first layer on the surface, as its advantage in terms of uniform distribution onto semiconductor surfaces with thicknesses of microns was shown earlier [35]. IgG molecules have a negative charge at pH 7.4 [36], which is favorable for their interaction with PEI. Since the groups of negatively charged PSS completely dissociate in an experimental setting and help to form stable protein-polyelectrolyte complexes [37,38], it was used as a second polyelectrolyte. We examined the simplest matrix, which contained one layer of PEI and one layer of PSS, in order to identify the thinnest modifying layer of the surface relevant to the desired Debye length.

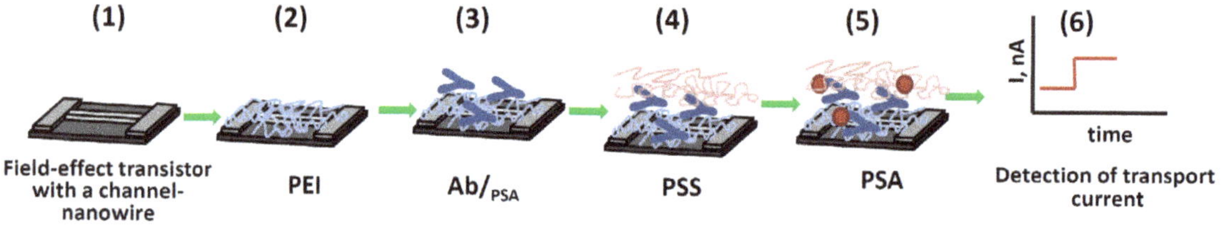

Figure 1. A scheme for determining PSA on a field-effect transistor with a nanowire using multilayers of charged polyelectrolytes. (**1**) Fabrication of a field-effect transistor with a nanowire channel; (**2**) The surface of the silicon nanowire is covered with positively charged polyelectrolyte PEI; (**3**) Antibodies to the analyte are adsorbed; (**4**) The surface is covered with negatively charged polyelectrolyte PSS; (**5**) PSA solution is passed through the nanowire; (**6**) A change in the transport current of the nanowire is registered.

The formation of a matrix of charged polyelectrolytes and antibodies involves several stages of spotting reagent solutions (PEI, PSA-specific antibodies, and PSS) on a silicon surface, drying, and several washings. As a result, we obtained a polyelectrolyte matrix in which the recognizing molecules of the immunosersor are located between the layers of oppositely charged polyelectrolytes.

Immune complexes are formed when the PSA solution is passed through a transistor with a nanowire channel. These complexes alter the conductivity of the transistor in response to changes in the electric charge on or near the nanowire's surface.

We examined the thickness of the polyelectrolyte layers and the formation of immune complexes when passing specific and non-specific proteins in order to investigate the preservation of antibody specificity and the reproducibility of the procedure.

3.2. Modification of Silicon Surface by Antibodies Adsorbed in a Matrix of Charged Polyelectrolytes and Study of the Formation of Immune Complexes

At the first stage, silicon samples (small pieces of silicon wafer) were utilized to study the polymer matrix's composition. The silicon surface was first coated with a layer of PEI, then covered with antibodies to PSA, and finally covered with a layer of PSS. In order to estimate the binding density of antibodies adsorbed in the matrix, we investigated the formation of sandwich immune complexes, composed of an antigen molecule and two molecules of antibodies, one immobilized on the modified silicon and the other labeled with AuNPs of 25 nm. Figure 2 illustrates the scheme for studying immune complex formation. To obtain sandwich immune complexes, modified silicon samples were successively incubated in a PSA solution and then in a solution of the second antibody to PSA with AuNPs. The conjugate of the second antibody was produced through covalent binding of AuNPs with their half-fragments containing thiol groups.

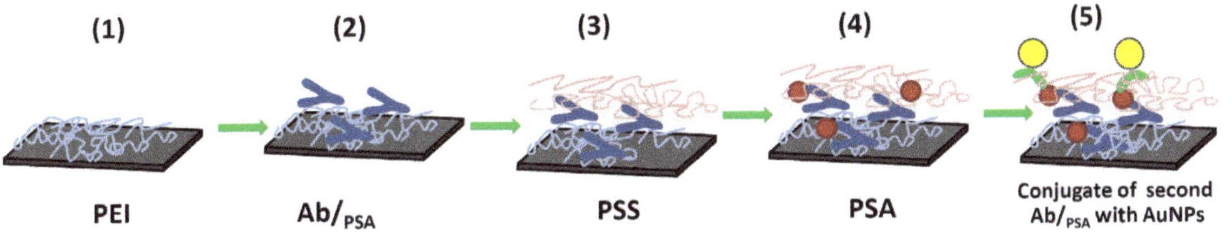

Figure 2. The scheme of silicon modification by the layers of polyelectrolytes and antibodies and the study of immune complex formation. (**1**) The surface of the silicon is covered with PEI; (**2**) Antibodies to PSA are adsorbed; (**3**) The surface is covered with PSS; (**4**) Incubation with PSA solution and washing; (**5**) Incubation with a conjugate of second antibodies labeled with gold nanoparticles and washing.

The thickness of the formed layers on the silicon surface as well as its morphology were investigated using SEM (a piece of silicon was split lengthwise and analyzed at an angle of at least 70°) and AFM (scanning of a silicon surface before and after making a scratch by applying 1 μN constant force along a 1 μm line). The combination of the two methods allowed us to determine the following parameters: the thickness of the PEI layer, the thickness of the matrix, the distribution of AuNPs on the surface, and their location inside the matrix.

The values of the matrix thickness on the silicon surface and the surface roughness are presented in Table 1. Native silicon is always covered with a thin layer of its oxide, which is characterized by a very small roughness of 0.5 nm. We changed the concentration of PEI to obtain a layer of minimal thickness (Figure S1). The thin layer of PEI that formed on the silicon surface had a thickness of 6 nm and was composed of half silicon oxide and half polymer. The optimal concentration of PEI for this process was 1 mg/mL (Figure S1). The surface of the layer has become notably rougher, but it has remained smooth overall. The slightly larger values of layer thicknesses obtained in AFM are apparently explained by the fact that AFM measured the depth of the scratch in unstressed layers, but SEM measured a slightly stressed (as a result of a splitting) edge of layers.

Table 1. Characteristics of the matrix on the silicon surface during sequential application of polyelectrolytes, antibodies, and antigen.

Components of the Matrix	Matrix Thickness Measured by SEM, nm	Matrix Thickness Measured by AFM, nm	Surface Roughness Measured by AFM, nm
No matrix (native silicon with a layer of silicon oxide)	3 ± 0.5	ND	0.5
PEI	6 ± 3	10 ± 2	3.7
PEI, $AB/_{PSA}$, PSS	22 ± 7	23 ± 5	3.7
(PEI, $AB/_{PSA}$, PSS) after interacting with PSA	27 ± 10	32 ± 5	7.8
(PEI, $AB/_{PSA}$, PSS) after interacting with PSA and $Ab/_{PSA}$–AuNPs	25 ± 3 *	35 ± 5	5.0
Control: (PEI, $AB/_{non\text{-}specific}$, PSS) after interacting with PSA and $Ab/_{PSA}$–AuNPs	20 ± 5 *	37 ± 5	5.3
Control: (PEI, $AB/_{non\text{-}specific}$, PSS) after interacting with PSA and $Ab/_{non\text{-}specific}$–AuNPs	20 ± 5 *	24 ± 5	3.3

* The thickness of the matrix layer was measured without accounting for nanoparticle size.

With good convergence of the two methods, the thickness of a matrix of two polyelectrolytes with adsorbed specific antibodies was increased up to about 22 nm, while the roughness remained relatively unchanged.

After interaction with the antigen, the layer's surface became more granular, and its thickness somewhat increased. A notable increase in the layer thickness value's standard deviation should be noted. This may be a sign of immune complexes formation in the matrix layer and is most likely the result of sufficiently large antibody and antigen molecules.

There was no discernible increase in layer thickness following interaction with the conjugate of the second antibody labeled with AuNPs (without accounting for nanoparticle size). The height of the matrix layer has also become more consistent at the same time. The determination of the PEI and the matrix thickness performed for different sample preparations (Figure S2) showed good reproducibility of the method.

Control experiments showed that the layer thickness was less when using non-specific antibodies and their conjugates compared to the use of specific antibodies. This suggests that formation of immune complexes results in a slight increase in matrix layer thickness and a decrease in density.

The layer height was obtained less in both control experiments using non-specific antibodies to PSA at the stage of polyelectrolyte matrix formation and as second antibodies than when using antibodies specific to PSA. At the same time, the layer thickness was lower compared to the formation of triple sandwich complexes from one PSA molecule and two antibody molecules. This is most likely due to the formation of fewer double complexes from one antigen molecule and one antibody molecule in control experiments compared to the number of triple sandwich complexes.

The processes that take place in the polymer matrix layer during the formation of double and triple sandwich immune complexes labeled with AuNPs were then visualized by applying SEM and AFM methods to a more detailed analysis of three silicon samples (Figures 3 and 4). The use of AuNPs as labels, reliably visualized by SEM and AFM, made it possible to compare the number of immune complexes formed of different antibody/antigen pairs and analyze their location relative to the surface of the nanowire. We compared the effectiveness of the formation of triple sandwich complexes from a PSA molecule and two molecules of specific antibodies, immobilized and labeled with AuNPs; double complexes from a PSA molecule and an antibody molecule labeled with AuNPs (control without immobilized specific antibodies); and non-specific complexes from a PSA molecule and non-specific antibodies labeled with AuNPs (control without immobilized and labeled specific antibodies). In all three studied combinations, we observed the formation of triple and double immune complexes, but their amounts differed significantly.

In order to calculate the number of complexes that were formed, the density of nanoparticles was counted (if each nanoparticle on the surface is part of an immune complex, then the total GNP number is proportional to the total number of complexes formed on the surface). Additionally, the location of the nanoparticles with respect to the silicon surface was examined by splitting the silicon sample.

When two antibody molecules and one antigen molecule combine to form sandwich complexes (Figure 3a), it is evident that AuNPs are "drawn" beneath the PSS layer. This suggests that second antibodies interact with the antigen by penetrating through the PSS layer. Sandwich complexes have a density of 188 AuNPs/μm^2 and are almost entirely embedded in a polyelectrolyte matrix. It is notable that AuNPs are located at an average distance of 3–5 nm from the silicon surface. As a result, AuNPs are sort of "pulled" beneath the PSS layer. The AFM image (Figure 4a) clearly shows the immersion of nanoparticles in polyelectrolyte matrices; only a small portion of the nanoparticle is still visible above the surface.

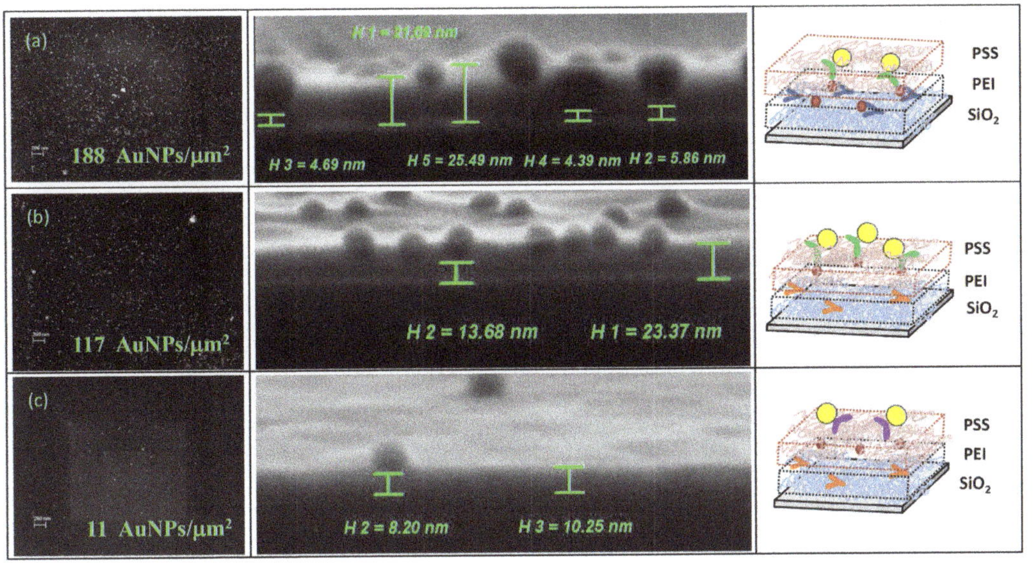

Figure 3. SEM images of silicon samples modified with polyelectrolytes with immune complexes of various compositions. (**a**) antibodies to PSA were adsorbed in the polyelectrolyte matrix, then a silicon sample was incubated in a solution of PSA and a conjugate of antibodies specific to PSA with AuNPs; (**b**) non-specific antibodies were adsorbed in the matrix, then a silicon sample was incubated in a solution of PSA and a conjugate of antibodies specific to PSA with AuNPs; (**c**) non-specific antibodies were adsorbed in the matrix, then a silicon sample was incubated in a solution of PSA and a conjugate of non-specific antibodies with AuNPs.

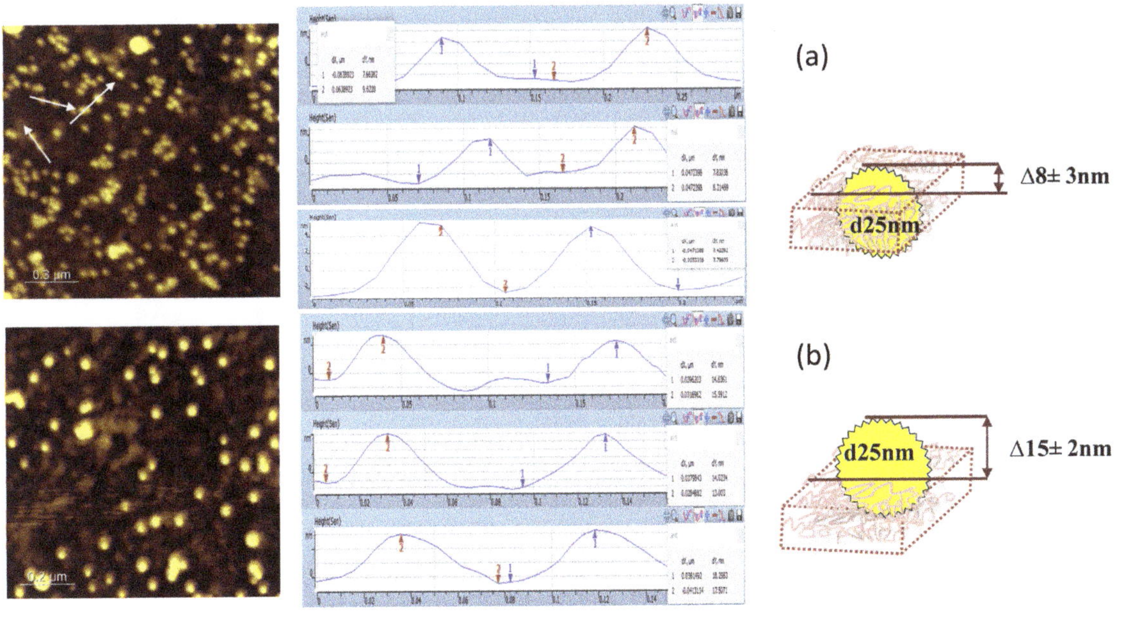

Figure 4. *Cont.*

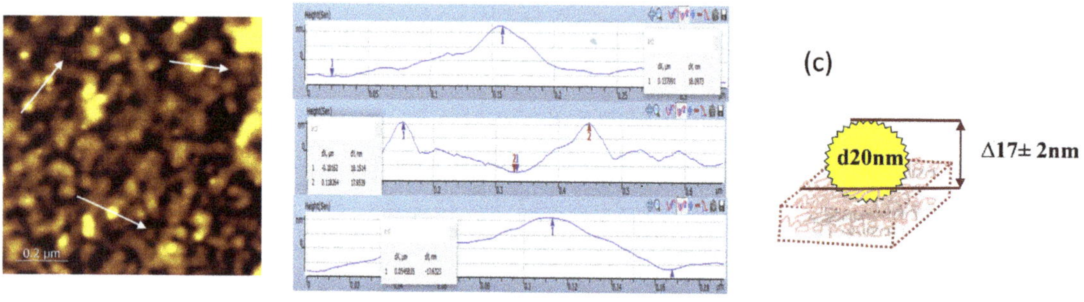

Figure 4. AFM images and surface profiles of silicon samples modified with polyelectrolytes with immune complexes of various compositions. (**a**) antibodies to PSA were adsorbed in the polyelectrolyte matrix, then a silicon sample was incubated in a solution of PSA and a conjugate of antibodies specific to PSA with AuNPs; (**b**) non-specific antibodies were adsorbed in the matrix, then a silicon sample was incubated in a solution of PSA and a conjugate of antibodies specific to PSA with AuNPs; (**c**) non-specific antibodies were adsorbed in the matrix, then a silicon sample was incubated in a solution of PSA and a conjugate of non-specific antibodies with AuNPs.

When PSA interacts with a matrix with non-specific antibodies (Figure 3b), its molecules enter the matrix and form double complexes of one antigen molecule and one antibody molecule labeled with a nanoparticle. When compared to triple sandwich complexes, we see a less complete partial immersion of AuNPs into the matrix (Figures 3b and 4b). The nanoparticle density (117 AuNPs/$\mu m^{2)}$ is lower than that for triple sandwich complexes, and they are farther away from the silicon surface (10–15 nm). In this case, the nanoparticles are not covered with a layer of PSS. Thus, the antigen penetrates into the matrix of polyelectrolytes but does not concentrate near the silicon surface. Apparently, it is hindered by a layer of adsorbed, nonspecific antibodies.

In a control experiment using non-specific antibodies and their conjugate with AuNPs, a certain level of non-specific binding was observed: the density of nanoparticles was 11 AuNPs/μm^2 (Figure 3c). This indicates that non-specific antibodies were adsorbed in small amounts on the surface of the matrix, did not penetrate into the polyelectrolyte layers, and the nanoparticles remained entirely on the surface of the upper PSS layer (Figure 4c).

Our investigation demonstrated the efficiency of antibody adsorption in the 3D matrix containing the layers of oppositely charged polyelectrolytes (PEI and PSS) on the silicon surface. It was discovered that this matrix was permeable to various biomolecules, including PSA, antibodies, and even 25 nm-sized AuNPs. The antigen interacts with antibodies that are specific to it by penetrating the matrix.

The distribution of nanoparticles inside the matrix showed that triple sandwich complexes are located near the surface of the nanowire, while the nanoparticles are drawn into the matrix. In the absence of immobilized specific antibodies, no such effect was observed; double immune complexes were formed at a greater distance from the surface, and the nanoparticles were slightly drawn into the matrix layer. In the control experiment with non-specific antibodies and a conjugate, a small number of nanoparticles remained on the surface of the matrix and were not drawn into it.

Thus, it can be concluded that PSA penetrates into the matrix of polyelectrolytes with specific antibodies and concentrates near the surface. Antibodies immobilized between the layers of polyelectrolytes are able to effectively form complexes with antigens due to the crowding effect, preserving the accessibility and mobility of their active centers. Due to the fact that only immune complexes localized in a thin, near-surface layer of the nanowire are detected by NWFETS, this information is crucial for detecting the antigen using this technique.

3.3. Characteristics of NWFETs with Antibodies Adsorbed in the Multilayers of Polyelectrolytes

NW FETs were fabricated as described in detail in the Supplementary Materials. The nanowires were then sequentially modified with layers of polyelectrolyte PEI, PSA-specific antibodies, and polyelectrolyte PSS. In order to improve NW sensitivity to external charges near their surface, the response of the immunosensors was registered in the sub-threshold region characterized by low bias voltage and transport current through the transistors. The regime of low currents is preferable for NW FETs both in terms of low self-noise (current shot noise) and low heating of the NWs [8,18,39].

The operation of the FET with a nanowire channel is based on a change in the conductivity of a semiconductor nanowire in an external electric field, which is formed by any charged particles (ions, molecules, etc.) located on or in the nearest vicinity of the nanowire surface. The surface of the immunosensor is a complex matrix formed by multilayers of charged polyelectrolytes and antibodies, which are also charged under experimental conditions. To test the sensitivity of the nanowire to changes in the electric field, at the first stage we studied its response to changes in proton (hydrogen ion) concentration with changes in the pH of the external environment. A change in pH leads to a change in the concentration of hydrogen ions on the nanowire surface and, as a result, a change in the nanowire conductivity, so the pH sensitivity is an indicator of the transistor sensitivity to any charged particles.

The conductivity of the NW channel depends on the local variation of the field or concentration of hydrogen ions (pH) connecting with the surface of modified silicon. Accordingly [40], the pH sensitivity of the NW FET can be estimated as:

$$\delta\Psi_0/\delta pH = \Delta U_{ref}/\Delta pH$$

where $\delta\Psi_0$ is the change of the insulator–electrolyte potential and U_{ref} is the voltage on the reference electrode.

In order to measure the pH sensitivity of the sensor, the optimal operating point of the NWFET corresponding to the high signal-to-noise ratio of the transistor and a low value of power dissipation in the nanowire channel [18,39] was determined by setting the voltages at the source drain (V_{sd} = 100 mV), the control electrode (V_g = 5 V), and the reference electrode (V_{ref} = 450 mV). When the pH value was changed, the voltage on the reference electrode (V_{ref}) was adjusted so that the transistor's current level returned to the value recorded at the prior pH value. Sensitivity to pH was determined as dV_{ref}/dpH.

The transistor's current response to a pH change is shown in Figure 5a. At the beginning of the experiment (section a1), the silicon nanowire covered with a PEI layer was placed in a buffer pH 7.5, and the reference electrode voltage was V_{ref} = 450 mV. A change in the pH of the environment led to a change in the conductivity of the nanowire channel and, as a result, to a change in the initial value of the transistor transport current. To determine the sensitivity to pH, the voltage at the reference electrode (V_{ref}) was adjusted so that the transistor's transport current level returned to its initial value.

The transistor's initial current level was approximately 0.5 nA. The current flowing through the transistor increased to about 1.5 nA when the buffer's pH dropped to 4.7 units (section a2). The voltage at the reference electrode was decreased from 450 mV to V_{ref} = 265 mV, which corresponds to the initial current level of I~0.5 nA (section a3). Therefore, a 2.8 unit change in pH corresponded to a 185 mV voltage change at the reference electrode, meaning that the pH sensitivity was 66 mV/pH.

Then, the buffer changed back to the original one with pH 7.5, and the level of current decreased to a value of I~0.1 nA (section a4). The voltage at the reference electrode was raised by dV_{ref} = 175 mV to bring it back to the starting level (section a5) and measure the sensitivity at the altered operating point. The sensitivity to pH in this case matched a value of 62 mV/pH.

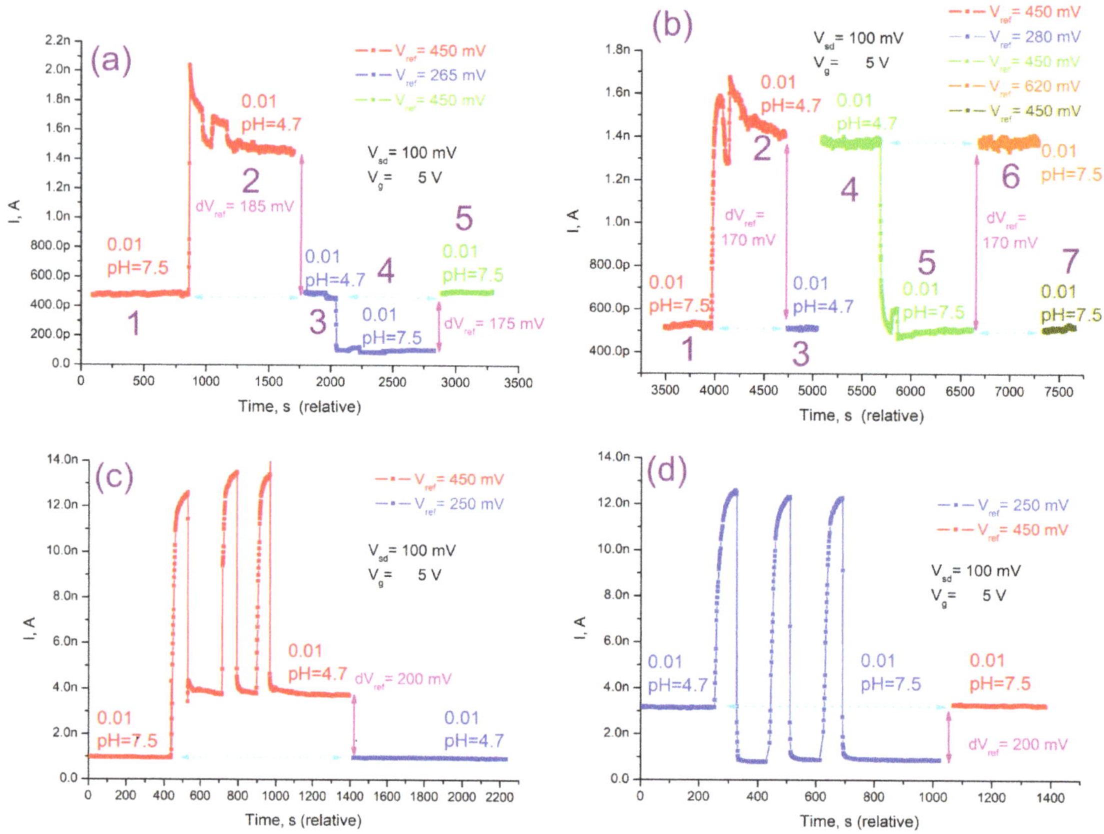

Figure 5. Responses of the NWFETs modified with the layers of polyelectrolyte PEI (**a**,**b**) and a matrix of PEI and PSS with antibodies included (**c**,**d**) to the change in pH. (**b**,**d**) Repeated measurements with a different method of changing pH and V_{ref} for (**a**,**c**), respectively. The current fluctuations (peaks) are caused by replacing a drop of buffer several times with a new one.

A slightly different approach was applied during the repeated measurements (Figure 5b). Sections b1, b2, and b3 were measured using the same methodology as in the prior case, but the dV_{ref} change was 170 mV by 2.8 pH units (61 mV/pH). Then, the voltage at the reference electrode returned to the initial level V_{ref} = 450 mV (section b4) and at this operating point the buffer changed to the initial one with pH = 7.5, which led to a decrease in the transistor current to the initial value I~0.5 nA (section b5). The voltage at the reference electrode was raised to V_{ref} = 620 mV in order to return to the signal level that corresponded to pH = 4.7 and V_{ref} = 450 mV (section b6). In this case, the dV_{ref} change was likewise 170 mV, which corresponded to a sensitivity to pH of 68 mV/pH. The current level was restored to the initial level of I~0.5 nA by lowering the V_{ref} in reverse to 450 mV (section b7). The error in the pH measurements, determined mainly by the drift of the current line, was ±3 mV/pH, so the measured values can be considered the same.

Measurements of the nanowire covered by PEI and PSS matrix with antibodies adsorbed between them (Figure 5c,d) were performed in the same manner; the change in V_{ref} (200 mV by 2.8 pH units) corresponds to a pH change of 71 mV/pH.

As a result, the pH sensitivity of the nanowires containing antibodies adsorbed in the polyelectrolyte matrix corresponds to a value of 71 ± 3 mV/pH, which is higher than the value of 60 mV/pH observed when silicon was treated with silanes [8].

3.4. Detection of PSA on the NWFETs

The determination of PSA on NW FETs is based on the direct binding of the antigen with specific antibodies immobilized in the complex matrix formed by multilayers of charged polyelectrolytes on the surface of a thin nanowire channel (Figure 1). During the formation of immune complexes of charged antibody and antigen molecules, electric charges are redistributed near the surface of the nanowire, which led to a change in the nanowire conductivity. The specificity and selectivity of PSA determination are ensured by the properties of monoclonal antibodies. The high specificity of this antibody clone against PSA was characterized earlier in the development of a standard ELISA method and the FETs with nanowires modified with gold nanoparticles using a collection of clinical samples [8].

The performance of the NWFETs at various PSA concentrations is shown in Figure 6. It can be seen that as the PSA concentration increases, the transport current increases as well. Even at a PSA concentration as low as 0.02 fg/mL, there is already a noticeable and substantial increase in the transport current (Figure 6a). About 10 s were needed to reach the maximum response to antigen solutions, and the signal remained steady for at least a few minutes. The analysis time per sample was 1 min, which included signal detection and washing. The control experiment with BSA (1 μg/mL) showed no detectable shift in the current of the NW FETs above the noise level.

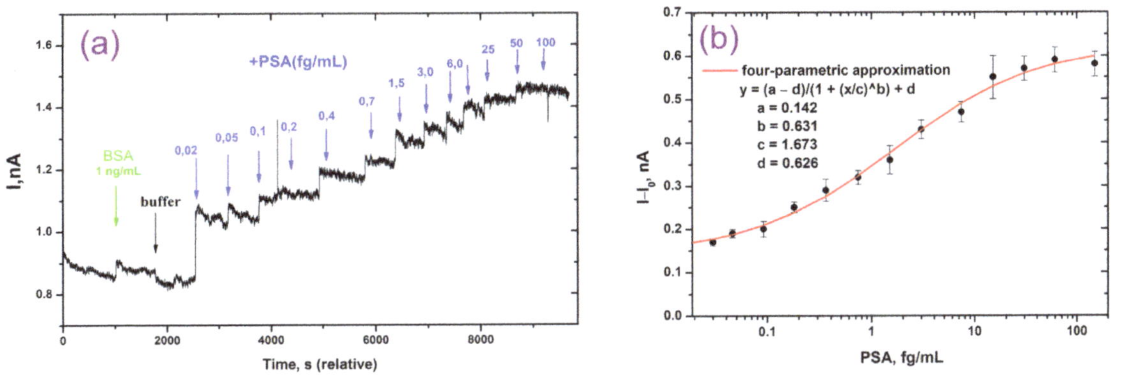

Figure 6. Determination of PSA on the NWFETs modified by a matrix of charged polyelectrolytes. (**a**) Transistor responses after adding buffer, BSA solution, and PSA solutions with different concentrations; (**b**) Calibration curve for the detection of PSA in 0.01 × PBS buffer.

The calibration curve for the detection of PSA in a 0.01 × PBS buffer is presented in Figure 6b. The calibration curve demonstrated a direct relationship between response and concentration of analyte. It was nonlinear, which is typical for immunoassay methods. In order to fit the curve, we used a 4-parameter mathematical model. The method is characterized by an ultra-low detection limit of 0.04 fg/mL (1.2 aM). For PSA determination in a buffer, the dynamic range is 0.08 fg/mL to 60 fg/mL.

Table 2 presents a comparison of the new approach with several electrochemical immunosensors for PSA detection that have been proposed recently. For comparison, biosensors based on NWFETs, FETs, and several variants of electrodes (screen-printed and gold) coated with films of various compounds, including polymers, were selected. Analysis of the analytical characteristics of biosensors shows that NWFETs are characterized by the lowest detection limit of PSA (this work, [41]). Between them, our technique with multilayer polyelectrolyte-modified NWs achieved an ultra-low PSA detection limit. Modification of imunosensors with different films, including polymers, improves their sensitivity [42–45]. Several immunosensors [42,43] showed comparable LOD with the NWFETs modified with APTES and glutaraldehyde [41]. It should be noted that for these immunosensors, one type of electrode coating was used. In contrast, in our immunosensor, antibodies were immobilized between the layers of oppositely charged polymers. As a result, we were able to reduce the detection limit of PSA by several orders of magnitude using monoclonal antibodies of the same clone compared to the NWFETs, in which silicon was modified by AuNPs [8]. This fact demonstrates how the polyelectrolyte matrix preserves the immunological characteristics of antibodies and has a positive impact on their properties.

Table 2. Analytical performances of different electrochemical immunosensors developed for quantitative determination of PSA.

Biosensor Type/Surface Modification	Limit of PSA Detection, fg/mL	Detection Range, fg/mL	Reference
Si NWFETs//matrix of polyelectrolytes PEI and PSS	0.04 (1.2 aM)	0.06–60	This study
Si NWFETs//GOPS-SH and AuNPs//the same clone of antibodies to PSA	23	23–5 × 10^8	[8]
n- and p-type Si NW FET arrays//APTES and glutaraldehyde	1.0	1.0–10^5	[41]
AlGaN/GaN HEMT//ethanolamine	1 (in PBS buffer)	1–10^8 (1 fg/mL to 100 ng/mL).	[42]
Electrochemical capacitive immunosensor/gold electrode modified with a thin film of isophthalic acid	3.3 (3.3 × 10^{-6} ng/mL)	1–10^8 (1 × 10^{-5}–100 ng/mL)	[43]
Immunosensor//screen-printed electrodes//AgNPs/ds-DNA/PEDOT	17	100–10^8 (0.1 pg/mL–100 ng/mL)	[44]
Electrochemical immunosensor with silica nanochannel films//GPTMS	100 0.1 pg/mL	10^3–10^8 (1 pg/mL to 100 ng/mL)	[45]
Graphene FET with coplanar electrodes//glutaraldehyde	10^3 (1 pg/mL in serum)	Up to 4 × 10^7 (up to 4 ng/mL in 100 mM phosphate buffer)	[46]
A dual-gate PEG-modified NR-ISFET nanoribbon-based ion-sensitive field-effect transistor (NR-ISFET)//polyethylene glycol	10 pM (in 100 mM phosphate buffer)	10 pM to 1 μM (in 100 mM phosphate buffer)	[47]
Si NW-Ionsensitive field-effect transistor (Si-NW ISFET)//DNA aptamer	ND *	10^3–10^9 (1 pg/mL to 1 μg/mL)	[48]

* Not determined.

At the same time, the dynamic range of detectable PSA concentrations by our immunosensor is three orders of magnitude and is not as wide as in other techniques. The possibility of dilution of the serum sample makes it possible to reduce the potential interference of biological samples. Despite the fact that the range of three orders of magnitude meets the diagnostic requirements, further research is needed to examine the range in biological samples.

The analysis of the applicability of various electrochemical immunosensor technologies for POC is of great practical interest because PSA detection is valuable for early screening of prostate cancer and monitoring recurrence after treatment. The criteria for POC systems involve rapid assay time, measuring to be performed at the place of the patient, minimal sample preparation, small sample volume, automation by a portable device, and fitting the sensitivity and concentration range [49]. At present, commercial POC systems for PSA detection are based on lateral-flow immunoassays with colorimetric detection. They require little sample volume and are quick, but they do not always fulfill the criteria for sensitivity, specificity, and measurement range. It is believed that electrochemical biosensors better match POC requirements. The immunosensors reported in the literature exhibit promising results and have good potential for ensuring low detection limits, small sample volumes, and automation capabilities. At the same time, only some immunosensors have been validated using real serum samples [44–46]. Our approach and a few other novel principles of PSA determination have only been evaluated with PSA samples in buffer systems. Further studies are required to investigate the long-term stability of the immunosensors, explore the need to expand the dynamic range, and validate them with serum samples.

4. Conclusions

In this work, we have demonstrated the advantages of coating silicon nanowires with layers of charged polyelectrolytes. Together, they form a three-dimensional matrix that is permeable to biomolecules and efficiently absorbs antibodies. Multilayers of polyelectrolytes provide a polyelectrolyte-protein "crowding" medium that promotes the preservation of antibody properties and contributes to the concentration of antigen in the near-surface layer. As a result, NWFETs with antibodies in a polyelectrolyte matrix show a high sensitivity to pH changes, which significantly exceeds the theoretical limit of Nernst. The immunosensor is characterized by the lowest PSA detection limit in buffer systems to date. Coating the NW surface with multiple polymer layers significantly simplifies the process of preparing the sensor surface compared to the chemical modification of silicon for the covalent attachment of antibodies. The sensor takes less than a minute, and the measurement is performed in real-time. This approach seems promising for further investigation of its applicability for early screening of prostate cancer and POC systems. Future research will focus on optimizing immunosensors for measuring PSA in biological samples, examining their long-term stability, and validating them with real serum samples.

Supplementary Materials: The following supporting information can be downloaded: https://www.mdpi.com/article/10.3390/polym16030332/s1, Figure S1: Dependence of the thickness of the PEI layer on the silicon surface on its concentration; Figure S2: Thickness of the PEI layer (a) and the matrix layer of polyelectrolytes with antibodies (b) for different silicon samples; Figure S3: Main fabrication steps of a transistor with a nanowire channel. (a)—formation of an aluminum mask; (b)—transfer of the mask pattern to the top layer of silicon using RIE; (c)—formation of metallic contact pads; (d)—isolation of current-carrying elements of the structure from contact with the liquid medium; Figure S4: SEM images of an aluminum mask on the top of the active silicon layer (a) and the structure of a nanowire transistor with contact pads covered by dielectric layers (b); Figure S5: Transistor transport current vs. main gate voltage V_g (top axis) and reference electrode voltage V_{ref} (bottom axis). References [33,34,39] are cited in the supplementary materials.

Author Contributions: Conceptualization, M.Y.R., E.V.S., O.V.S. and D.E.P.; methodology, G.V.P., A.S.T., M.M.U., I.I.T., O.V.S. and E.V.S.; validation and formal analysis, G.V.P., M.Y.R., V.A.K. and D.E.P.; visualization, G.V.P. and D.E.P.; writing—original draft preparation, G.V.P., M.Y.R., V.A.K. and D.E.P.; writing—review and editing, G.V.P., M.Y.R., V.A.K. and D.E.P.; supervision, M.Y.R. and D.E.P.; project administration and funding acquisition, O.V.S. All authors have read and agreed to the published version of the manuscript.

Funding: The work was conducted with the support of the MSU Program.

Institutional Review Board Statement: Not applicable.

Data Availability Statement: Data are contained within the article and Supplementary Materials.

Acknowledgments: The authors acknowledge the use of the research infrastructure of the "Educational and Methodical Center of Lithography and Microscopy" at Lomonosov Moscow State University.

Conflicts of Interest: The authors declare no conflicts of interest. The funders had no role in the design of the study, in the collection, analysis, or interpretation of data, in the writing of the manuscript, or in the decision to publish the results.

References

1. Sadighbayan, D.; Hasanzadeh, M.; Ghafar-Zadeh, E. Biosensing based on field-effect transistors (FET): Recent progress and challenges. *Trends Analyt. Chem.* **2020**, *133*, 116067. [CrossRef]
2. Cui, Y.; Lieber, C.M. Functional nanoscale electronic devices assembled using silicon nanowire building blocks. *Science* **2001**, *291*, 851–853. [CrossRef] [PubMed]

3. Salfi, J.; Savelyev, I.G.; Blumin, M.; Nair, S.V.; Ruda, H.E. Direct observation of single-charge detection capability of nanowire field-effect transistors. *Nat. Nanotech.* **2010**, *5*, 737–741. [CrossRef] [PubMed]
4. Clement, N.; Nishiguchi, K.; Dufreche, J.F.; Guerin, D.; Fujiwara, A.; Vuillaume, D. A silicon nanowire ion-sensitive field-effect transistor with the elementary charge sensitivity. *Appl. Phys. Lett.* **2011**, *98*, 014104. [CrossRef]
5. Wang, Z.; Lee, S.; Koo, K.-I.; Kim, K. Nanowire-Based Sensors for Biological and Medical Applications. *IEEE Trans. NanoBiosci.* **2016**, *15*, 186–199. [CrossRef] [PubMed]
6. Panahi, A.; Sadighbayan, D.; Forouhi, S.; Ghafar-Zadeh, E. Recent Advances of Field-Effect Transistor Technology for Infectious Diseases. *Biosensors* **2021**, *11*, 103. [CrossRef] [PubMed]
7. Malsagova, K.A.; Ivanov, Y.D.; Pleshakova, T.O.; Kaysheva, A.L.; Shumov, I.D.; Kozlov, A.F.; Archakov, A.I.; Popov, V.P.; Fomin, B.I.; Latyshev, A.V. A SOI-nanowire biosensor for the multiple detection of D-NFATc1 protein in the serum. *Anal. Methods* **2015**, *7*, 8078–8085. [CrossRef]
8. Presnova, G.; Presnov, D.; Krupenin, V.; Grigorenko, V.; Trifonov, A.; Andreeva, I.; Ignatenko, O.; Egorov, A.; Rubtsova, M. Biosensor based on a silicon nanowire field-effect transistor functionalized by gold nanoparticles for the highly sensitive determination of prostate specific antigen. *Biosens. Bioelectron.* **2017**, *88*, 283–289. [CrossRef]
9. Zhao, W.; Hu, J.; Liu, J.; Li, X.; Sun, S.; Luan, X.; Zhao, Y.; Wei, S.; Li, M.; Zhang, Q.; et al. Si nanowire Bio-FET for electrical and label-free detection of cancer cell-derived exosomes. *Microsyst. Nanoeng.* **2022**, *8*, 57. [CrossRef]
10. Seo, G.; Lee, G.; Kim, M.J.; Baek, S.H.; Choi, M.; Ku, K.B.; Lee, C.S.; Jun, S.; Park, D.; Kim, H.G.; et al. Rapid Detection of COVID-19 Causative Virus (SARS-CoV-2) in Human Nasopharyngeal Swab Specimens Using Field-Effect Transistor-Based Biosensor. *ACS Nano* **2020**, *14*, 5135–5142. [CrossRef]
11. Kim, K.; Park, C.; Kwon, D.; Kim, D.; Meyyappan, M.; Jeon, S.; Lee, J.S. Silicon nanowire biosensors for detection of cardiac troponin I (cTnI) with high sensitivity. *Biosens. Bioelectron.* **2016**, *77*, 695–701. [CrossRef] [PubMed]
12. Espinosa, F.M.; Uhlig, M.R.; Garcia, R. Molecular Recognition by Silicon Nanowire Field-Effect Transistor and Single-Molecule Force Spectroscopy. *Micromachines* **2022**, *13*, 97. [CrossRef] [PubMed]
13. Hu, Q.; Chen, S.; Solomon, P.; Zhang, Z. Ion sensing with single charge resolution using sub–10-nm electrical double layer–gated silicon nanowire transistors. *Sci. Adv.* **2021**, *7*, eabj6711. [CrossRef] [PubMed]
14. Wasfi, A.; Awwad, F.; Gelovani, J.G.; Qamhieh, N.; Ayesh, A.I. COVID-19 Detection via Silicon Nanowire Field-Effect Transistor: Setup and Modeling of Its Function. *Nanomaterials* **2022**, *12*, 2638. [CrossRef] [PubMed]
15. Abidin, W.A.B.Z.; Nor, N.M.M.; Arshad, M.K.M.; Fathil, M.F.M.; Parmin, N.A.; Sisin, N.A.H.T.; Ibau, C.; Azlan, A.S. Femtomolar Dengue Virus Type-2 DNA Detection in Back-gated Silicon Nanowire Field-effect Transistor Biosensor. *Curr. Nanosci.* **2022**, *18*, 139–146. [CrossRef]
16. Zafar, S.; D'Emic, C.; Jagtiani, A.; Kratschmer, E.; Miao, X.; Zhu, Y.; Mo, R.; Sosa, N.; Hamann, H.; Shahidi, G.; et al. Silicon Nanowire Field Effect Transistor Sensors with Minimal Sensor-to-Sensor Variations and Enhanced Sensing Characteristics. *ACS Nano* **2018**, *12*, 6577–6587. [CrossRef]
17. Kesler, V.; Murmann, B.; Soh, H.T. Going beyond the Debye Length: Overcoming Charge Screening Limitations in Next-Generation Bioelectronic Sensors. *ACS Nano* **2020**, *14*, 16194–16201. [CrossRef] [PubMed]
18. Presnov, D.E.; Bozhev, I.V.; Miakonkikh, A.V.; Simakin, S.G.; Trifonov, A.S.; Krupenin, V.A. Local sensor based on nanowire field effect transistor from inhomogeneously doped silicon on insulator. *J. Appl. Phys.* **2018**, *123*, 054503. [CrossRef]
19. Rajan, N.K.; Routenberg, D.A.; Reed, M.A. Optimal signal-to-noise ratio for silicon nanowire biochemical sensors. *Appl. Phys. Lett.* **2018**, *98*, 264107. [CrossRef]
20. De Vico, L.; Sorensen, M.H.; Iversen, L.; Rogers, D.M.; Sorensen, B.S.; Brandbyge, M.; Nygard, J.; Martinez, L.; Jensen, J.H. Quantifying signal changes in nano-wire based biosensors. *Nanoscale* **2011**, *3*, 706–717. [CrossRef]
21. Li, J.; Zhang, Y.; To, S.; You, L.; Sun, Y. Effect of nanowire number, diameter, and doping density on nano-FET biosensor sensitivity. *ACS Nano* **2011**, *5*, 6661–6668. [CrossRef] [PubMed]
22. Stern, E.; Wagner, R.; Sigworth, R.J.; Breaker, R.; Fahmy, T.M.; Reed, M.A. Importance of the Debye screening length on nanowire field effect transistor sensors. *Nano Lett.* **2007**, *7*, 3405–3409. [CrossRef] [PubMed]
23. Awsiuk, K.; Budkowski, A.; Psarouli, A.; Petrou, P.; Bernasik, A.; Kakabakos, S.; Rysz, J.; Raptis, I. Protein adsorption and covalent bonding to silicon nitride surfaces modified with organo-silanes: Comparison using AFM, angle-resolved XPS and multivariate ToF-SIMS analysis. *Colloids Surf. B Biointerfaces* **2013**, *110*, 217–224. [CrossRef] [PubMed]
24. Ghobaei Namhil, Z.; Kemp, C.; Verrelli, E.; Iles, A.; Pamme, N.; Adawi, A.M.; Kemp, N.T. A label-free aptamer-based nanogap capacitive biosensor with greatly diminished electrode polarization effects. *Phys. Chem. Chem. Phys.* **2019**, *21*, 681–691. [CrossRef] [PubMed]
25. Gao, N.; Zhou, W.; Jiang, X.; Hong, G.; Fu, T.-M.; Lieber, C.M. General Strategy for Biodetection in High Ionic Strength Solutions Using Transistor-Based Nanoelectronic Sensors. *Nano Lett.* **2015**, *15*, 2143–2148. [CrossRef] [PubMed]
26. Becker, A.L.; Henzler, K.; Welsch, N.; Ballauff, M.; Borisov, O. Proteins and polyelectrolytes: A charged relationship. *Curr. Opin. Colloid Interface Sci.* **2012**, *17*, 90–96. [CrossRef]
27. Lee, A.A.; Kostinski, S.V.; Brenner, M.P. Controlling Polyelectrolyte Adsorption onto Carbon Nanotubes by Tuning Ion–Image Interactions. *J. Phys. Chem. B* **2018**, *122*, 1545–1550. [CrossRef] [PubMed]
28. Stekolshchikova, A.A.; Radaev, A.V.; Orlova, O.Y.; Nikolaev, K.G.; Skorb, E.V. Thin and Flexible Ion Sensors Based on Polyelectrolyte Multilayers Assembled onto the Carbon Adhesive Tape. *ACS Omega* **2019**, *4*, 15421–15427. [CrossRef]

29. Zhao, H.; Ibrahimova, V.; Garanger, E.; Lecommandoux, S. Dynamic Spatial Formation and Distribution of Intrinsically Disordered Protein Droplets in Macromolecularly Crowded Protocells. *Angew. Chem. Int. Ed.* **2020**, *59*, 11028–11036. [CrossRef] [PubMed]
30. Baldina, A.A.; Nikolaev, K.G.; Ivanov, A.S.; Nikitina, A.A.; Rubtsova, M.Y.; Vorovitch, M.F.; Ishmukhametov, A.A.; Egorov, A.M.; Skorb, E.V. Immunochemical Biosensor for Single Virus Particle Detection Based on Crowding Molecular Polyelectrolyte System. *J. Appl. Polym. Sci.* **2022**, *139*, 52360. [CrossRef]
31. Frens, G. Controlled nucleation for the regulation of the particle size in monodisperse gold suspensions. *Nat. Phys. Sci.* **1973**, *241*, 20–22. [CrossRef]
32. Green, C.P.; Lioe, H.; Cleveland, J.P.; Proksch, R.; Mulvaney, P.; Sader, J.E. Normal and torsional spring constants of atomic force microscope cantilevers. *Rev. Sci. Instr.* **2004**, *75*, 1988–1996. [CrossRef]
33. Tsiniaikin, I.I.; Presnova, G.V.; Bozhev, I.V.; Skorik, A.A.; Rubtsova, M.Y.; Kamalov, A.A.; Matskeplishvili, S.T.; Snigirev, O.V.; Krupenin, V.A.; Presnov, D.E. A sensor system based on a field-effect transistor with a nanowire channel for the quantitative determination of thyroid-stimulating hormone. *Mosc. Univ. Phys. Bull.* **2020**, *75*, 645–656. [CrossRef]
34. Presnova, G.V.; Tcinyaykin, I.I.; Bozhev, I.V.; Rubtsova, M.Y.; Shorokhov, V.V.; Trifonov, A.S.; Ulyashova, M.M.; Krupenin, V.A.; Presnov, D.E. Thyroglobulin detection by biosensor based on two independent Si NWFETs. *Proc. SPIE-Int. Soc. Opt. Eng.* **2019**, *11022*, 110220Z. [CrossRef]
35. Soranno, A.; Koenig, I.; Borgia, M.B.; Hofmann, H.; Zosel, F.; Nettels, D.; Schuler, B. Single-molecule spectroscopy reveals polymer effects of disordered proteins in crowded environments. *Proc. Natl. Acad. Sci. USA* **2014**, *111*, 4874–4879. [CrossRef]
36. Filoti, D.I.; Shire, S.J.; Yadav, S.; Laue, T.M. Comparative study of analytical techniques for determining protein charge. *J. Pharm. Sci.* **2015**, *104*, 2123–2131. [CrossRef]
37. Füzik, T.; Formanov, P.; Ružek, D.; Yoshii, K.; Niedrig, M.; Plevka, P. Structure of tick-borne encephalitis virus and its neutralization by a monoclonal antibody. *Nat. Commun.* **2018**, *9*, 436. [CrossRef] [PubMed]
38. Straeten, A.V.; Bratek-Skicki, A.; Jonas, A.M.; Fustin, C.-A.; Dupont-Gillain, C. Integrating Proteins in Layer-by-Layer Assemblies Independently of their Electrical Charge. *ACS Nano* **2018**, *12*, 8372–8381. [CrossRef] [PubMed]
39. Trifonov, A.S.; Presnov, D.E.; Bozhev, I.V.; Evplov, D.A.; Desmaris, V.; Krupenin, V.A. Non-contact scanning probe technique for electric field measurements based on nanowire field-effect transistor. *Ultramicroscopy* **2017**, *179*, 33–40. [CrossRef] [PubMed]
40. Van Hal, R.E.G.; Eijkel, J.C.T.; Bergveld, P. A novel description of ISFET sensitivity with the buffer capacity and double-layer capacitance as key parameters. *Sens. Actuators B* **1995**, *24*, 201–205. [CrossRef]
41. Gao, A.; Lu, N.; Dai, P.; Fan, C.; Wang, Y.; Li, T. Direct ultrasensitive electrical detection of prostate cancer biomarkers with CMOS-compatible n- and p-type silicon nanowire sensor arrays. *Nanoscale* **2014**, *6*, 13036–13042. [CrossRef]
42. Gu, Z.; Wang, J.; Miao, B.; Zhao, L.; Liu, X.; Wu, D.; Li, J. Highly sensitive AlGaN/GaN HEMT biosensors using an ethanolamine modification strategy for bioassay applications. *RSC Adv.* **2019**, *9*, 15341–15349. [CrossRef]
43. Mwanza, D.; Adeniyi, O.; Tesfalidet, S.; Nyokong, T.; Mashazi, P. Capacitive Label-Free Ultrasensitive Detection of PSA on a Covalently Attached Monoclonal Anti-PSA Antibody Gold Surface. *J. Electroanal. Chem.* **2022**, *927*, 116983. [CrossRef]
44. Liu, Y. Highly sensitive sensing detection of prostate-specific antigen based on point-of-care electrochemical immunosensor. *Int. J. Electrochem. Sci.* **2023**, *18*, 100028. [CrossRef]
45. Ma, K.; Zheng, Y.; An, L.; Liu, J. Ultrasensitive Immunosensor for Prostate-Specific Antigen Based on Enhanced Electrochemiluminescence by Vertically Ordered Mesoporous Silica-Nanochannel Film. *Front. Chem.* **2022**, *10*, 851178. [CrossRef] [PubMed]
46. Mandal, N.; Pakira, V.; Samanta, N.; Das, N.; Chakraborty, S.; Pramanick, B.; RoyChaudhuri, C. PSA Detection using Label Free Graphene FET with Coplanar Electrodes Based Microfluidic Point of Care Diagnostic Device. *Talanta* **2021**, *222*, 121581. [CrossRef] [PubMed]
47. Ma, S.; Li, X.; Lee, Y.K.; Zhang, A. Direct Label-free Protein Detection in High Ionic Strength Solution and Human Plasma Using Dual-Gate Nanoribbon-based Ion-Sensitive Field-Effect Transistor Biosensor. *Biosens. Bioelectron.* **2018**, *117*, 276–282. [CrossRef] [PubMed]
48. Rani, D.; Pachauri, V.; Madaboosi, N.; Jolly, P.; Vu, X.T.; Estrela, P.; Chu, V.; Conde, J.P.; Ingebrandt, S. Top-Down Fabricated Silicon Nanowire Arrays for Field-Effect Detection of Prostate-Specific Antigen. *ACS Omega* **2018**, *3*, 8471–8482. [CrossRef] [PubMed]
49. Garg, S.; Sachdeva, A.; Peeters, M.; McClements, J. Point-of-Care Prostate Specific Antigen Testing: Examining Translational Progress toward Clinical Implementation. *ACS Sens.* **2023**, *8*, 3643–3658. [CrossRef]

polymers

MDPI

Article

Effects of Beech Wood Surface Treatment with Polyethylenimine Solution Prior to Finishing with Water-Based Coating

Tanja Palija [1,*], Milica Rančić [1], Daniela Djikanović [2], Ksenija Radotić [2], Marko Petrič [3], Matjaž Pavlič [3] and Milan Jaić [1]

[1] Faculty of Forestry, University of Belgrade, 11030 Belgrade, Serbia; milica.rancic@sfb.bg.ac.rs (M.R.); milan.jaic@sfb.bg.ac.rs (M.J.)

[2] Institute for Multidisciplinary Research, University of Belgrade, 11108 Belgrade, Serbia; danielle@imsi.rs (D.D.); xenia@imsi.rs (K.R.)

[3] Biotechnical Faculty, University of Ljubljana, 1000 Ljubljana, Slovenia; marko.petric@bf.uni-lj.si (M.P.); matjaz.pavlic@bf.uni-lj.si (M.P.)

* Correspondence: tanja.palija@sfb.bg.ac.rs

Abstract: The surfaces of beech wood samples were treated with polyethylenimine (PEI) solutions at three different concentrations—0.5%, 1% and 2%—and two molecular weights—low molecular weight (LMW) and high molecular weight (HMW). The effects of PEI surface treatment of wood were characterized by FT-IR spectroscopy, the penetration depth of PEI (EPI fluorescence spectroscopy), the bonding position of PEI (by SEM), the wetting and surface energy, and the water uptake. After PEI treatment, the samples were coated with a water-based transparent acrylic coating (WTAC). The dry film thickness, the penetration depth of the coating, the adhesion strength and the surface roughness of the coated wood surface were evaluated. EPI fluorescence and SEM micrographs showed that PEI HMW chains were deposited on the surface, in contrast to PEI LMW, which penetrates deeper into layers of the wood cells. Treatment with a 1% PEI HMW solution resulted in a 72% reduction in water uptake of the wood (compared to untreated samples after 5 min of applying water droplets to the surface) and a 23.2% reduction in surface energy (compared to untreated samples) while maintaining the adhesion strength of the applied WTAC. The lower water uptake of the treated wood samples reduced the roughness of the coated surface, which is particularly important when the wood surface is finished with water-based coatings.

Keywords: wood; polyelectrolyte; polyethyleneimine; water-based coating; wood-coating interface; wood surface finishing

1. Introduction

Wood as a carbon-neutral natural material and innovative wood products have attracted great interest in recent years due to increasing environmental awareness. However, in order to extend the service life of wood products, it is necessary to protect their surfaces from environmental influences. Surface protection is usually achieved by coating. The main disadvantage of conventional coatings based on organic solvents is the emission of volatile organic compounds (VOCs). In recent years, the coatings industry has faced increasing pressure to reduce its overall environmental impact [1]. For this reason, environmentally friendly materials for the surface protection of wood, such as water-based coating formulations, have a growing share of the coatings market. In the wood industry, the use of water-based coatings has increased significantly in recent decades [1]. Today, water-based resins for coating formulations are in many ways equivalent to or better than their solvent-based counterparts [2]. Due to lower VOC exposure, water-based coatings are less toxic and flammable and easy to clean [3]; they have higher water vapor permeability and water uptake than their solvent-based counterparts [4]. New trends in water-based coatings are mostly based on further reducing greenhouse gas emissions by using bio-based materials that are sustainable both in terms of the production process and the raw materials used [5]. Given that wood is a hygroscopic material, the main disadvantage of using water-based coatings is that it swells upon contact with water, which manifests itself in a raising of the wood fibers and an increase in the roughness of the coated surface [6,7]. In addition, the swelling of the wood leads to a narrowing of the coating's path of movement into the interior of the wood tissue, which can affect the depth of penetration, which is a prerequisite for achieving adequate adhesion between the coating and the wood substrate [8].

Previous studies have shown that it is possible to utilize the electrical potential of wood, which is due to the dissociation of functional groups of the basic polymeric structures, especially the carboxyl groups [9], for the electrostatic binding of various materials to the wood surface. Polyelectrolytes (PEs) are charged macromolecules with a positive or negative net charge depending on the pH value of the environment. Most of them are water-soluble [10] and can dissociate in polar solvents, such as water, producing counterions that are dispersed in the solution. PE coatings can change the charge, surface energy and mechanical properties of various substrates. Furthermore, functionalized PE coatings are considered a potential solution in the context of green chemistry [11]. Since cellulose, like wood, has a negative electrical surface potential in moderately acidic or neutral environments, cationic polyelectrolytes can be used to bind efficiently to negatively charged wood surfaces. In addition, the water from the polyelectrolyte solution has a positive effect on increasing the electrical conductivity of the wood (a 10^{10} to 10^{13} fold increase in electrical conductivity when the moisture content of the wood increases from near zero to fiber saturation, according to Simpson and TenWolde (2007) in [12]).

In the previous study, the influence of the application of commercially available cationic polyelectrolyte solutions (twelve different types) to samples of oak (*Quercus robur*) and Scots pine (*Pinus sylvestris*) on the extent of water uptake was investigated [13]. The adsorbed polyelectrolytes hindered the penetration of water into the pore system of the wood, according to the results of the relative water uptake, which indicates the percentage ratio of water uptake to total water consumption. However, with the different polyelectrolyte solutions, the oak and pine samples showed a different degree of decrease in water uptake. Compared to untreated oak and pine samples, treatment with a 1% PEI solution reduced the relative water uptake of the treated samples by 41.1% and 36.9%, respectively. The concentration and molecular weight of the aqueous polyelectrolyte solutions proved to be key in maximizing the modification effect. In the study investigating the degree of adsorption of poly(AM-co-DADMAC) (copolymers of acrylamide and diallyldimethylammonium chloride) in the cell wall of oxygen-bleached, never-dried kraft pulp, only polyelectrolytes with a charge density and a molecular weight below a critical value could be adsorbed in the cellulose fiber cell wall [14]. When investigating the influence of the molecular weight of poly(AM-co-DADMAC) (8×10^3 g/mol; 1.1×10^5 g/mol; 4.0×10^5 g/mol) on the adsorption of cellulose fibers, the polyelectrolyte with the lowest molecular weight showed a significantly higher degree of adsorption due to its penetration into the cell walls. In another study [15], it was found that pretreatment of beech wood with a 1% high-molecular-weight PEI solution resulted in a reduction in the roughness of the wood coated with a water-based coating, especially when the initial roughness was higher. This result shows that when using water-based coatings for the surface treatment of wood, surface treatment with a polyelectrolyte solution can effectively replace more intensive sanding (in terms of the number of sanding steps and grit) to produce a coated surface with lower roughness. In addition, the PEI surface treatment had no effect on the adhesion strength of the water-based acrylic coating on the treated samples. In testing the adhesion strength of three different types of coatings for exterior application on redwood, PEI and aminosilanes were used for chemical bonding of the coating to the substrate by the grafted molecular brushes [16]. The highest adhesion strength of the coatings (measured by visual assessment of the delamination effect using the cross-cut method) was observed after the aforementioned treatment, which was preceded by surface oxidation with corona discharge and/or oxidation.

PEI is considered an environmentally friendly substance that is safe and easy to use [17]. It is used as a component in the production of environmentally friendly adhesive without formaldehyde emissions [18,19], as well as for the formation of a coating on the wood surface through a layer-by-layer system [20] and polyelectrolyte complexes [21] to achieve flame-retardant wood. One of the main advantages of PEI is the ability to form hydrogen bonds [17], which is due to the large number of amino groups, whereby each nitrogen-bonded hydrogen atom within the amino group can be involved in the formation of hydrogen bonds.

The optimal parameters of the PEI solution for surface treatment must be determined in relation to the properties of the final product, i.e., the wood coated with a water-based coating. However, it is not known how the ability of PEI to bind water affects the behavior of water-based coatings that come into contact with the treated wood surface in terms of wetting, penetration and bonding. In this context, it is important to know the effects of PEI absorption on chemical, structural and physical changes in wood surfaces. In order to exploit the potential of PEI in the field of wood finishing with water-based coatings, it is important to understand the indirect effects of the parameters of the polyelectrolyte solution on the properties of the finished wood.

Therefore, the main objective of this study is to investigate the possible effects of using the weak polycation PEI as a pretreatment in the finishing of wood surfaces with water-based coatings. The effects of different concentrations and molecular weights of PEI solutions on the properties of the treated wood surface were investigated. It was then analyzed how the pretreatment of wood with PEI affects the key properties of the coated wood when using water-based coatings. The low-molecular-weight PEI and the high-molecular-weight PEI were used to modify the penetration effects of the polyelectrolytes into the wood surface.

2. Materials and Methods

2.1. Preparation of Wood Samples

Samples of beech wood (*Fagus moesiaca* C.) with dimensions of 225 mm × 30 mm × 10 mm were obtained by cutting from boards and then conditioned for two months in a climate chamber with controlled environmental parameters ($t = 15 \pm 1$ °C and $\varphi = 50 \pm 2\%$). All samples were prepared from quarter-sawn boards (radial cut) to exclude the influence of the anatomical direction on the test results, taking into account its influence on the differences in swelling of the wood in contact with water. After the conditioning phase, the average wood moisture content was 9.53% (measured according to ISO 13061-1:2014), which corresponds to the percentage of moisture content of the wood in wood end-use products. After conditioning, the samples were sanded on a wide-belt sanding machine (LSM 8, Heesemann, Germany) in a four-stage system with the following sanding belt grits: P60, P80, P100 and P150. The sanding speed was 20 m/s and the conveyor belt speed was 16.5 m/min. Before applying the polyelectrolyte solution, the surface of the samples was sanded manually using a hand sander and the abrasive belt grit size P150 to obtain a fresh surface. Smaller test samples were prepared with the dimensions: 90 mm × 30 mm × 10 mm; 50 mm × 30 mm × 10 mm and 30 mm × 30 mm × 10 mm, for various subsequent tests.

Aqueous PEI solutions of low and high molecular weight with similar pH values were used for the surface treatment of the samples. The specifications of the polyelectrolytes used are listed in Table 1.

Table 1. Specification of used polyelectrolytes.

Polyelectrolyte Type	Manufacturer	Molecular Weight, g/mol	pH	Abbreviation	Concentration, %
Polyethyleneimine (PEI)	Sigma-Aldrich, St. Lous, MO, USA	1800	12.0	PEI LMW	0.5 1 2
	MP Biomedicals, Illkirch, France	50,000–100,000	10.5–11.0	PEI HMW	0.5 1 2

Deionized water with an electrical conductivity of approx. 0 µS/cm was used to prepare the PEI solutions. The diluted solutions were homogenized with a magnetic stirrer for 2 h at a speed of 1000 rpm. The chosen concentrations are consistent with previous tests on cellulose fibers and veneer surfaces [9,14]. The polyelectrolyte solutions were manually applied to the surface of the samples using a sponge-coated roller.

After drying the samples under controlled ambient conditions (t = 23 ± 2 °C and φ = 50 ± 5%), WTAC ("AQUAL—basecoat for furniture", Pitura, Belgrade, Serbia) was applied to the surface using a manual applicator with a fixed gap height at a theoretical wet film thickness of 200 µm. The basic properties of the applied liquid coating were tested and are shown in Table 2.

Table 2. Properties of liquid WTAC.

Density, g/mL (EN ISO 2811-1: 2016) [22]	Solids Content, % (EN ISO 3251: 2019) [23]	Flow Time, s (EN ISO 2431: 2012) [24]	Apparent Viscosity, mPa × s (EN ISO 2555: 2018) [25]	pH Value	Surface Tension, N/m (ISO 304: 1985) [26]
1.024	31.3	55.5	3460	7.34	0.02815

2.2. Test Methods for Wood Treated with PEI Solution

2.2.1. Determination of the Chemical Characteristics of the Wood Surface After Treatment with Polyelectrolyte Solution

The changes in the chemical properties of the wood surface were investigated using FT-IR spectroscopy (MB-102 FTIR, BOMEM Michelfan, Frankfurt, Germany spectrophotometer in transmission mode). FT-IR images of untreated and PEI solution-treated wood samples were obtained by homogeneously mixing a small amount of finely dispersed wood flour with potassium bromide (KBr) and pressing it into a thin round plate. The wood powder that was used to form the pastille was obtained by sanding the surface layer of untreated and treated samples.

2.2.2. Determination of Penetration Depth of PEI into Wood

The position of the polyelectrolyte in the wood tissue after application was determined with an EPI fluorescence microscope (Axio Observer Z1, Carl Zeiss Microscopy, Jena, Germany) by staining the polyelectrolyte with a fluorescent agent. Rhodamine B (Centrohem, Stara Pazova, Serbia) was used as a fluorescent staining agent due to its ability to form covalent bonds with PEI molecules. Staining of the polyelectrolytes with a fluorescent dye was performed according to a procedure similar to that described in the literature [27]. In total, 0.042 g of rhodamine B was dissolved in 1 mL of deionized water; then, 0.035 g of PEI LMW dissolved in 3 mL of deionized water was added to the solution. In this way, a solution with a concentration of rhodamine five times higher than that of PEI was obtained. The pH of the solution was adjusted to 9 by adding hydrochloric acid solution. The solution was stirred for 24 h with a magnetic stirrer. Excess rhodamine B was removed by dialysis using a dialysis device (Float-a-Lyzer, Spectrapor, Massachusetts, CA, USA) with a dialysis membrane whose pore size corresponds to the MWCO (Molecular Weight Cut Off) of the polyelectrolyte of 500–1000 Da for two weeks, with deionized water being exchanged every 24 h as a buffer (Figure 1). During the dialysis process, the buffer solution changes color due to the release of excess unbound dye. Dialysis of PEI HMW was performed using dialysis cassettes (Slide-a-LyzerTM, Thermo Scientific, Waltham, MA, USA) with a MWCO pore size of 10 kDa. Due to the larger difference in the size of the membrane aperture and the size of the molecules of the samples, PEI HMW dialysis was performed for 48 h, with a buffer change after 4 h, 8 h, and 16 h.

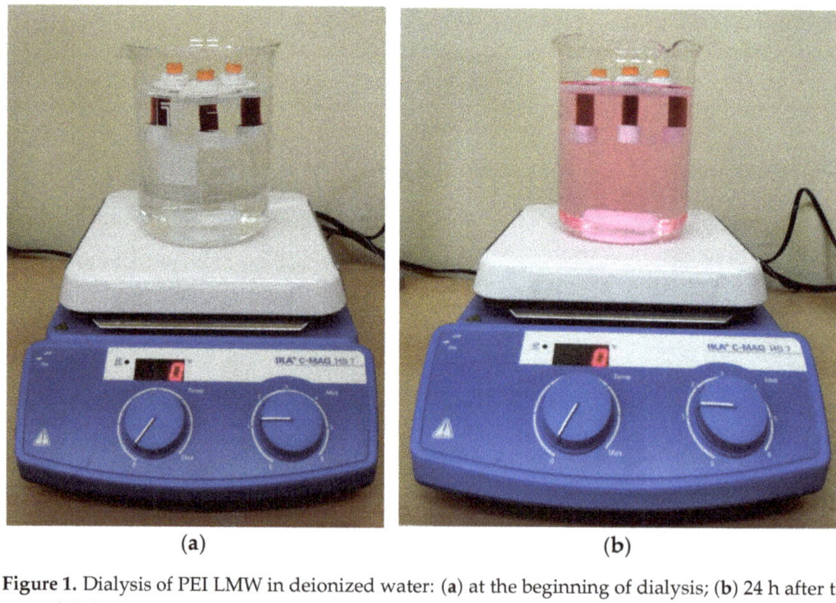

(a) (b)

Figure 1. Dialysis of PEI LMW in deionized water: (**a**) at the beginning of dialysis; (**b**) 24 h after the start of dialysis.

After completion of dialysis, aqueous solutions of polyelectrolytes were prepared at the desired concentrations and applied to wood samples measuring 50 mm × 30 mm × 10 mm. After drying, the surface of the samples was covered with self-adhesive tape and then coated with WTAC to prevent damage to the surface of the treated samples′ during preparation of the microtome samples. For better visual observation of the polyelectrolytes in the wood tissue cells, three filters were used to image the samples: 49 DAPI (excitation at

358 nm; emission at 461 nm), FAM (excitation at 490 nm; emission at 520 nm) and DsRED (excitation at 563 nm; emission at 581 nm). The application of the DAPI filter leads to autofluorescence of the wood cell wall, which is primarily due to the fluorescence of lignin as the main fluorescent component [28], which occurs at around 445 nm with an excitation of 360 nm [29]. The individual images taken with these three filters were combined into a single photo (using Photoshop). Finally, photos were taken in RBG mode with the red color assigned to the DsRED filter, the blue color to the FAM filter, and the green color to the DAPI filter.

2.2.3. Determination of the Binding Position of Polyelectrolytes in the Wood Structure After Adsorption

A scanning electron microscope (SEM) (Tescan Vega TS 5130MM, Brno, Czech Republic) equipped with a back-scattered electron (BSE) detector was used to qualitatively evaluate the penetration depth of the polyelectrolytes. The SEM images were taken on representatives of PEI samples with a 1% concentration of different molecular weights (LMW and HMW).

The SEM images were taken on previously prepared microtome samples. The preparation of the samples for SEM analysis consisted of gluing the microtome samples to metal carriers with conductive double-sided carbon adhesive tape. The glued supports are coated with a very thin layer of gold (approx. 10–20 nm) by means of argon ionization under high voltage. The impact of argon ions on the cathode, i.e., the gold-coated chamber shell, causes gold atoms to break off and coat the sample. Due to the higher energy compared to secondary electrons, "backscattered" electrons can penetrate deeper into the surface of the sample. To investigate the effect of the polyelectrolyte treatment, an image was therefore generated by detecting bounced elastic electrons.

2.2.4. Determination of Wetting Ability of Wood Surface Treated with PEI Solution

To assess the wettability of the treated surface of the samples with a water-based coating, the contact angle was measured using the drop method according to EN 828: 2009 ("sessile-drop test") of water and WTAC droplets and then the surface energy of the wood was determined on the basis of the measured contact angles of liquids with known surface tension—water, formamide and diodomethane (according to Chibowski and Perea-Carpio (2002) in [30]). Liquid droplets with a volume of approx. 2 μL were dropped from a low height relative to the substrate surface. For each group, 5 test specimens (dimensions 90 mm × 30 mm × 10 mm) were selected on which four drops of the tested liquids were placed. The contact angle was measured in the direction of the wood fibers using the images taken with a video camera through an optical microscope (Olympus SZH, Olympus, Tokyo, Japan) with an objective (Olympus DF PLAN, Olympus, Tokyo, Japan) with a magnification of 1.5. The change in contact angle compared to the initial position, i.e., the zero position, was tracked for 25 s for water and water-based coating.

2.2.5. Determination of Water Uptake of Wood Surface Treated with PEI Solution

The degree of water uptake was determined by placing water droplets on the surface of test samples measuring 30 mm × 30 mm × 10 mm. For each group of samples (control, treated with PEI LMW, and treated with PEI HMW), nine samples were used, and the narrower sides of the test tubes were coated with a coating to ensure water uptake only from the wide side of the samples. Five droplets of water (total approximate mass 0.2 g) were placed on each sample, using a syringe (TROJECTOR-3, TROGE, Hamburg, Germany) with an injection needle (0.4 mm × 19 mm). Excess water was removed from the surface of the sample with absorbent material. The degree of water uptake was expressed as a percentage of the increase in the mass of the sample after a certain time concerning the total mass of applied water droplets, according to the equation:

$$a = \frac{m_0 + m_{aw}}{m_0 + m_w} \cdot 100\ [\%] \quad (1)$$

where a—degree of water uptake; m_0—the initial mass of the sample [g]; m_{aw}—the mass of water absorbed by the sample [g]; m_w—the mass of water applied to the surface of the samples [g].

Three test samples were used for each group to determine water intake at different time points: after 1 min, after 2, and after 5 min.

2.3. Methods of Testing the Properties of Dry Film Coating on the Surface of Wood Treated with PEI Solution

2.3.1. Dry Film Thickness of Coating

The dry film thickness (DFT) of WTAC was measured using an ultrasonic gauge (according to EN ISO 2808: 2019 [31]). For each group of samples, the DFT was measured at 30 positions (Figure 2).

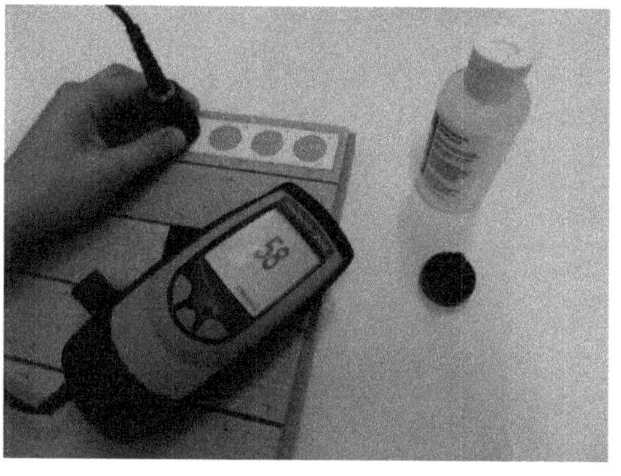

Figure 2. Dry film thickness measurement of water-based transparent acrylic coating.

2.3.2. Coating Penetration Depth

To determine the penetration depth of WTAC, the technique of superimposing images taken with an EPI fluorescence microscope with different filters was used. (The same method was used to determine the penetration depth of polyelectrolytes). The quantitative determination of the coating penetration depth was carried out with the computer program Image J (version 1.46). For each sample group, 30 microtome samples were prepared and photographed with EPI fluorescence microscopy. The total number of photomicrographs was 210.

The maximum penetration depth of the coating (D_{max}) and the average penetration depth of the coating (D_{av}) were measured on each micrograph. For the calculation of the penetration depth parameters, the coating penetration surface (PS) is defined by the initial penetration plane passing through the highest point of the wood section (according to Van den Bulcke et al. (2010) in [32]) and the plane parallel to a previous plane passing through the lowest point where the coating is visible (Figure 3). Within the PS of each micrograph, the area of filled cell lumens (A_{LF}) and the area of available cell lumens

(A_{LA}) were calculated. For the calculation of D_{av}, each photomicrograph was divided into 30 sections of equal width for which the penetration depths were measured. D_{av} was calculated as the mean of the 30 results for each photomicrograph.

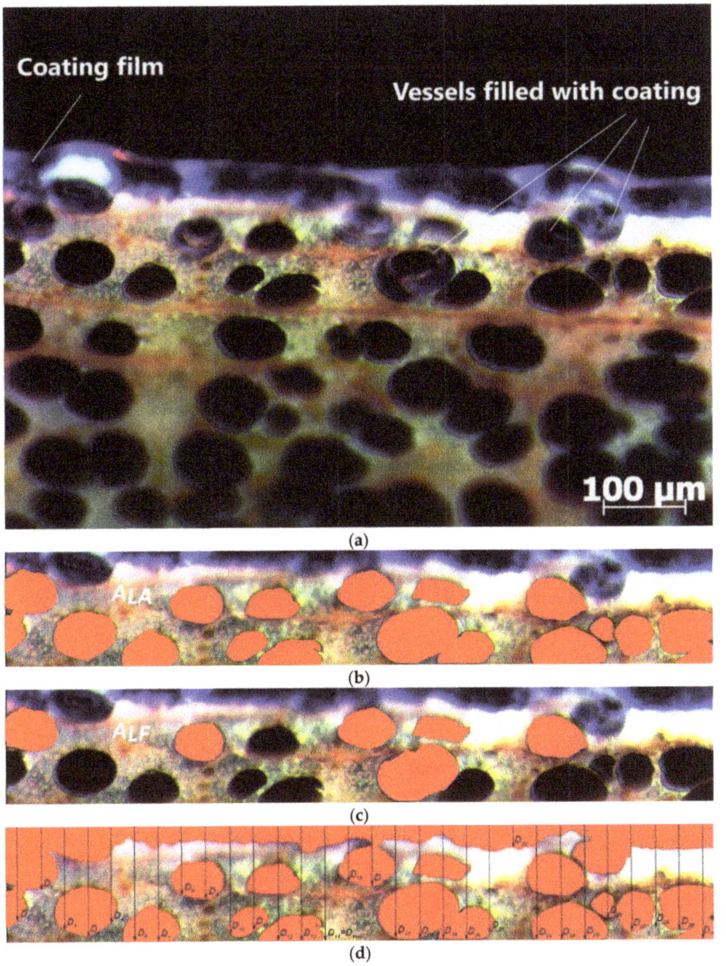

Figure 3. Calculation of coating penetration parameters: (**a**) snapshot of microtome sample; (**b**) area of available cell lumens (A_{LA}) in the penetration surface (PS); (**c**) area of filled cell lumens (A_{LF}) in the penetration surface (PS); (**d**) coating penetration depth at 30 positions and maximum coating penetration depth (D_{max}) in the penetration surface (PS).

For each recording, the filling of the available cell lumens (LF) with the coating within the coating penetration zone was calculated according to the formula:

$$LF = \frac{A_{LF}}{A_{LA}} \cdot 100\ [\%] \quad (2)$$

2.3.3. Adhesion Strength of Coating

The adhesive strength of WTAC was measured by the pull-off test according to EN ISO 4624: 2023 [33] using an automatic device with a hydraulic pump (POSITest AT-A, DeFelsko,

Ogdensburg, NY, USA). Dollies with a diameter of 20 mm were glued to the coated surface of the samples using a 2K epoxy adhesive (Epoxy Universal, Bison, Zoetermeer, The Netherlands). The adhesive cured for 24 h under room climate conditions. After the adhesive had cured, the coating was cut around the circumference of the dollies which were then pulled in a vertical plane at a speed of 0.7 MPa/s until breakage (Figure 4). The adhesion strength of WTAC was measured for each sample group on 5 samples measuring 225 mm × 50 mm × 10 mm at six positions, i.e., a total of 210 measurements. The results of adhesion strength with the failure mode outside the adhesive coating plane were not considered in the statistical analysis.

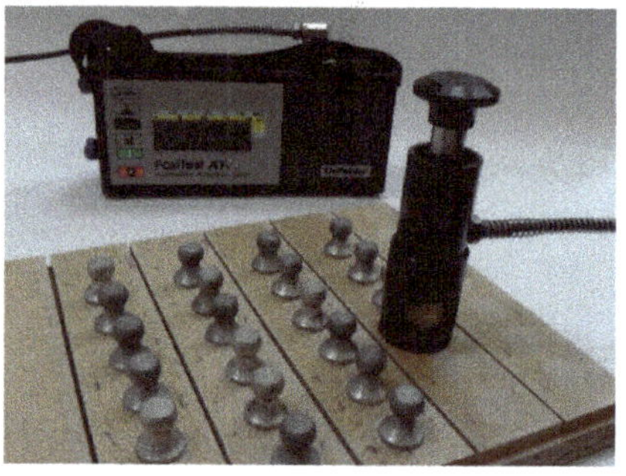

Figure 4. Determination of WTAC adhesion strength by the pull-off test.

2.3.4. Roughness of the Coated Surface

A stylus contact tester (TimeSurf TR200, Beijing TIME High Technology Ltd., Beijing, China) was used to determine the surface roughness of coated samples (Figure 5). The device is equipped with a diamond-tipped needle (diameter 2 μm) that touches the surface with a force of 4 mN. The following surface roughness parameters were used to characterize the coated surface: R_a (arithmetic mean deviation of the profile), R_z (maximum profile height) and R_t (total height of the profile) according to ISO 4287: 1997 [34], with a selected sampling length of 2.5 mm. The roughness was measured on samples with the dimensions of 225 mm × 50 mm × 10 mm at six positions on five samples, i.e., at a total of 30 measuring positions for each sample group.

Figure 5. Measuring the roughness of the wood-coated surface.

2.4. Statistical Analysis

The results of the parameters DFT, adhesion strength, surface roughness and WTAC penetration dept were analyzed with SPPS IBM Statistics 20. These results were compared using analysis of variance (ANOVA) to determine whether there was a statistically significant difference between the result groups. The basis of ANOVA is the F-statistic, which determines the relationship between the variability within the variables (also known as error) and the variability between the variables being compared (also known as systematic variance). If the assumption of homogeneity of variances was met (by the Levine test), the ANOVA was performed. If the ANOVA revealed a statistically significant difference between the sample groups, a post hoc analysis was performed (Tukey's HSD) to clarify between which groups significant differences were found. In cases where the assumption of homogeneity of variances was not met (different variances), the test results were checked using the Welch test. A post hoc test (Games–Howell) was used to determine which group differed from the others if the Welch test showed a statistically significant difference in the results between the groups. All tests were performed with a confidence level of 95% ($p < 0.05$).

3. Results and Discussion

3.1. FTIR Analysis

FT-IR spectra show the presence of a broad absorption band in the range of 3600–3000 cm^{-1} (Figure 6a), which is attributed to the vibrations of hydroxyl groups (–OH) [21] and water adsorption [27]. In samples treated with polyelectrolyte solutions, the intensity of this band was reduced compared to untreated control samples. The decrease in intensity can be explained by the breaking of hydrogen bonds between the cellulose and hemicellulose chains during the penetration of PEI into the cell walls. Hydrogen bonds occur between neighboring cellulose chains as well as between hemicellulose molecules and cellulose chains in the amorphous region. According to Gibson [29], during the sorption and desorption process, water molecules move rapidly through the material from one free-binding site to another, followed by the formation of new hydrogen bonds and the breaking of old ones. During the swelling of wood, there is a spatial rearrangement of neighboring cellulose molecules, while during desorption, the tension created by the loss of water molecules reduces the distance between the newly formed regions. During water loss, the reduction in the distance between the cellulose and hemicellulose chains enables the formation of stronger hydrogen bonds between them.

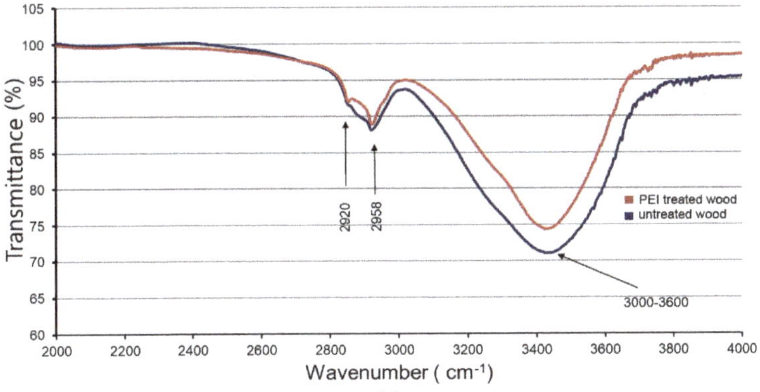

(a)

Figure 6. *Cont.*

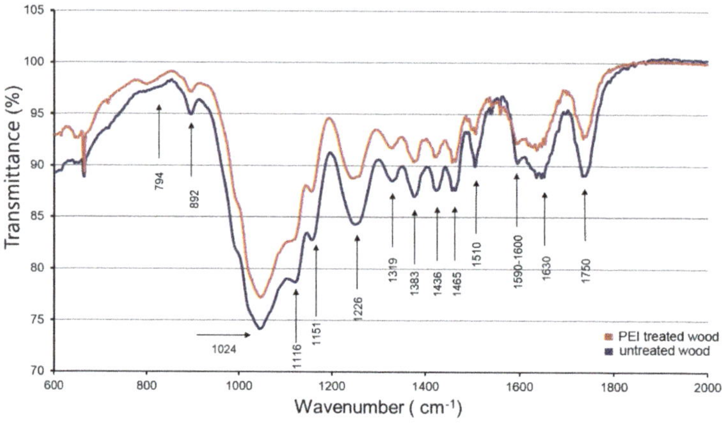

(**b**)

Figure 6. FT-IR spectra of untreated and samples treated with PEI solution in the range of 2000–4000 cm^{-1} (**a**) and from 600 to 2000 cm^{-1} (**b**).

In the samples treated with PEI solution, new hydrogen bonds are probably formed between the electronegative atom N and the protons of the wood cell wall components. In addition, hydrogen bonds can be formed between cellulose molecules in the amorphous region (intermolecular hydrogen bonds) and between parallel cellulose chains in the crystalline region (intramolecular hydrogen bonds). The vibration of the absorption band around 3335 cm^{-1} indicates intramolecular hydrogen bonds (according to Liang and Marchessault (1959) in [30]). The formation of hydrogen bonds between parallel cellulose chains in the crystalline region is possible due to the separation of the planes of parallel cellulose chains in an alkaline medium, whereby the crystal lattice is not disturbed but only transformed [32]. The high pH of the PEI solution (pH = 10) during treatment has an effect on the reorganization of the crystal structure and consequently on the swelling of the cell wall.

The absorption bands at 2958 cm^{-1} and 2920 cm^{-1} show characteristic peaks, which are attributed to asymmetric and symmetric stretching vibrations of CH and CH_2 groups, respectively [21]. A large number of clearly defined peaks can be observed in the 1800–1600 cm^{-1} range. The intensity of the vibrational band at 1750 cm^{-1}, which is due to the stretching of the non-conjugated C=O bond in the carboxyl groups present in xylan [35], is lower in the samples treated with PEI solution, indicating a reduction in the content of carboxyl groups (Figure 6b). On the other hand, the appearance of a peak at about 1650–1580 cm^{-1} corresponds to NH bending in amide bonds (type II), indicating the chemical bonding of carboxyl groups of wood and charged nitrogen atoms of PEI. The formation of amide bonds was observed in the reaction of cellulose with PEI in a molar ratio of 4:1 [36].

The bands below 1600 cm^{-1} show no significant changes (Figure 6b) after treatment with PEI. The weak FT-IR band at 790 cm^{-1}, which is attributed to the out-of-plane vibration of the NH ion, shows that PEI is only involved to a certain extent in the chemical bonding with the carboxyl groups of cellulose.

Analysis of the FT-IR spectra led to the conclusion that PEI molecules have an affinity for hydrogen bond formation due to the presence of nitrogen atoms, as well as the possibility of chemical bonding with the carboxyl groups of the wood. Consequently, there is not only the possibility of forming electrostatic interactions with oppositely charged atoms but also the ability to form bonds outside the electrostatic field.

3.2. Penetration Depth of PEI Solution

Figure 7 shows photographs of the penetration of PEI LMW and PEI HMW solutions of different concentrations obtained by computer overlay of images taken with three different filters (DAPI, FAM and DsRED). In the PEI-treated samples, a higher concentration of polyelectrolytes in the surface layers of the wood was observed in higher molecular weight samples (Figure 7d–f), compared to samples treated with a low-molecular-weight solution of this polyelectrolyte (Figure 7a–c). These results are consistent with previous studies on the influence of polymer molecular weight on penetration into the wood cell wall, in which it was found that only low-molecular-weight PEI (800 Da) could sufficiently penetrate the cell walls, while low- and high-molecular-weight PEI (800 Da and 750,000 Da, respectively) coated the lumens at high density [35]. It was observed (Figure 7) that the penetration depth of the PEI solution increased with increasing concentration regardless of the molecular weight. A lower penetration depth of the polyelectrolyte was observed in samples where the cell rays were located on the surface itself or in several rows of cells below the surface. This phenomenon can be explained by the transport paths of the matter through the wood tissue.

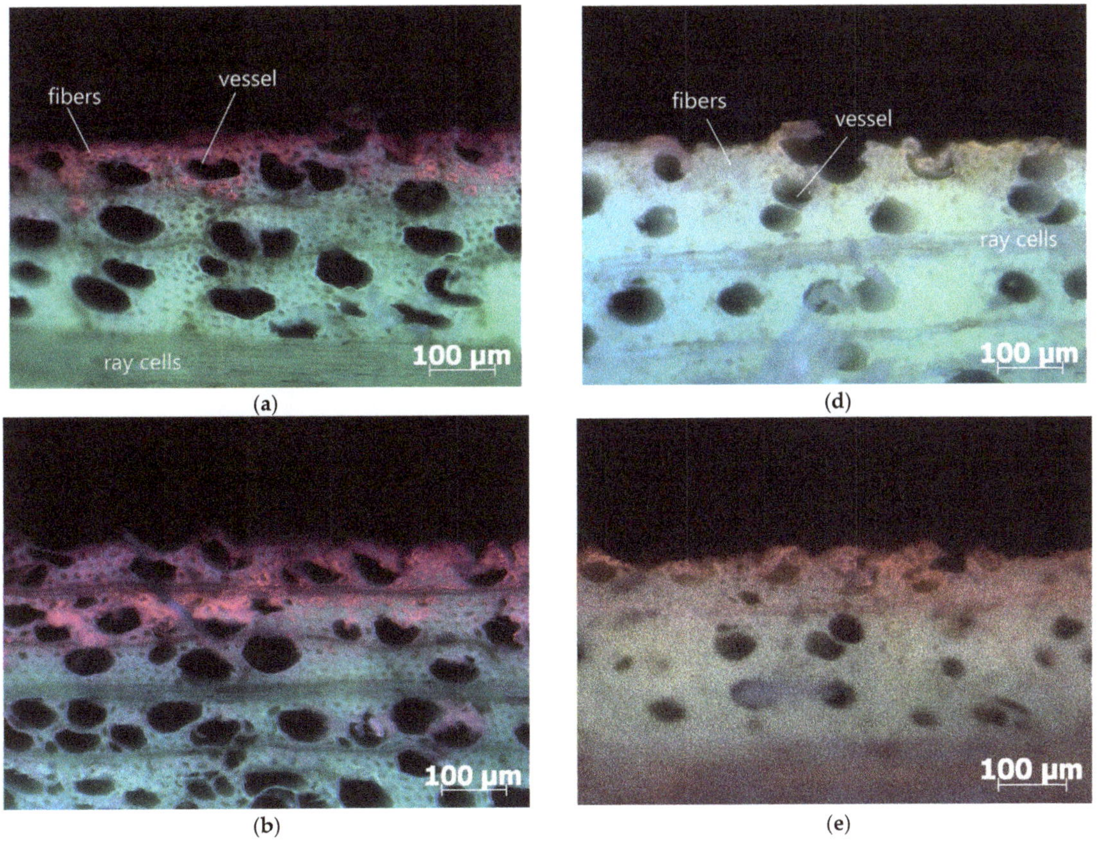

Figure 7. *Cont.*

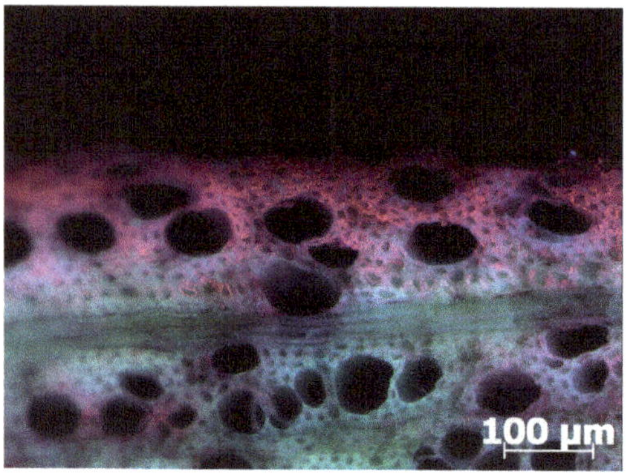

(c)

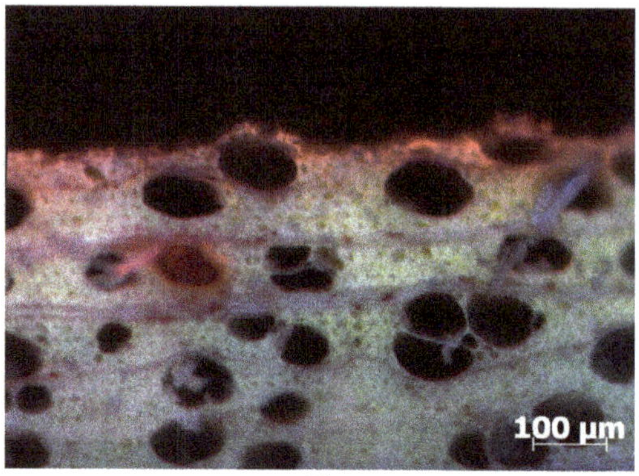

(f)

Figure 7. Images of penetration of low- (LMW) and high-molecular-weight (HMW) PEI solutions of different concentrations, stained with rhodamine B, into the surface layer of wood, obtained by overlaying images with DAPI, FAM, and DsRED filters from an epi-fluorescence microscope: (**a**) 0.5% PEI LMW; (**b**) 1% PEI LMW; (**c**) 2% PEI LMW; (**d**) 0.5% PEI HMW; (**e**) 1% PEI HMW; (**f**) 2% PEI HMW.

On the SEM image of untreated beech wood (Figure 8a), two main components can be distinguished: larger cells (vessels) with large lumens and thin walls and smaller cells (fibers) with thick walls. Fluid transport between the cells takes place through pits that connect neighboring individual cells. According to studies of the porosity of beech wood, the pore size of the fiber lumens is in the order of 0.5 to 2 µm [36]. The untreated samples (Figure 8a) and the samples treated with 1% PEI LMW or PEI HMW solutions (Figure 8b or Figure 8c) differ significantly in the thickness of their fiber walls. The change in the conformation of the cellulose molecules can be seen on a microscopic level by the swelling of the wood cell wall, which is confirmed by SEM images (Figure 8b,c). In addition, a "smoother texture" of the wood fibers can be seen in the surface layer of the treated wood samples. This phenomenon indicates that the cell walls were filled (bulking) with polyelectrolytes, which led to an increase in the volume of the walls themselves. The swelling of the fiber walls can be observed not only when the samples were treated with PEI LMW (Figure 8b), but also when they are treated with PEI HMW (Figure 8c). This result indicates that the swelling of the fibers is related to the water uptake from the aqueous PEI solution through hydrogen bonding with the hydroxyl groups of the wood cell wall polymers [37]. The samples treated with polyelectrolyte solutions (Figure 8b,c) showed a clear difference in the cell structure of the wood fibers in the form of thicker walls and narrower lumens in the surface layers of the wood compared to those in the deeper layers of the sample. This result confirms that PEI penetrates into the wood, as shown in the EPI images (Figure 7a–f).

Regarding the penetration of the WTAC coating into the beech wood samples, the filling of the vessels is only visible in individual cells of the first row of the untreated and treated samples. This result is consistent with other studies in which the penetration depth of water-based acrylic coatings into wood surfaces is relatively low compared to solvent-based coatings [38,39]. In the study on beech wood, the penetration depth of water-based acrylic coatings was limited to the surface layer of the wood cells (a few tens of micrometers); only individual lumens of larger vessels below the surface on the investigated surfaces were filled with coating [40]. The low penetration depth of the coating can be attributed to the high viscosity of the liquid coating due to the application technique used, which affects the spreading rate and uniformity of the applied coating on the surface [41].

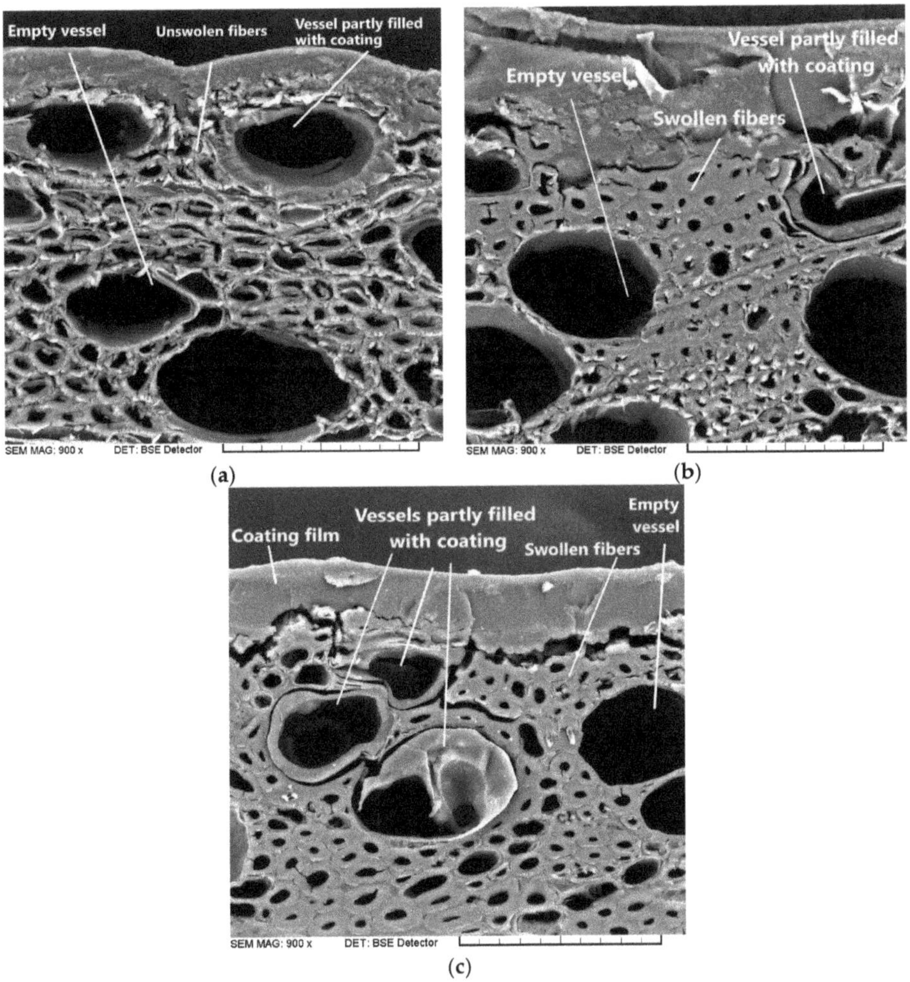

Figure 8. SEM images of wood samples protected by a coating layer (900× magnification): (**a**) untreated wood; (**b**) samples treated with a solution of 1% PEI LMW; (**c**) samples treated with a solution of 1% PEI HMW.

3.3. Characterization of the Surface of the Treated Samples

The results of the change in the contact angle of the water droplet over time on the surface of the untreated samples and samples treated with PEI solutions are shown in the graph (Figure 9).

In general, the value of the contact angle decreases with time, which is due to the spreading of the water droplets and the uptake of the water in the wood [42,43]. According to a previous study in which the water uptake of oak veneers treated with different polyelectrolyte solutions was investigated, the decrease in the water contact angle of the treated samples over time is to a greater extent due to the evaporation of water from the surface of the samples and to a lesser extent due to the uptake of water into the wood [13].

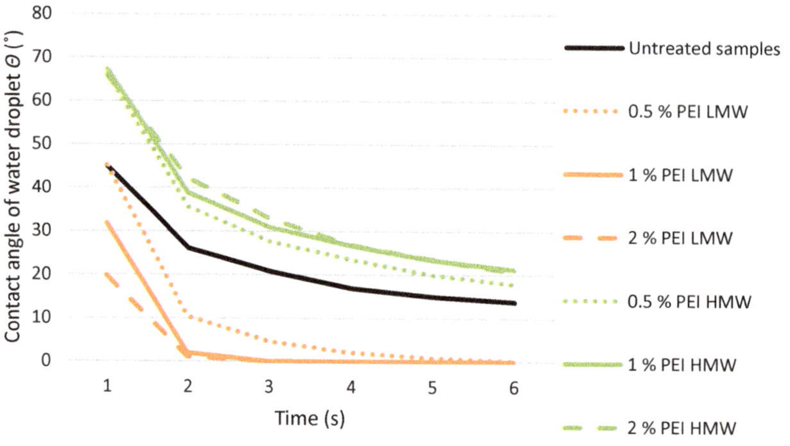

Figure 9. The contact angle of a water droplet on the surface of untreated samples and samples treated with PEI solution of low (LMW) and high (HMW) molecular weight of different concentrations, during the observation period from 1 to 25 s.

The water-wetting of the surface of samples treated with a polyelectrolyte solution depends on the molecular weight of the dissolved PEI polyelectrolytes. Samples treated with a PEI LMW solution showed lower contact angle values and thus better wetting with water. One of the factors determining the wetting of liquid droplets on the wood surface is the surface structure of the substrate [43,44]. After treating the wood with an aqueous solution of polyelectrolytes, the roughness of the wood surface increases, which probably influenced the reduction in the contact angle, as it is known that the roughness of the wood promotes the wetting process [45].

Higher water contact angle values were measured on the samples treated with HMW solution compared to the untreated samples, which can be explained by differences in the penetration depth and location of PEI accumulation with higher molecular mass in the wood surface layer. This finding confirms the results of previous studies that an increase in PEI molecular weight reduces the wettability of biofilms on agar substrate [46]. The higher values of the water contact angle on the surface of veneer samples of white pine (*Pinus sylvestris*) and oak (*Quercus robur*) treated with different polyelectrolyte solutions were attributed to the change in the wood surface [13].

The results of the change in the contact angle of the WTAC droplet as a function of time are shown in the graph (Figure 10).

In contrast to the results of wetting with water, wetting with WTAC had almost the opposite effect on the samples treated with PEI solutions of different molecular weights. When comparing all treated sample groups, the samples treated with 1% PEI HMW solution had the lowest values of WTAC droplet contact angle on the wood surface, which at the same time had the highest values of water droplet contact angle (both at the beginning and after 5 min of observation). Even with untreated wood, the contact angle of the coating droplet remains larger than the contact angle of the water droplet during the observation period. This result is unexpected considering that the surface tension of water (0.072 N/m) is significantly higher than the surface tension of the coating (0.02815 N/m). It is possible that the density and viscosity of the undiluted WTAC used are the reason for the high value of the contact angle of the coating droplet, as well as the fact that the evaporation rate of the water from the coating droplet is significantly slower than that of the water droplet. These results confirm the conclusion from previous studies that the wetting of the wood surface with water should not be regarded as an indicator of the quality of the wetting of the wood surface with water-based coatings [47].

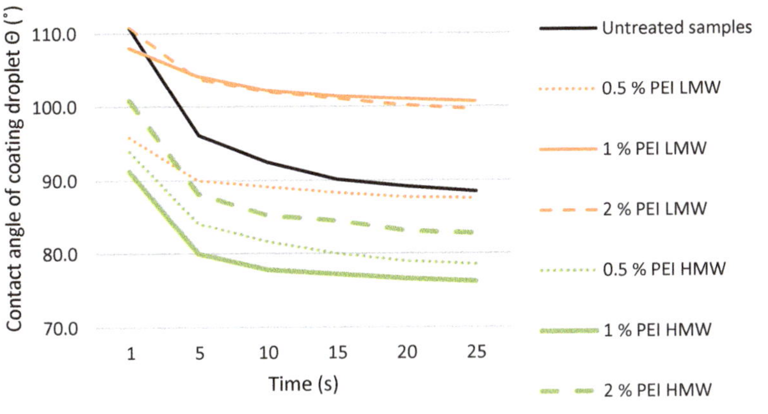

Figure 10. The contact angle of the WTAC droplet on the surface of untreated samples and samples treated with PEI solution of low (LMW) and high (HMW) molecular weight of different concentrations, during the observation period from 1 to 25 s.

The surface energy results are given in Table 3.

Table 3. Results of total surface energy (γ) of untreated samples and samples treated with PEI solutions of different MW and concentration.

Group of Samples		Concentration, %	Surface Energy (γ), N/m
Untreated samples		/	0.0587
Treated samples	PEI LMW	0.5	0.0576
		1	0.0575
		2	0.0550
	PEI HMW	0.5	0.0489
		1	0.0451
		2	0.0416

The samples treated with the PEI HMW solution showed a significant reduction in total surface energy compared to the untreated wood (16.7, 23.2% and 29.1% for concentrations of 0.5%, 1% and 2%, respectively). The highest values of the contact angle of the water droplets were also calculated for the same sample groups. Considering that the method for determining the surface energy is based on measuring the contact angle of liquid droplets with known surface tension, it is clear that lower values of surface energy can be explained by higher values of the contact angle of water droplets on the surface of the treated samples. In previous research [46], polyelectrolyte coating is described as a simple and efficient method to control surface energy, but our results show that this is only true for high-molecular-weight polyelectrolytes.

The results of water uptake are shown in the graph (Figure 11).

All samples treated with polyelectrolyte solution showed a lower degree of water uptake compared to the untreated samples. It is expected that during sample preparation by cutting, milling and especially sanding, new fibrillar structures are opened in the surface layer of the wood, forming new pores between the fibril sections [3]. When water evaporates from the swollen wood fibers, it is possible that the newly formed pores between individual fibers close, which reduces water uptake when rewetting with water from a subsequent water-based coating. The occurrence of irreversible pore closure is associated with water evaporation during the initial drying of wood in the green moisture state (after cutting) [48] but has also been observed during the removal of water from the cell walls of wood fibers in the papermaking process [49]. In addition to a possible pore closure, the swollen wood fibers in the surface layer of the wood physically block the penetration of water into the deeper layers of the wood tissue, which certainly influences the degree of water uptake [3].

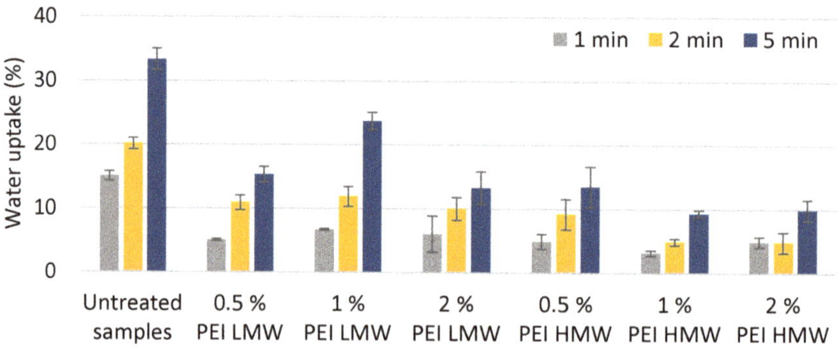

Figure 11. Water uptake of untreated samples and surface treated samples with PEI LMW and PEI HMW solutions of different concentrations (0.5%, 1%, and 2%).

The analysis of the effect of the treatment of polyelectrolyte solutions with different molecular weights on the degree of water uptake showed that when PEI HMW is used, lower water uptake of the treated samples can be obtained than in samples treated with the same type of polyelectrolyte but with a low molecular weight. This result is consistent with the conclusions of previous studies [13], according to which the water uptake of pine veneers treated with polyelectrolytes was more pronounced with polyelectrolytes of higher molecular weight. Since high-molecular-weight polyelectrolytes can only penetrate the cell wall to a limited extent (Figure 8c), it is assumed that high-molecular-weight polyelectrolytes coat the lumens of the cell walls and form a barrier for the penetration of water molecules into the interior of the wood tissue, which is also confirmed by epifluorescence microscope images.

The results of the degree of water uptake showed no clear correlation between the concentration of the polyelectrolyte in the solution and the degree of water uptake of the samples treated with the selected solution. The lowest water uptake was observed in the samples treated with 1% PEI HMW (78.7% and 72% lower uptake of water droplets compared to untreated samples after 1 and 5 min of observation), confirming that the concentration of the polyelectrolyte solution is a key parameter for optimizing the polyelectrolyte solution for minimal water uptake on wood surfaces of certain species [13].

The treatment with PEI LMW reduced the water uptake of the beech wood samples only at the highest concentration in the solution of 2% when observing the long-term effect (after 5 min observation time). In addition, it was found that a 2% concentration of PEI solution resulted in a more stable increase in the mass of the samples in contact with water, regardless of the MW of PEI.

Reduced water uptake can be of particular importance when using wood outdoors, where the resistance of the coated surface to liquid water determines the use category of the product (stable, semi-stable or unstable end product according to EN 927-2: 2014 [50]).

3.4. Results of the Dry Coating Film on the Surface of Wood Treated with PEI Solution

The results of the dry film thickness WTAC on untreated and treated samples are shown in Table 4.

Table 4. Dry film thickness of WTAC of untreated samples and samples treated with PEI of different MW and concentration.

Dry Film Thickness (DFT) of WTAC *, μm						
Untreated samples	Treated samples					
	PEI LMW			PEI HMW		
	Concentration, %			Concentration, %		
	0.5	1	2	0.5	1	2
57.80 (5.27)	64.57 (4.17)	58.07 (6.05)	55.63 (7.50)	61.80 (4.50)	59.73 (3.11)	69.83 (6.35)

* Values in parenthesis are standard deviations.

A one-way ANOVA showed that there was a statistically significant difference in DFT between untreated and PEI LMW and PEI HMW treated samples ($F(6,203) = 23.65$ at $p < 0.5$). The Games–Howell post hoc analysis revealed that the samples treated with 2 % PEI HMW had significantly higher DFT compared to the other sample groups. These results are consistent with epifluorescence microscope images showing that high-molecular-weight PEI is retained on the wood surface and forms a barrier to coating penetration (Figure 7d–f). The DFT of the samples treated with a 0.5% PEI solution with both molecular weights was slightly lower, but still significantly higher than the DFT of the untreated samples and all other groups of treated samples. These results indicate that at the low concentration, the water from the polyelectrolyte solution had a preeminent effect on the PEI, resulting in a higher DFT due to the narrowing of the penetration pathway and grain raising of the wood fibers on the surface.

3.5. Penetration Parameters of WTAC

Table 5 shows the basic parameters of WTAC coating penetration for untreated samples and samples treated with PEI of different MW and concentrations.

One-way ANOVA showed significant differences between the D_{max} of WTAC between untreated samples and samples treated with PEI of different MW and concentrations ($F(6,195) = 11.246$ for $p < 0.5$). The post hoc Tukey test revealed that the group of samples treated with PEI HMW of all concentrations and with 2% PEI LMW had a statistically lower penetration depth compared to the untreated samples.

The groups of treated samples that had a lower maximum penetration depth also had a lower average penetration depth ($F(6,195) = 14.237$ at $p < 0.5$). As shown in Table 4, the untreated samples with the highest penetration depth had the lowest DFT. The lower penetration depth of the water-based coating in the treated samples can be attributed to the swelling of the cell walls of the wood and the associated narrowing of the penetration paths, which was confirmed on the SEM images (Figure 8b,c). The average penetration depth (D_{av}) of WTAC decreased by 36% when the wood surface was treated with 1% PEI HMW. When the samples were treated with lower concentrations of PEI LMW (0.5 and

1%), the penetration depth (both D_{max} and D_{av}) was not statistically lower compared to the untreated samples.

Table 5. Penetration parameters of WTAC untreated and samples treated with PEI solution of different concentrations: Maximum penetration depth (D_{max}); average penetration depth (D_{av}) and lumen filling (*LF*).

Samples		Concentration, %	Penetration Parameters of WTAC		
			D_{max} *, µm	D_{av} *, µm	*LF* *, %
Untreated samples		/	84.18 (19.40)	43.31 (11.67)	37.6 (18.8)
Treated samples	PEI LMW	0.5	80.05 (18.17)	48.33 (13.26)	29.5 (16.4)
		1	68.65 (21.60)	37.55 (11.17)	37.8 (19.6)
		2	56.95 (16.71)	32.20 (8.36)	55.3 (21.3)
	PEI HMW	0.5	57.56 (11.59)	33.21 (5.86)	40.3 (22.1)
		1	66.12 (19.27)	30.13 (8.37)	32.3 (16.5)
		2	62.35 (13.44)	35.46 (6.53)	47.7 (15.8)

* Values in parenthesis are standard deviations.

With regard to the filling of the available cell lumens (parameter LF), there was no difference between untreated and polyelectrolyte solution-treated samples ($F(6,163) = 1.007$ at $p = 0.423$). The statistical homogeneity of the LF values confirms that the possibilities of penetration of the coating into the wood depth were uniform in all sample groups so the influence of structural elements on the penetration depth can be excluded.

3.6. The Adhesion Strength of WTAC Coating

Table 6 shows the results of the adhesion strength of the WTAC in the untreated and treated samples.

Table 6. The adhesion strength of WTAC on the surface of untreated and samples treated with PEI of different MW and concentration.

Samples		Concentration, %	Adhesion Strength *, MPa
Untreated samples		/	3.54 (0.58)
Treated samples	PEI LMW	0.5	3.44 (0.59)
		1	3.10 (0.43)
		2	3.28 (0.47)
	PEI HMW	0.5	3.15 (0.38)
		1	3.55 (0.51)
		2	3.50 (0.48)

* Values in parenthesis are standard deviations.

A one-way ANOVA with Tukey's HSD had shown that treatment of the samples with 1% PEI LMW and 0.5% PEI HMW resulted in significantly lower adhesion strength of WTAC to the wood surface compared to untreated samples ($F(6,193) = 4.714$ for $p < 0.5$). All other groups of treated samples showed lower values for the adhesion strength of WTAC, but the difference was not statistically significant. The SEM images showed that the treatment with polyelectrolyte solutions leads to a filling of the cell walls, which can lead to a weakening of the mechanical properties of the wood in the surface layer [51], as the number of microfibrils per unit area decreases. Since the adhesion of the coating depends on the mechanical properties of the substrate [52], the filling of the cell walls can reduce the adhesion of the coating and the wood. On the other hand, treatment with polyelectrolyte solutions led to an increase in roughness in the surface layer of the wood, so that the contact zone between the coating and the wood is larger. Tests have shown that mechanical bonding is one of the most important adhesion mechanisms of water-based coatings [53]. It is possible that the increase in roughness and thus the increase in the contact zone between the coating and the wood in the treated samples has reduced the losses in the adhesive bonds that are to be expected due to the decrease in the mechanical properties of the cells in the surface layer of the wood.

3.7. The Surface Roughness of the Wood Coated Surface

Table 7 shows the roughness parameters R_a, R_z and R_t of the wood-coated surface.

Table 7. The surface roughness parameters R_a, R_z and R_t of the coated surface of untreated samples and samples treated with PEI of different MW and concentration.

Group of Samples		Concentration, %	The Roughness of the Varnished Wood Surface *, µm		
			R_a	R_z	R_t
Untreated samples		/	6.659 (0.752)	38.84 (5.42)	54.20 (9.66)
Treated samples	PEI LMW	0.5	5.508 (0.653)	32.91 (4.74)	43.44 (5.98)
		1	3.949 (0.418)	24.55 (2.21)	32.41 (4.22)
		2	4.253 (0.546)	26.14 (2.98)	34.25 (3.51)
	PEI HMW	0.5	4.106 (0.537)	26.71 (3.50)	38.80 (6.73)
		1	3.318 (0.287)	21.14 (2.90)	30.93 (4.02)
		2	3.997 (0.344)	24.35 (2.87)	33.86 (6.04)

* Values in parenthesis are standard deviations.

All samples treated with PEI solutions of different MW and concentrations show an improvement in surface quality in the form of a reduction in surface roughness. The best result in terms of the roughness of the coated surface was shown by the samples treated with a 1% PE HMW solution, where the roughness of the coated surface was reduced by 44.5, 40.8, and 45.0% for the parameters R_a, R_z, and R_t, respectively (compared to untreated samples). A statistically significant difference was found between the untreated samples and the samples treated with PEI LMW solutions and PEI HMW solutions with respect to the parameters R_a (F (6,203) = 140.40 for $p < 0.5$), R_z (F (6,203) = 80.96 for $p < 0.5$) and the parameter R_t (F (6,203) = 55.10 for $p < 0.5$).

Within the treated samples, the lowest reduction in surface roughness in terms of R_a and R_z was observed in the samples treated with 0.5% PEI LMW, while the highest value for R_t was obtained in the samples treated with 0.5% PEI of both MW, with no statistically significant difference between these two groups.

4. Conclusions

The aim of this study was to investigate the influence of treating the wood with weak cationic polyelectrolyte, PEI, on the adhesion of the water-based transparent acrylic coating to the wood surface. The conditions for the deposition of the polyelectrolyte were varied by using PEI with two different molecular weights—low-molecular-weight PEI (PEI LMW) and high-molecular-weight PEI (PEI HMW).

The results of this study suggest that by treating the surface of beech wood prior to coating, water uptake can be reduced by 72% when the surface is treated with a 1% solution of PEI HMW (compared to untreated samples after 5 min of applying water droplets to the surface). This result is illustrated by the significant reduction in surface energy of the treated surface by 23.2% compared to the untreated surface. Treatment with a 1% PEI HMW solution resulted in the lowest WTAC wetting angles and the highest water droplet wetting angles compared to other combinations of molecular weight and concentration of PEI solution.

PEI LMW solutions had a greater effect on reducing sample surface water uptake only at the highest solution concentration of 2%. The reduction in water uptake of the treated wood samples had a positive effect on the roughness of the coated surface, which is of particular importance when water-based coatings are used for the surface treatment of wood.

It was found that the molecular weight has an influence on the penetration depth of the PEI solution at all concentrations (0.5, 1 and 2%), so that PEI HMW should be used if the polyelectrolyte network is to be retained in the outer layer. If, on the other hand, deeper penetration of PEI into the interior of the wood is required, PEI LMW solutions should be used. These results were confirmed by epifluorescence and SEM microscopy.

Treatment with a PEI solution of different concentrations and MW did not reduce the adhesion strength of WTAC, which allows the introduction of this process into industrial production without compromising the protective properties of the coated surface. This exception was observed in the samples treated with 1% PEI LMW and 0.5% PEI HMW. In industrial practice, the surface treatment of wood with 1% PEI HMW can be used before the wood is finished with water-based coatings to improve the surface properties in terms of roughness. This is particularly important for small and medium-sized businesses that do not have intensive and fast-working dryers, because a straightforward surface treatment, can improve the quality of the finished product. Including the 1% PEI HMWsurface treatment in the wood surface finishing would require industrial application techniques (e.g., spraying), which would change the viscosity of the coating and possibly its penetration depth. The higher penetration depth expected due to the lower viscosity of the coating could even increase the adhesion strength of the water-based coating to the wood surface. In the meantime, according to our findings, the 1% PEI HMW treatment of the wood maintains the ability of the water-based coating to sufficiently adhere to the treated surfaces.

Author Contributions: Conceptualization, T.P.; methodology, T.P., M.R., D.D., K.R. and M.J.; software, T.P., M.P. (Matjaž Pavlič) and D.D.; validation, M.P. (Marko Petrič), K.R. and M.J.; formal analysis, T.P., M.P. (Matjaž Pavlič), D.D. and M.R.; investigation, T.P. and D.D.; resources, T.P., K.R. and M.P. (Marko Petrič); data curation, T.P., M.P. (Matjaž Pavlič), D.D. and M.R.; writing—original draft preparation, T.P.; writing—review and editing, M.R.; visualization, T.P. and D.D.; supervision, M.P., K.R. and M.J.; project administration, T.P.; funding acquisition, T.P. All authors have read and agreed to the published version of the manuscript.

Funding: This research was funded by the Ministry of Science, Technological Development and Innovation of the Republic of Serbia, grant number 451-03-65/2024-03/200169 and grant number 451-03-66/2024-03/200053.

Institutional Review Board Statement: Not applicable.

Data Availability Statement: The original contributions presented in the study are included in the article, further inquiries can be directed to the corresponding authors.

Conflicts of Interest: The authors declare no conflicts of interest.

References

1. Landry, V.; Boivin, G.; Schorr, D.; Mottoul, M.; Mary, A.; Abid, L.; Carrère, M.; Laratte, B. Recent Developments and Trends in Sustainable and Functional Wood Coatings. *Curr. For. Rep.* **2023**, *9*, 319–331. [CrossRef]
2. Ifijen, I.H.; Maliki, M.; Odiachi, I.J.; Aghedo, O.N.; Ohiocheoya, E.B. Review on Solvents Based Alkyd Resins and Water Borne Alkyd Resins: Impacts of Modification on Their Coating Properties. *Chem. Afr.* **2022**, *5*, 211–225. [CrossRef]
3. Palija, T.B. *The Impact of Polyelectrolyte on Interaction between Wood and Water-Borne Coatings*; University of Belgrade Faculty of Foresty: Belgrade, Serbia, 2015.
4. Gibbons, M.; Nikafshar, S.; Saravi, T.; Ohno, K.; Chandra, S.; Nejad, M. Analysis of a Wide Range of Commercial Exterior Wood Coatings. *Coatings* **2020**, *10*, 1013. [CrossRef]
5. Sarcinella, A.; Frigione, M. Sustainable and Bio-Based Coatings as Actual or Potential Treatments to Protect and Preserve Concrete. *Coatings* **2023**, *13*, 44. [CrossRef]
6. Rowell, R.M. Understanding Wood Surface Chemistry and Approaches to Modification: A Review. *Polymers* **2021**, *13*, 2558. [CrossRef]
7. Shukla, S.; Pandey, K.K.; Kishan Kumar, V.S. Chemical and Application Properties of Some Solvent and Water Based Coatings on Wooden Substrate. *Drv. Ind.* **2019**, *70*, 107–114. [CrossRef]
8. de Meijer, M.; Militz, H. Wet Adhesion of Low-VOC Coatings on Wood A Quantitative Analysis. *Prog. Org. Coat.* **2000**, *38*, 223–240. [CrossRef]
9. Zhou, Y. *Nanoscale Surface Modification of Wood Veneers for Adhesion*; Faculty of Virginia Polytechnic Institute and State University: Blacksburg, VA, USA, 2008.
10. Delgado, M.Z.; Aranda, F.L.; Hernandez-Tenorio, F.; Garrido-Miranda, K.A.; Meléndrez, M.F.; Palacio, D.A. Polyelectrolytes for Environmental, Agricultural, and Medical Applications. *Polymers* **2024**, *16*, 1434. [CrossRef]
11. Gîfu, I.C.; Ianchiș, R.; Nistor, C.L.; Petcu, C.; Fierascu, I.; Fierascu, R.C. Polyelectrolyte Coatings—A Viable Approach for Cultural Heritage Protection. *Materials* **2023**, *16*, 2873. [CrossRef]
12. Simpson, W.; TenWolde, A. Physical Properties and Moisture Relations of Wood. In *The Encyclopedia of Wood*; Skyhorse Publishing, Inc.: New York, NY, USA, 2007; pp. 3–21.
13. Frenzel, R.; Swaboda, C.; Petzold, G.; Emmler, R.; Simon, F. Controlling the Water Uptake of Wood by Polyelectrolyte Adsorption. *Prog. Org. Coat.* **2011**, *72*, 88–95. [CrossRef]
14. Horvath, A.E. The Effects of Cellulosic Fiber Charges on Polyelectrolyte Adsorption and Fiber-Fiber Interactions. Ph.D. Thesis, Royal Institute of Technology, Stockholm, Sweden, 2006.
15. Palija, T.; Rančić, M.; Pavlič, M.; Petrič, M.; Jaić, M. Water-Borne Coatings: Reducing the Impact of Wood Grain Raising. In Proceedings of the Final COST FP1006 Meeting "Advances in Modified and Functional Bio-Based Surfaces", Thessaloniki, Greece, 8 April 2015.
16. Gutowski, V.; Li, S.; Kuys, B.; Filippou, C.; Russell, L. Bonding Connections. *Eur. Coat. J.* **2015**, *2*, 18–22.
17. Treusch, O.G.; Petutschnigg, A. A Combination of Polyethylenimine and Phenolic Resin as an Adhesive for Wood-Based Panels. *BioResources* **2012**, *7*, 789–798. [CrossRef]
18. Gao, S.; Qi, J.; Jiang, S.; Wu, T.; Wang, W.; Cai, Y.; Ma, C.; Zhang, B.; Huang, J.; Yan, Y. Green Wood Adhesives from One-Pot Coacervation of Folic Acid and Branched Poly (*Ethylene imine*). *ACS Appl. Bio Mater.* **2021**, *4*, 7314–7321. [CrossRef] [PubMed]
19. Peng, W.; Dong, C.; An, J.; Zhang, G.; Wang, P.; Xie, Y. A Novel Formaldehyde-Free Wood Adhesive Synthesized by Straw Soda Lignin and Polyethyleneimine. *BioResources* **2023**, *18*, 3123–3143. [CrossRef]
20. Dong, S.; Jiang, H.; Hou, B.; Du, Y.; Jin, C.; Yan, Y. High Efficient Fire-Retardant Coatings on Wood Fabricated by Divalent Metal Ion Cross-Linked Multilayer Polyethyleneimine and Ammonium Polyphosphate Polyelectrolytes. *Polym. Degrad. Stab.* **2022**, *204*, 110112. [CrossRef]
21. Soula, M.; Samyn, F.; Duquesne, S.; Landry, V. Innovative Polyelectrolyte Treatment to Flame-Retard Wood. *Polymers* **2021**, *13*, 2884. [CrossRef]
22. *EN ISO 2811-1*; Paints and Varnishes. Determination of Density Pycnometer Method. ISO: Geneva, Switzerland, 2016.
23. *EN ISO 3251*; Paints, Varnishes and Plastics-Determination of Non-Volatile-Matter Content. ISO: Geneva, Switzerland, 2019.
24. *EN ISO 2431*; Paints and Varnishes-Determination of Flow Time by Use of Flow Cups. ISO: Geneva, Switzerland, 2012.
25. *EN ISO 2555*; Plastics-Resins in the Liquid State or As Emulsions or Dispersions-Determination of Apparent Viscosity Using a Single Cylinder Type Rotational Viscometer Method. ISO: Geneva, Switzerland, 2018.
26. *ISO 304*; Surface Active Agents-Determination of Surface Tension by Drawing Up Liquid Films. ISO: Geneva, Switzerland, 1985.
27. Sunoqrot, S.; Bae, J.W.; Jin, S.-E.; M Pearson, R.; Liu, Y.; Hong, S. Kinetically Controlled Cellular Interactions of Polymer-Polymer and Polymer-Liposome Nanohybrid Systems. *Bioconjug. Chem.* **2011**, *22*, 466–474. [CrossRef]

28. Đikanović Golubović, D. Structural Investigation of Plant Cell Walls and Lignins of Different Origin. Ph.D. Thesis, Faculty of Physical Chemistry University of Belgrade, Belgrade, Serbia, 2013.
29. Albinsson, B.; Li, S.; Lundquist, K.; Stomberg, R. The Origin of Lignin Fluorescence. *J. Mol. Struct.* **1999**, *508*, 19–27. [CrossRef]
30. Chibowski, E.; Perea-Carpio, R. Problems of Contact Angle and Solid Surface Free Energy Determination. *Adv. Colloid. Interface Sci.* **2002**, *98*, 245–264. [CrossRef]
31. *EN ISO 2808*; Paints and Varnishes-Determination of Film Thickness. ISO: Geneva, Switzerland, 2019.
32. Van den Bulcke, J.; Boone, M.; Van Acker, J.; Van Hoorebeke, L. High-Resolution X-Ray Imaging and Analysis of Coatings on and in Wood. *J. Coat. Technol. Res.* **2010**, *7*, 271–277. [CrossRef]
33. *EN ISO 4624*; Paints and Varnishes-Pull-Off Test for Adhesion. ISO: Geneva, Switzerland, 2023.
34. *ISO 4287*; Geometrical Product Specifications (GPS)-Surface Texture: Profile Method-Terms, Definitions and Surface Texture Parameters. ISO: Geneva, Switzerland, 1997.
35. DorvelL, B.; Boopalachandran, P.; Chen, I.; Bowling, A.; Kerry, W.; Steve, K. Probing the Effect of Polymer Molecular Weight on Penetration into the Wood Cell Wall Using Polyethylenimine (PEI) as a Model Compound. *J. Microsc.* **2018**, *270*, 121–128. [CrossRef] [PubMed]
36. Vitas, S.; Segmehl, J.S.; Burgert, I.; Cabane, E. Porosity and Pore Size Distribution of Native and Delignified Beech Wood Determined by Mercury Intrusion Porosimetry. *Materials* **2019**, *12*, 416. [CrossRef] [PubMed]
37. He, M.; Xu, D.; Li, C.; Ma, Y.; Dai, X.; Pan, X.; Fan, J.; He, Z.; Gui, S.; Dong, X.; et al. Cell Wall Bulking by Maleic Anhydride for Wood Durability Improvement. *Forests* **2020**, *11*, 367. [CrossRef]
38. den Bulcke, J.V.; Rijckaert, V.; Acker, J.V.; Stevens, M. Quantitative Measurement of the Penetration of Water-Borne Coatings in Wood with Confocal Lasermicroscopy and Image Analysis. *Eur. J. Wood Prod.* **2003**, *61*, 304–310. [CrossRef]
39. Xie, Y.; Krause, A.; Militz, H.; Mai, C. Coating Performance of Finishes on Wood Modified with an N-Methylol Compound. *Prog. Org. Coat.* **2006**, *57*, 291–300. [CrossRef]
40. Žigon, J.; Kovač, J.; Petrič, M. The Influence of Mechanical, Physical and Chemical Pre-Treatment Processes of Wood Surface on the Relationships of Wood with a Waterborne Opaque Coating. *Prog. Org. Coat.* **2022**, *162*, 106574. [CrossRef]
41. Palija, T.; Jaić, M.; Džinčić, I.; Šućur, A.; Dobić, J. Variability of Dry Film Thickness of a Coating Applied by Roller Coater on Wood in a Real Industrial Process. *Drewno* **2018**, *61*, 153–164. [CrossRef]
42. Kúdela, J. Wetting of Wood Surface by a Liquids of a Different Polarity. *Wood Res.* **2014**, *59*, 11–24.
43. Shi, S.Q.; Gardner, D.J. Dynamic Adhesive Wettability of Wood. *Wood Fiber Sci.* **2001**, *33*, 58–69.
44. Qin, Z.; Gao, Q.; Zhang, S.; Li, J. Surface Free Energy and Dynamic Wettability of Differently Machined Poplar Woods. *BioResources* **2014**, *9*, 3088–3103. [CrossRef]
45. de Moura, L.F.; Hernández, R.E. Evaluation of Varnish Coating Performance for Two Surfacing Methods on Sugar Maple Wood. *Wood Fiber Sci.* **2005**, *37*, 355–366.
46. Ryzhkov, N.V.; Nikitina, A.A.; Fratzl, P.; Bidan, C.M.; Skorb, E.V. Polyelectrolyte Substrate Coating for Controlling Biofilm Growth at Solid–Air Interface. *Adv. Mater. Interfaces* **2021**, *8*, 2001807. [CrossRef]
47. Miklečić, J.; Lončarić, A.; Veseličić, N.; Jirouš-Rajković, V. Influence of Wood Surface Preparation on Roughness, Wettability and Coating Adhesion of Unmodified and Thermally Modified Wood. *Drv. Ind.* **2022**, *73*, 261–269. [CrossRef]
48. Stone, J.; Scallan, A. A Structural Model for the Cell Wall of Water-Swollen Wood Pulp Fibres Based on Their Accessibility to Macromolecules. *Cell Chem. Technol.* **1968**, *2*, 343–358.
49. Fahlén, J. The Cell Wall Ultrastructure of Wood Fibres-Effects of the Chemical Pulp Fibre Line. Ph.D. Thesis, KTH Royal Institute of Technology, Stockholm, Sweden, 2005.
50. *EN 927-2*; Paints and Varnishes-Coating Materials and Coating Systems for Exterior Wood-Part 2: Performance Specification. ISO: Geneva, Switzerland, 2014.
51. Bjurhager, I. *Effects of Cell Wall Structure on Tensile Properties of Hardwood*; KTH Chemical Science and Engineering: Stockholm, Sweden, 2011.
52. Jaić, M.; Palija, T. Uticaj Vrste Drveta i Sistema Površinske Obrade Na Adheziju Premaza. *Zaštita Mater.* **2012**, *53*, 299–303.
53. Jaić, M.; Palija, T.; Đorđević, M. The Impact of Surface Preparation of Wood on the Adhesion of Certain Types of Coatings. *Zaštita Mater.* **2014**, *55*, 163–169. [CrossRef]

polymers

MDPI

Review

Polyelectrolytes: From Seminal Works to the Influence of the Charge Sequence

Nam-Kyung Lee [1], Min-Kyung Chae [2], Youngkyun Jung [3], Albert Johner [4,*] and Jean-Francois Joanny [5]

1 Department of Physics and Astronomy, Sejong University, Seoul 05006, Republic of Korea; lee@sejong.ac.kr
2 National Institute for Mathematical Sciences, Daejeon 34047, Republic of Korea; mkc@nims.re.kr
3 Supercomputing Center, Korea Institute of Science and Technology Information, Daejeon 34141, Republic of Korea; yjung@kisti.re.kr
4 Institut Charles Sadron CNRS-Unistra, 6 rue Boussingault, 67083 Strasbourg, France
5 Institut Curie, Physique des cellules et Cancer, Collège de France Soft Matter and Biophysics Chair, 11 , PSL University, Place Marcelin-Berthelot, 75231 Paris, France; jean-francois.joanny@college-de-france.fr
* Correspondence: albert.johner@ics-cnrs.unistra.fr

Abstract: We propose a selected tour of the physics of polyelectrolytes (PE) following the line initiated by de Gennes and coworkers in their seminal 1976 paper. The early works which used uniform charge distributions along the PE backbone achieved tremendous progress and set most milestones in the field. Recently, the focus has shifted to the role of the charge sequence. Revisited topics include PE complexation and polyampholytes (PA). We develop the example of a random PE in poor solvent forming pearl-necklace structures. It is shown that the pearls typically adopt very asymmetric mass and charge distributions. Individual sequences do not necessarily reflect the ensemble statistics and a rich variety of behaviors emerges (specially for PA). Pearl necklaces are dynamic structures and switch between various types of pearl-necklace structures, as described for both PE and PA.

Keywords: polyelectrolytes; polyampholytes

1. Introduction

Modern polymer science started with the 1920 paper by Staudinger [1], who introduced the concept of long molecules comprising many repeating units held together by ordinary covalent chemical bonds. A nice account of Staudinger's manifesto (which has never been translated) was given in 2020 for its 100th anniversary [2]. The emergence of the concept of covalent polymer is contemporary to the early development of quantum mechanics. Considering the scale of science, it is a young field. Staudinger himself was inspired by natural rubber and later also studied biopolymers, foreshadowing the extended scope of the field as it stands. Polymer science including polymer chemistry, polymer physics, biological aspects and applications grew to an extent which precludes the project of an exhaustive review.

Water, which may still be termed abundant, is an ecologically harmless solvent, in contrast to most organic solvents. Unfortunately, polymer backbones are typically insoluble in water. Some polymer backbones, which can establish hydrogen bonds, are nonetheless soluble in water, among which the most popular are Polyethylene Oxide (PEO) [3] and Poly(N-isopropylacrylamide) PNIPAM. Various theoretical descriptions for uncharged water-soluble polymers were proposed: de Gennes suggested that the Flory–Huggins type theory with three body attractions can lead to closed-loop solubility diagrams [4], and detailed lattice (Flory–Huggins type) theories were developed [5,6], which generated induced three-body attractions [6]; water-soluble polymers were later described as annealed hydrated/dehydrated copolymers [7]. The solubility of these polymers in water decreases with increasing temperature (the loop diagram is usually not closed), in contrast to most polymers in organic solvents. PNIPAM changes solubility in water at about body temperature and attracted considerable academic attention [8,9]. PEO is also of considerable practical interest [10].

Besides these important exceptions, most water-soluble polymers carry charged monomers whose charge is compensated by counterions. They are generically called polyelectrolytes (PEs) and are the topic of this paper (see Figure 1). We try to provide a consistent minimal path through the literature on PEs, allowing to embrace the field in its actual state. Many original important contributions are hence not mentioned and are replaced by subsequent works on similar topics; this partially reflects the taste of the authors.

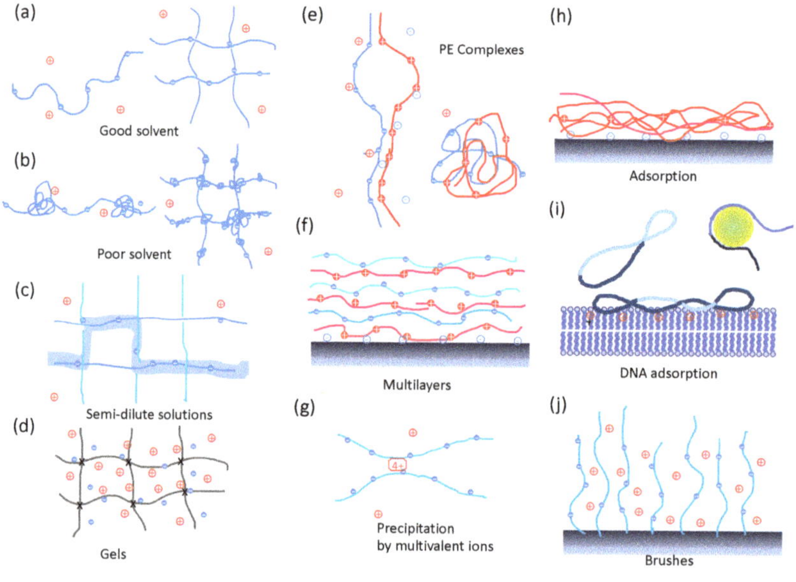

Figure 1. (**a–c**) PE solutions. (**a**) Left—Single PE chain in good solvent which is stretched by electrostatics with a few counterions. Right—PE solution in good solvent where the correlation length corresponding to typical distances between chains is represented as the mesh size, without physical crosslinks. (**b**) Left—Overall stretched pearl-necklace structure of a single PE chain in poor solvent. Right—PE solution in poor solvent where meshes interconnected by pearls appear. (**c**) Semi-dilute solution: The depicted polyelectrolyte configuration (2D illustration) gains entropy by selecting orientations at each 'crossing'. In the case of an intrinsically flexible polyelectrolyte, the correlation length, illustrated as the distance between chains, can be qualitatively equated to the electrostatic persistence length. The thick shadow line represents a conformation of a test chain. (**d**) PE gel. Crosses (×) indicate permanent (chemical) crosslinks. The osmotic pressure of trapped counterions tends to swell the gel. (**e**) PE Complexes. Left—Ladder structure with an open loop. Right—Scrambled-egg structure. (**f**) Multilayers. Sketch of a periodic multilayer structure. (**g**) PE precipitation mediated by multivalent ions. (**h**,**i**) PE adsorption on a substrate. (**h**) Intrinsically flexible PE. (**i**) rigid PE. Left—DNA plasmid on a planar substrate. Right—ds-DNA on a colloid/histone. (**j**) PE brushes. PEs are grafted to a solid substrate by one end and form a so-called brush. The brush of flexible PEs is swollen by the osmotic pressure of the trapped counterions.

The objective of this review is to provide the development/history of the field though essential papers that have substantially improved our understanding. Besides, we push forward the point of view that the recent trend focuses on sequence-dependent properties, bringing the field somewhat closer to biophysics, although this evolution is of considerable interest for material science, too. Generic PE properties based on a description with the charge as a continuous variable, smeared out along the polymer are presented first. Then, we proceed to more recent investigations on PE systems, where the charge sequence plays an important role. The last part describes some of our own recent works on the structure of random PE sequences in poor solvent, with a few results on random polyampholytes (PAs) for comparison. The final discussion also briefly mentions approaches and topics that deserve further attention.

From the perspective of electrical charges, we may distinguish between polymers that carry charges of both signs and polymers that carry charges of the same sign. Polymers carrying charges of both signs along their backbone are called polyampholytes (PAs) [11]. Most proteins are PAs. Proteins, unlike random synthetic PAs, generally have a rugged energy landscape and a well-defined ground state, not ruled by electrostatics only, which determines their biological activity. An important class of proteins known as Intrinsically Disordered Proteins (IDPs) [12] which does not have a well-defined ground state shares many features of synthetic PAs. This PA character of IDPs is well documented experimentally [13,14] and has attracted a lot of attention recently, including from theorists [15,16]. Continuous attention was given to the phase behavior of IDPs or proteins with disordered protein regions (IDPRs) and other associative biomacromolecules in water [17], sometimes in close relation with membrane-less organelles [18,19]. For clarity, we call polyelectrolyte (PE) a polymer carrying charges all of the same sign as opposed to PAs, from here on.

Throughout, a continuous description of water is used, which considers water as a dielectric fluid with a dielectric constant $\epsilon_r = 80$. This is fair in the dilute and semidilute polymer solutions considered below. The contrast in dielectric constant between the polymer and the water may become relevant in the dense PE phase [20]. Because of its high dielectric constant, water serves as a good solvent for ions. Typically, it is energetically more advantageous to distribute counterions within water rather than creating a shared (ionic) structure that is locally neutralized with the polymer. The entropy of released counterions is driving the dispersion of the counterions and polymer in water. The charged polymer backbone stretches out under the influence of electrostatic repulsions. Added salt screens the electrostatic interaction, so adding salt is an easy way to tune the properties of PE solutions.

2. Generic Descriptions Using Smeared Out Charge Distributions

Among the essential properties of single chains, two prevalent phenomena are the distribution of counterions around the chain and the stiffening of the chain due to electrostatic repulsion. The question as to which extent counterions are released by the polymer into the solution was raised early on by osmotic pressure measurements, and concepts of charge regulation and charge condensation were introduced, as discussed in the book by Oosawa [21]. This very classical question obtained an improved solution (beyond mean-field) more recently; see, for example, reference [22] and references therein. The case of strong coupling of the counterions, mainly of higher valency, to the polymer was considered in the same work [22]. The condensed counterions form a more-or-less rigid crystal-type structure on the cylinder, mimicking the rigid PE backbone.

The bending rigidity is of special importance for biopolymers such as double-stranded DNA (ds-DNA) and is typically characterized in terms of the persistence length, which qualitatively separates small scales, where the backbone behaves as stiff, and large scales where it behaves as totally flexible. The dependence of the persistence length upon added salt was well characterized for intrinsically rigid PEs like ds-DNA and understood [23]. Measurements of the bending rigidity as a function of ionic strength on other PEs were roughly split into two categories: part of the measurements were consistent with the variation observed for ds-DNA and with ref. [23], others, where the persistence is mainly due to electrostatic repulsion, were in conflict [24]. This puzzle regarding the variation in the persistence length with salt attracted considerable efforts from theory [24–27] and simulations [28]. Simulations mainly support the model of Khokhlov and Katchatourian of a rod of electrostatic blobs [29]. Quantitative theories [24,26] are of variational type.

Reference [24] predicts two regimes with an unusually complex crossover. The subsequent ref. [26] explains the weakness of ref. [24] and identifies the previous Khokhlov and Katchaturian behavior, in accord with simulations. These theoretical works [24,26] and simulations [28] utilize the Debye–Hückel screened interaction potential rather than explicit salt ions. On the other hand, explicit salt ions raise questions about the very interaction between the salt ions and the PE also beyond electrostatics. The emerging picture is conceptually much simpler than that presented in ref. [24], but the question of the persistence length remains, to some extent, open. The iconic example of ds-DNA benefited from a recent advanced theory [30] which allowed us to refine the measured value of its intrinsic rigidity. The two limitations of the standard theory, namely the weak charge hypothesis and the vanishing chain thickness, are typically not fulfilled in real systems. The authors relax both limitations in a relatively simple way, leading to a compact expression for the persistence length [30].

The understanding of the behavior of a single polymer chain is a good starting point for comprehending the behavior of solutions in the same solvent. This is, to some extent, true for PEs, too. The seminal paper by de Gennes, Pincus, Velasco and Brochard [31], published in 1976, describes correlations in PE solutions and gives a theoretical description of PE solutions from a physical point of view. Several important points are made in their work. It is shown that the single PE (dilute regime) is stretched with an extension proportional to the number of monomers and the electrostatic blob concept is introduced; the upper critical dimension 6 is given; and the idea of a Newton-type equation for stretched (weakly fluctuating) conformations is sketched; this was discussed later (essentially one decade later) as the classical limit of the Schroedinger-type Edwards–Lifshiz equation for the propagator [32]. It is argued that the semi-dilute solution remains isotropic (for flexible PE backbones) and one chain in the solution follows large-scale Gaussian statistics with the electrostatic persistence length equal to the correlation length. Predictions are made for the scattering function and the viscosity. The careful and inspired conclusion calls for experiments. SAXS and fluorescence emission experiments by Essafi et al. showed qualitatively different behaviors for PEs with hydrophilic and hydrophobic backbones [33]. Hydrophobic PEs do not follow predictions of ref. [31]. Earlier, Khokhlov [34] predicted the influence of the solvent quality on the structure of a dilute PE and a semi-dilute polyelectrolyte solution. The dense PE solution was subsequently described quantitatively in the framework of RPA by Borue and Erukhimovitch [35] and independently by Leibler and Joanny [36]. In ref. [34], Khokhlov introduced the cylindrical, cigar-shaped, single-PE conformation as a result of local collapse due to hydrophobicity imposing the internal density of the cylinder and global elongation under the long-range electrostatic repulsion. Although this model was recognized as unstable due to the instability of the charged cylinder demonstrated by Rayleigh [37], some fairly elongated cylindrical sections can nonetheless be found mainly in random PAs under special conditions as discussed in Section 4. The pearl-necklace conformation was introduced by Dobrynin, Rubinstein and Obukhov [38] for PEs and by Kantor and Kardar for PAs [39]. (See Figure 2.) The authors of Ref. [38] also proposed an elegant description of the semi-dilute PE solution based on the same pearl-necklace model. As a result, PEs in poor solvent (with a hydrophobic backbone) and PE in good solvent (with a hydrophilic backbone) need to be considered separately.

A series of reviews [40–42] marks out a period of intense (theoretical) activity roughly ending around the year 2005.

We concentrated here on progress in our fundamental understanding of PE where the single chain and solutions are key. Besides solutions (Figure 1a–c), other PE-based systems were considered: PE gels [43] (Figure 1d), complexes of polyanions and polycations [44] (Figure 1e), PE multilayers [45] (Figure 1f), precipitation or folding of PEs by multivalent counterions [46–48] (Figure 1g), PE adsorbed at an interface [41,49] (Figure 1h,i), and grafted PEs [50,51] (Figure 1j), to cite just a few. These systems are sketched in Figure 1 with an extended caption. Compacted PE complexes in short COPECS, sometimes also called saloplastics, obtained via (ultra)centrifugation or extrusion of well-selected PE complexes under high salt, exhibit enhanced mechanical properties and represent a promising variant of PE complexes for applications [52–55].

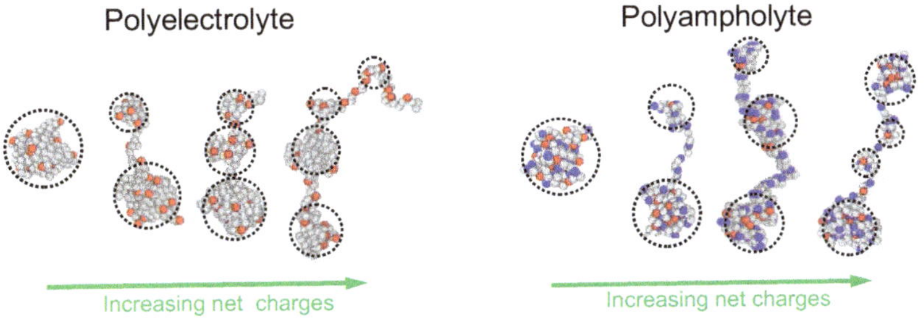

Figure 2. Pearl-necklace conformations of PE in poor solvent conditions (**left**) and pearl necklaces of random PA sequences in weakly poor solvent conditions (**right**). The necklace structure consists of pearls (indicated by dashed circle) and connecting strings. The number of pearls increases with increasing net charges. The conformations are obtained from MD simulations for PA and PE chains consisting of $N = 202$ monomers in which every 3rd monomers can carry charges ($p = 3$). All charge sequences are created by unbiased Markovian processes. Blue and red colors represent different types of charges. The net charges of the chains increase from left to right: $Q_{PE} = 16$, 22, 28 and 34 for PE chains and $Q_{PA} = 12$, 16, 20 and 24 for PA. (See, Supplementary Materials for simulation model).

3. Charge Sequence Effects

The considerable progress in our generic understanding of PEs achieved in the seminal papers cited earlier is solely based on a smeared-out charge description, accompanied by the idea that the importance of the charge sequence is marginal due to the long-range character of the electrostatic interaction (except for the case of somewhat extreme charge correlations). In contrast, smearing out the charge of a PA does not make sense and the charge correlations are essential [56,57]. It turns out that the sequence does matter for polyanion/polycation complexation. Indeed, complexation was shown to depend on the charge correlation along the PEs: charge blockiness favors complexation [58]. A comprehensive account of recent progress on complex coacervation can be found in the recent reviews by Rumyantsev [59] and by Sing and Perry [60]. In the same spirit, the influence of the sequence on self-complexation of PAs [56,57], where charges of both signs are located on the same polymer, was re-examined recently [61,62]. A complete phase diagram was also given for IDPs [15]. As to the structure of a single PE (or PA) chain, the sequence is specially important for partially collapsed structures like pearl necklaces. Figure 2 illustrates various conformations of pearl necklaces composed of pearls connected by strings. It is energetically favorable to form strings between pearls out of overcharged subsequences which link uneven pearls, provided the smaller pearls carry a higher charge density [63,64]. In other words, it is favorable to localize a relative charge excess on average in strings than in pearls [62,64], and also in smaller pearls as compared to larger ones [65]. This is easier for annealed charges (say, pH-dependent charges) where the actual sequence can adapt to the structure, but is impossible for quenched regular sequences. How easily these inhomogeneous structures can be achieved for quenched PEs depends on the detailed charge sequence [64].

In the following section, we explore the structure of PE pearl necklaces featuring quenched charge sequences. The theoretical picture is supported by Molecular Dynamics (MD) simulations. The uncorrelated random sequences and correlated sequences, generated by unbiased Markovian processes, are explored in our previous publication [65]. Here, we primarily focus on presenting analysis on the charge sequences with positive correlations, incorporating previously undisclosed data. The results are consistent with our previous findings and are explained using the same physical approach. We invite the reader to compare the results presented below with those in Ref. [65].

4. Hydrophobic Polyelectrolytes: Influence of the Charge Sequence

The standard pearl-necklace model for hydrophobic PE describes regular charge sequences and can be considered as a mean-field model for random PE. However, recent research has revealed that random PEs do not tend to adopt configurations with an equal distribution of pearls, as previously suggested by mean-field theory [65]. The structure of random PEs results from the interplay between the disorder along the charge sequence and thermal fluctuations. Moderately charged random PEs exhibit very asymmetric mass distributions, as depicted in Figure 2. MD simulations and energy considerations demonstrate that moderately charged random PEs primarily concentrate their mass and charge within the pearls. (As mentioned earlier, this behavior contrasts with very blocky sequences.) Moreover, random PEs do not tend to divide into evenly sized pearls; instead, they form pearl necklaces characterized by significant asymmetry. The disorder in the sequence also disrupts the correlations between adjacent pearls to some degree, resulting in the occurrence of two large pearls in close proximity (see Figure 2).

To gain further insight, we describe the energy landscape of the n-pearl states, taking into account the distribution of mass and charge among the pearls. The key question is how a PE with a quenched PE charge sequence can visit energetically favorable regions.

A pearl can be considered as a drop of dense polymer phase (of almost vanishing osmotic pressure) with a surface tension associated with the polymer/water(solvent) interface. With M_p and Q_p the total mass and charge belonging to pearls, the retained energy E comprises the electrostatic self-energy of the n-pearls and their surface energy. At this point, the energy of strings between pearls and the interaction among pearls and among strings and pearls are neglected. It is suggested by the standard pearl-necklace model where, for the necklace to be stable at all, the self-energy of the pearls must dominate. Consistently, the order of the pearls along the sequence is not considered [65]. This simplified model helps to structure the parameter space and to provide clarity when interpreting the data clouds generated by simulations. In particular, the model predicts the location of the energy extrema and their stability, which, in turn, provide a rationale for the global distribution of data points obtained from the samples (See, also Supplementary Materials). The model is expected to be less suitable when dealing with highly blocky statistics.

The configuration with mass distribution $\{m_i\}$ and charge distribution $\{q_i\}$ among the pearls is characterized by a set of asymmetry parameters for the mass distribution $\{x_i\}$ and charge distribution $\{y_i\}$. These parameters are defined by $m_i/M_p = 1/n + x_i$ and $q_i/Q_p = 1/n + y_i$, where x_i is the relative excess mass of i-th pearl referenced to the uniform mass distribution $m_i = M_p/n$ and y_i is similarly defined for the charge distribution. When all pearls have equal mass and charges, $x_i = 0, y_i = 0$. The main physics is captured by the following expression of the energy:

$$\frac{E}{E_S} = \sum_{i=1}^{n}(1/n + x_i)^{2/3} + \chi\sum_{i=1}^{n}\frac{(1/n + y_i)^2}{(1/n + x_i)^{1/3}} - 1 - \chi. \tag{1}$$

The conservation of mass and charge imposes the two constraints.

$$\sum_{i=1}^{n} x_i = 0; \ \sum_{i=1}^{n} y_i = 0. \tag{2}$$

The single globule state is chosen as the reference state and its surface energy E_s sets the overall energy scale, while the control parameter χ measures the ratio of the electrostatic energy to the surface energy in the uni-globular reference state. The first term on the right-hand side carries a $2/3$ power and represents the surface energy of actual bead i, while the second term stands for its electrostatic self-energy. The subsequent additive terms are non-essential and adjust the energy of the reference state to zero.

Then, the n-pearl structure can be pictured in a $2(n-1)$ dimensional space. The uniform pearl distribution corresponds to the zero vector in which all components $\{x_i\}, \{y_i\}$ are zeros. An analysis of Equation (1) reveals that, in the case of moderate charge ($\chi < n$), the state with evenly sized pearl is not the energy minimum but instead the highest energy saddle point. The configuration with $n-1$ evenly large pearls complemented by one small pearl represent the next highest energy saddle point, and so forth. The absolute energy minimum is reached with only one large pearl complemented by $n-1$ evenly sized small pearls. Note that in the mean-field description, where the condition $\chi = n/2 < n$ holds, the asymmetric structure is generally expected, except for instances with unusually small numbers of pearls due to rare fluctuations. Increasing χ while keeping n constant results in a flatter energy landscape. Conversely, for a fixed χ value, for a fixed sequence, a realization with fewer pearls exhibits greater fluctuations within a smoother landscape. This feature also influences kinetic properties, as demonstrated later in this section.

A comprehensive analysis of the energy described in Equation (1) can be found in reference [65]. This reference explores the energy extrema associated with a specific number of pearls across the entire range of χ values. Figure 2 in ref. [65] displays the energies of non-symmetric states for systems composed of three beads. The symmetric state has the lowest energy when χ exceeds n ($\chi > n$), while the non-symmetric states cease to exist beyond a critical value χ_c, which is somewhat larger.

For MD simulations, we consider PE and PA chains consisting of $N = 202$ monomers, where every third monomer ($p = 3$) can bear a charge. All charge sequences are generated by unbiased Markovian processes. In the case of PE, every third monomer can carry a charge (1) or remain neutral (0). Correlations between charges can be positive, negative, or absent. In uncorrelated sequences, the probability of having a charge is $1/2$ for each charge site, leading to the global unbiased statistics $\langle Q \rangle \approx N/(2p)$ for PE. Note that there are variations in net charge from sequence to sequence. We may choose a sub-sample that retains specified net charges, differing from the global average. The global sample is generated with a prescribed correlation following a Markov process. If we select a large sub-sample with an imposed (net) charge, its statistics are well-defined, but its correlation can vary. To take an extreme example: selecting a sub-sample where all sites are charged (1) results in a single correlated block, irrespective of the initial Markov process used, although such sequences may be scarce in the global sample). Positively correlated sequences tend to favor blocks of either zero or one, while negatively correlated sequences tend to alternate between charges zero and one. Below, we consider charge sequences with positive correlations which have an average block length of 4, and uncorrelated random sequences which have an average block length of 2. The former is referred as a "blocky" sequence. Among these, sequences that satisfy specific net charge conditions are selected for further analysis.

In the case of PA, our focus is on uncorrelated PA chains where either positive (+1) or negative (−1) charges are randomly assigned with a probability of $1/2$ at every third monomer. The ensemble of PA chains results in a unbiased global statistics of vanishing average charge $\langle Q \rangle = 0$. There are variations in net charges from sequence to sequence. The net charge of each chain is determined by the difference between the majority charge and the minority charge. Detailed simulation descriptions can be found in the Supplementary Materials.

Figure 3 depicts the density of states in the space of $\{x, y\}$ for PE chains when they adopt two-pearl structures. Here, 'x' and 'y' represent the relative excess mass and charge of a pearl, defined as $x = m/M_p - 1/2$ and $y = q/Q_p - 1/2$ for two pearl states. The line $x = y$ corresponds to the case in which each pearl has charge in proportion to its "share" of mass (Here and below, "share" refers to the amount of charge in proportion to the mass of the pearl). For large pearls, their mass exhibits a positive excess value ($x > 0$), and their charge distribution lies below the $x = y$ line. This implies that the larger pearls carry fewer charges than their "share," while the smaller pearls ($x < 0$) carry more charges than their "share".

To better illustrate how charges are distributed between pearls and strings, we may define the reduced mass and charge at strings as $m_s = N_s/N$ and $q_s = Q_s/Q$, respectively, where N_s and Q_s refer total mass and charge allocated at strings. In Figure 4, we evaluate the density of states $p(m_s, q_s)$ of blocky PE sequences with n = three pearls. With increasing net charges, more mass is allocated to the string. A greater population above the line $m_s = q_s$ indicate that strings carry more charges than what corresponds to their shares.

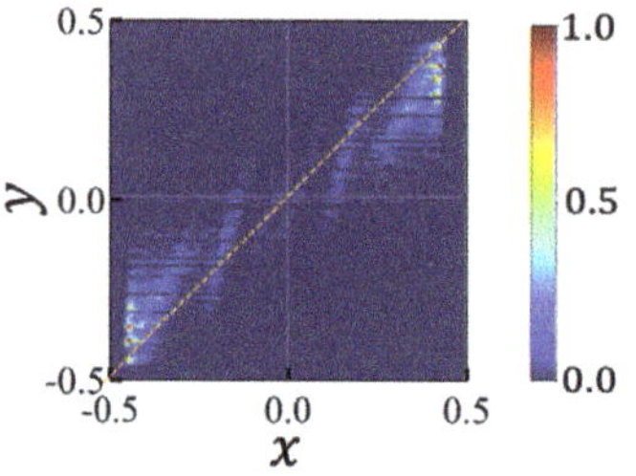

Figure 3. Density of states $p(x, y)$ of a 2-pearl structure of PE chain consisting of N = 202 monomers and net charge Q = 22. Charge sequences have positive charge correlation with average block length 4. The color bar in [0,1] scale represents the density of visited states realized in MD simulations. Relative excess mass $x = m/M_p - 1/2$ and excess charge $y = q/Q_p - 1/2$ measure the asymmetry between the two pearls. The region where $x > 0$ corresponds to larger pearls, i.e., x_1, while the region where $x < 0$ corresponds to smaller pearls, i.e., x_2. The line $x = y$ represents the uniform charge distribution where each pearl has a charge proportional to its mass. States below (above) the line indicate undercharged (overlarged) large (small) pearl states. Note that the data points exhibit point symmetry with respect to the origin, a property given by the constraints, as shown in Equation (2).

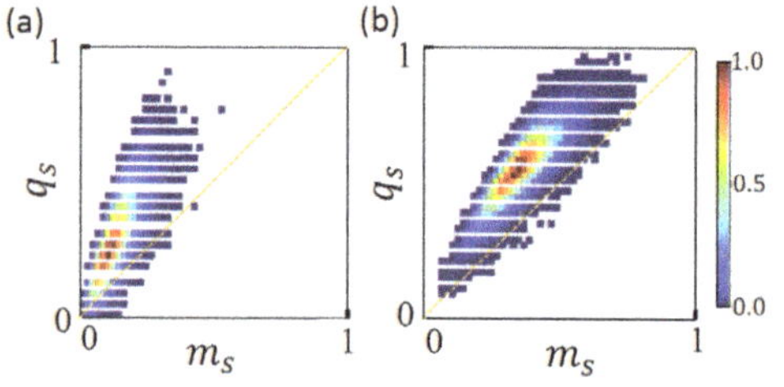

Figure 4. Density of states $p(m_s, q_s)$ of PE conformations in pearl necklaces where the number of pearls is n = 3. The states are evaluated according to the relative mass and charge allocated at strings, $m_s = N_s/N$ and $q_s = Q_s/Q$. Ensemble of PE chains consisting of N = 202 monomers and carrying net charges of (**a**) Q = 22 and (**b**) Q = 34 are considered. Each charge sequence has charge–charge correlations with the average block length of four. The color bar in [0,1] scale represents the density of visited states realized in MD simulations. With increasing net charges, more mass is allocated to the string. A greater population above the line $m_s = q_s$ indicates that strings carry more charges than what corresponds to their shares.

Since we are mainly interested in the pearl size distributions, we project the parameter space to the $(n-1)$-dimensional sub-space spanned by the mass parameters $\{x_i\}$ only. After projection, the parameter space reduces to the $(n-1)$-simplex defined by the x_i constraint (Equation (2)) and the ranking conditions on the pearl masses (The order of the pearls along the structure is not retained). For the n pearls, we exclude the smallest pearl, whose mass is set by the conservation laws but must still be positive. Additionally, there are $n-1$ ranking conditions of the retained masses (x_i's). Together, these n linear inequalities establish the boundaries of the interior region of the $(n-1)$-simplex. Turning the inequalities into equalities defines the n sides of the $(n-1)$-simplex. The inequalities defining the interior of the simplex and the position of the vertices are explicitly given in the Supplementary Materials, together with equations determining the location of the energy extrema. Figure 5 shows the simplex representations of MD simulation data for two, three, and four pearls, respectively: two pearl states are represented in a one-simplex, which is a segment; three pearl states are in a two-simplex, which is a triangle; and four pearl states are shown in a three-simplex, which is a tetrahedron. Note that a face of a $(n-1)$-simplex is a $(n-2)$-simplex. In particular, the face that does not contain the origin embeds the $(n-1)$-pearl states. It is hence, in principle, possible (but not so appealing) to represent all states with up to n pearls in the $(n-1)$-simplex.

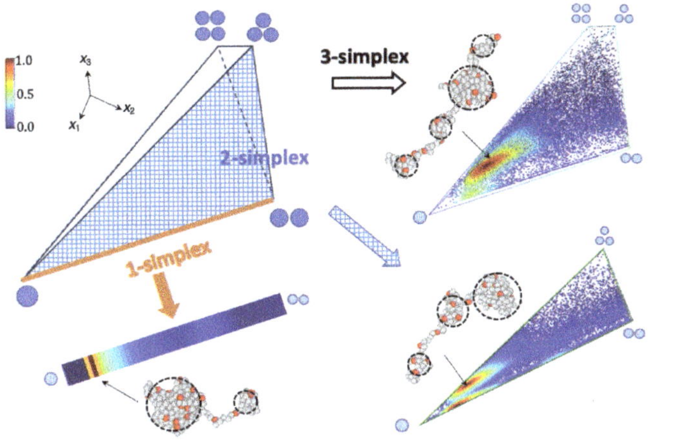

Figure 5. Simplex representations for PE with blocky charge sequences, $Q = 22$, $N = 202$. Mass asymmetry among n-pearl states are shown on $(n-1)$ simplex representation: 1-simplex representation for two pearl states, 2-simplex representation for three pearl states, and 3-simplex representations for four pearl states. The color bar in [0,1] scale represents the density of visited states realized in MD simulations. Representative conformations are shown for each $(n-1)$ simplex. Weakly charged PE occupies the conformational space near the vertex $V_{1,n-1}$ where one large pearl is dominant. In the presence of positive charge correlations, the asymmetry of the pearl structure is further accentuated compared to uncorrelated sequences, as demonstrated in Ref. [65]. Mass asymmetry of 4-pearl states can be shown as internal points in tetrahedron. A shaded triangular face of the tetrahedron corresponds to a 2-simplex. The thick orange line in a 2-simplex corresponds to 1-simplex. Each vertex V_{n_l,n_s} of tetrahedron is identified with number n_l of large pearls (shown as blue disks).

Each extremum of the energy Equation (1) is expressed by the number of evenly large pearls n_l and the number of evenly small pearls n_s. We use two subscripts n_l and n_s to denote each energy extremum E_{n_l,n_s} and the corresponding vertex V_{n_l,n_s} (see Supplementary Materials for the coordinates of vertices and energy extrema in simplex representation). Which regions of the simplex will be more densely populated is governed by the interaction energy and the allowed topology. Two important points to note are: (1) The positions of energy extrema E_{n_l,n_s} coincide with the vertices V_{n_l,n_s} if the PE does not carry any charges. This means that in the limit of vanishing charge, the energy extrema approach is applicable to the vertices of simplex. (2) When the electrostatic self energy is moderate $(\chi < n)$, the ground state conformation at $E_{1,n-1}$ is highly asymmetric and the symmetric states $\vec{0}$ $(E_{n,0})$ are only saddle points. State $\vec{0}$ is the energy minimum only for large χ, where pearl numbers larger than χ prevail and this symmetric state is hardly observed. In the case of strongly charged polyelectrolytes (PE) where χ approaches n, conformations tend to concentrate significantly around the vertex $V_{n,0}$. In the case of strongly charged PE sequences where nearly all sites are charged, the charge distribution becomes nearly regular, albeit with a few defects. This effect is somewhat analogous to what is observed in poly(styrene sulfonate) (PSS), which exhibits solubility at high sulfonation rates (greater than 2/3). Charge regulation of pearls accompanied by counterion condensation on strings is expected in this limit.

When positive charge correlations are present, charge sequences tend to be blocky. In blocky sequences, the distribution of pearls remains asymmetric even when there is large net charge Q (note that typically, strings carry more charges than their share). The most densely populated area in a simplex does not change significantly with increasing Q, as compared to uncorrelated random sequences.

The pearl necklace is a very dynamic structure. Figure 6a shows a series of snapshots which depict fluctuation of small pearls accompanied by one large pearl. In the multi-simplex representation, the trajectory generally remains near the vertex $V_{1,n-1}$, indicating that large pearls are rather robust against thermal fluctuations. Small pearls tends to nucleate on strings, and the typical lifetime of a small pearl is about $\sim 10\tau$ (with τ the MD time unit). The large pearl state with n = 2 has a longer lifetime of $\sim 50\tau$. Most random PE sequences, considered in a rather poor solvent, tend to have a well-defined number of large pearls. This contrasts with pearl necklaces formed by PAs in marginally poor solvent [66], where the number of large pearls also shows fast fluctuations, as depicted in Figure 6b.

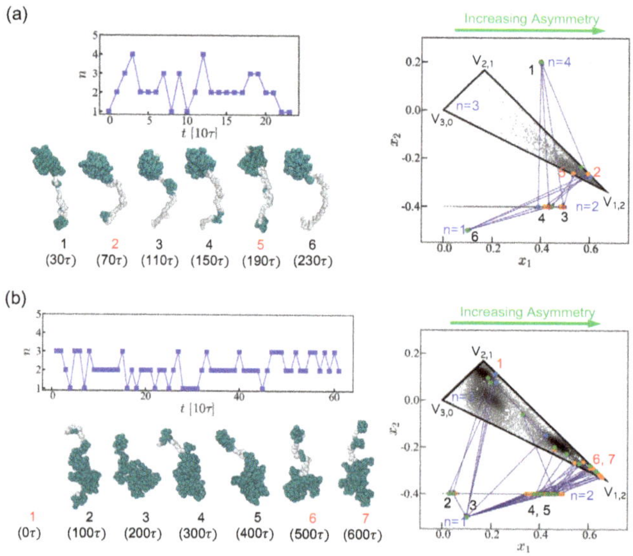

Figure 6. Structural changes accompanying changes in pearl number are shown for (**a**) PE and (**b**) PA. We retain the sequences from the sub-ensembles with the given net charges. (**a**) Fluctuation of PE with Q = 22 and N = 202. The conformation of pearl-necklaces is monitored with time interval 10τ and the change of the number of pearls $n(t)$ over 230τ is shown in the upper left panel. Small pearls appear and disappear on the tail/string. Snapshots are taken every 40τ (τ is the MD time unit.). (**b**) PA pearl-necklaces with Q = 20 (44 majority charges and 24 minority charges) and N = 202. The change in the number of pearls $n(t)$ is shown in the upper left panel over 600τ. Snapshots are taken every 100τ from MD simulations. In multiple simplexes (right panels), 1-pearl states ($n = 1$) are represented as a point, the 2-pearl states ($n = 2$) are depicted as a 1-simplex $x_1 \in [0, 0.5)$, 3-pearl states ($n = 3$) are shown as triangular 2-simplex $\{x_1, x_2\}$. The states with a larger number of pearls ($n \geq 4$) are shown as a point above the triangular 2-simplex for convenience. The chronologically numbered PE and PA forms are displayed in the corresponding simplex according to the number of pearls in the multiple simplexes. If the pearl number in the previous state is 1, 2, 3, or 4, the points on the simplex are displayed in blue, orange, green, and red, respectively. PA switches from the vicinity of $V_{2,1}$ to the vicinity of $V_{1,2}$ along a complex path through different number of pearls. A direct path inside the triangle (3-pearl states) is largely suppressed due to the very sparsely populated intermediate states within the triangle.

The polarization energy primarily provides the cohesive part of energy for PAs. For PAs in a marginally poor solvents, the boundaries of pearls are less well defined due to the weaker surface tension. It is remarkable that a few random PA sequences can adopt multiple shapes identified by different numbers of large pearls. Figure 6b shows a series of fluctuating PA shapes with a varying number of large pearls. In the triangular 2-simplex representation, the trajectory spans both vicinities of $V_{1,2}$ and $V_{2,1}$ vertices. The time scale of these switches can be as short as $\sim 1000\tau$. The trajectories typically go through a two-pearl-state with a single large pearl where small pearls merge into and split from.

5. Conclusions

We tried to draw a consistent minimal route through the physics of polyelectrolytes. We privileged the tangible and intuitive route initiated by de Gennes and Pincus, which was extended in early works by Khokhlov, and we did not aim to be exhaustive. Another review could start from the Flory school of thought [67], and could include work of the Dutch group in Wageningen [68] or the numerous contributions of Muthukumar [69,70]. The development of PE physics is presented as a two-step evolution: the seminal works using charge distributions homogeneously smear out along the polymer backbone, followed by recent developments emphasizing the role of the charge sequence along the backbone.

We primarily focus on single PE chains and their behaviors in solutions. While more complex PE systems are mentioned, along with some references, a comprehensive knowledge of these systems is based on understanding single PE chains in the solution. Systems involving both polyanions and polycations, such as PE complexes, need more specific attention and are partially related to PAs.

Recently, there has been increasing emphasis on the role of charge sequences in the context of polyelectrolyte (PE) complexation, polyampholytes (PAs), and PEs in poor solvents. The statistics of the charge distribution along the backbone are typically chosen as Markovian, and it is found that blocky statistics favor complexation. In the related context of PA pearl necklaces, it is observed that as the statistics become more blocky, a higher number of charges are allocated to the strings. Those works resort to ensemble averages and to the physics of disordered systems. Chemical synthesis of PEs often shows more complex statistics in the absence of specific control. Despite this complexity, predictions regarding complexation has been successfully verified. [58].

We developed the example of random PE in poor solvent based on our own recent research. It shows that pearl necklaces are highly irregular structures showing a small number of large pearls accompanied by a majority of small pearls. In contrast to the regular sequence, interactions between pearls do not dictate the arrangement of the pearls along the necklace and neighboring large pearls occur.

Large shape fluctuations are also predicted for PAs. Individual sequences do not always reproduce the global statistics and the structure and dynamics are very sequence-dependent. This similarity to conformational plasticity, a crucial concept in Intrinsically Disordered Proteins (IDPs), is striking. While being structured in pearls, the polymer can readily adapt to external stimuli and functions as a multitasker.

One of the most relevant criterion determining properties of a given sequence is its "blockiness". The length and number of blocks (mainly minority charge blocks when there is finite net charge) in the sequence significantly impact the behavior. Low blockiness, as also is the case for the charge sequence of IDPs, favors large shape fluctuations and labile aggregates from the identical sequences.

The control of the sequence opens the route to the control of self-complexation. Sequence specific studies seeking for sequences with peculiar properties, such as highly fluctuating structures or the control of self-aggregation of similar PAs, have indeed been conducted recently and may lead to high-tech applications. This idea is familiar to biochemists synthesizing de novo polypeptides for a specific purpose.

At the end of this study, we are left with the idea that pearl-necklace structures fluctuate significantly, although this tendency might be somewhat reduced for larger polymers, such as those with over a thousand monomers. This observation does not detract from the merits of the scaling pictures often developed for polymer research. The scaling pictures should not be taken too strictly as deterministic. They remain very useful in defining general trends and guide/interpret experiments.

Among the topics that we did not mention, dynamical aspects are likely the most important. A description of PE dynamics can be found in the review [40]. See also refs. [47,71,72] for collapse dynamics and ref. [73] on unfolding of polyelectrolyte condensates in electric fields. Recently, the occurrence of long-lived metastable knots in PAs has been investigated in relation with proteins [74].

The interaction of PEs with (biological) membranes and surfactants (phospholipids) in general is ubiquitous. See, for example, the less specifically biology-oriented review by Lindman [75]. Beyond the strict scope of PEs, there are a number of systems mainly brought along by biophysics which start to benefit from the knowledge of electrostatics accumulated in soft matter physics. Typically, additional physics enters the related theories besides electrostatics. Proteins are PEs and electrostatics is an important component of protein science [76]. Recently, the structure of viruses where electrostatic interactions play an important role received increasing attention. The review [77] describes recent advances in understanding viral structures. Biophysics puts forward questions which are sometimes raised in the surfactant community regarding short-distance electrostatics, typically close to an interface [42,78] or dense charged media [55]. Membraneless organelles are often seen as PE coacervates. Sometimes biophysics introduces brand new concepts into physics, such as active matter. More complex models for membraneless organelles involving concepts beyond passive matter were proposed recently [79]; for a review, see [80].

Supplementary Materials: The following supporting information can be downloaded at: https://www.mdpi.com/article/10.3390/polym15234593/s1, Molecular dynamics simulation description, Rules for constructing simplexes [64–66,81,82].

Author Contributions: Conceptualization, N.-K.L., A.J. and J.-F.J.; formal analysis and investigation, M.-K.C. and Y.J.; writing, N.-K.L., A.J. and J.-F.J.; visualization, M.-K.C. and Y.J. All authors have read and agreed to the published version of the manuscript.

Funding: Our work is supported by National Research Foundation grants provided by Korean government (2020R1A2B5B01002041, RS-2023-00208734 (N.-K.L.) and 2021R1F1A1064098 (Y.J.)) and the National Supercomputing Center (TS-2022-RE-0023).

Institutional Review Board Statement: Not applicable.

Data Availability Statement: Data are contained within the article and supplementary materials.

Conflicts of Interest: The authors declare no conflict of interest.

References

1. Staudinger, H. Uber Polymerisation. *Ber. Dtsch. Chem. Ges.* **1920**, *53*, 1073–1085. [CrossRef]
2. Frey, H.; Johann, T. Celebrating 100 years of "polymer science": Hermann Staudinger's 1920 manifesto. *Polym. Chem.* **2020**, *11*, 8–14. [CrossRef]
3. Vitz, J.; Majdanski, T.C.; Meier, A.; Lutz, P.J.; Schubert, U.S. Polymerization of ethylene oxide under controlled monomer addition via a mass flow controller for tailor made polyethylene oxides. *Polym. Chem.* **2016**, *7*, 4063–4071. [CrossRef]
4. de Gennes, P.G. A second type of phase separation in polymer solutions. *C. R. Acad. Sci.* **1991**, *313*, 1117–1122.
5. Matsuyama, A.; Tanaka, F. Theory of Solvation-Induced Reentrant Phase Separation in Polymer Solutions. *Phys. Rev. Lett.* **1990**, *65*, 341–344. [CrossRef] [PubMed]
6. Bekiranov, S.; Bruinsma, R.; Pincus, P. Solution behavior of polyethylene oxide in water as a function of temperature and pressure. *Phys. Rev. E* **1997**, *11*, 577–585. [CrossRef]
7. Jarkova, E.; Johner, A.; Maresov, E.A.; Semenov, A.N. Globules of annealed amphiphilic copolymers: Surface structure and interactions. *Eur. Phys. J. E* **2006**, *21*, 371–386. [CrossRef]
8. Wu, C.; Zhou, S. First Observation of the Molten Globule State of a Single Homopolymer Chain. *Phys. Rev. Lett.* **1996**, *77*, 3053.[CrossRef]
9. Wu, C.; Zhou, S. Thermodynamically Stable Globule State of a Single Poly(N-isopropylacrylamide) Chain in Water. *Macromolecules* **1995**, *28*, 8388–8390. [CrossRef]

10. Al Harbi, L. Tuning optical, dielectric, and electrical properties of Polyethylene oxide/Carboxymethyl cellulose doped with mixed metal oxide nanoparticles for flexible electronic devices. *J. Mol. Struct.* **2023**, *1272*, 134244. [CrossRef]
11. Candau, F.; Joanny, J.-F. *Polyampholytes*; Salamone, J.C., Ed.; CRC Press: Boca Raton, FL, USA, 1996; Volume 7, pp. 5462–5476.
12. Uversky, V.N. Intrinsically disordered proteins from A to Z. *Intl. J. Biochem. Cell Biol.* **2011**, *43*, 1090–1103. [CrossRef] [PubMed]
13. Müller-Späth, S.; Soranno, A.; Hirschfeld, V.; Hofmann, H.; Rüegger, S.; Reymond, L.; Nettels, D.; Schuler, B. Charge interactions can dominate the dimensions of intrinsically disordered proteins. *Proc. Natl. Acad. Sci. USA* **2010**, *107*, 14609–14614. [CrossRef] [PubMed]
14. Bianchi, G.; Longhi, S.; Grandori, R.; Brocca, S. Relevance of Electrostatic Charges in Compactness, Aggregation, and Phase Separation of Intrinsically Disordered Proteins. *Int. J. Mol. Sci.* **2020**, *21*, 6208. [CrossRef] [PubMed]
15. McCarty, J.; Delaney, K.T.; Danielsen, S.P.; Fredrickson, G.H.; Shea, J.-E. Complete phase diagram for liquid–liquid phase separation of intrinsically disordered proteins. *J. Phys. Chem. Lett.* **2019**, *10*, 1644–1652. [CrossRef] [PubMed]
16. Johner, A.; Joanny, J.-F. Translocation of polyampholytes and intrinsically disordered proteins. *Eur. Phys. J. E* **2018**, *41*, 78. [CrossRef] [PubMed]
17. Pappu, R.V.; Cohen, S.R.; Dar, F.; Farag, M.; Kar, M. Phase transitions of associative biomacromolecules. *Chem. Rev.* **2023**, *123*, 8945–8987. [CrossRef] [PubMed]
18. Brangwynne, C.P.; Tompa, P.; Pappu, R.V. Polymer physics of intracellular phase transitions. *Nat. Phys.* **2015**, *11*, 899–904. [CrossRef]
19. Wei, M.T.; Elbaum-Garfinkle, S.; Holehouse, A.S.; Chen, C.C.-H.; Feric, M.; Arnold, C.B.; Priestley, R.D.; Pappu, R.V.; Brangwynne, C.P. Phase behaviour of disordered proteins underlying low density and high permeability of liquid organelles. *Nat. Chem.* **2017**, *9*, 1118–1125. [CrossRef]
20. Nakamura, I.; Wang, Z.G. Effects of dielectric inhomogeneity in polyelectrolyte solutions. *Soft Matter* **2013**, *9*, 5686–5690. [CrossRef]
21. Oosawa, F. *Polyelectolytes*; Marcel Dekker: New York, NY, USA, 1971.
22. Mallarino, J.P.; Tellez, G.; Trizac, E. Counterion Density Profile around Charged Cylinders: The Strong-Coupling Needle Limit. *J. Phys. Chem. B* **2013**, *117*, 2702–12716. [CrossRef]
23. Odijk, T. Polyelectrolytes near the rod limit. *J. Polym. Sci.* **1997**, *15*, 477–48. [CrossRef]
24. Barrat, J.L.; Joanny, J.F. Persistence length of polyelectrolyte chains. *EPL* **1993**, *24*, 333. [CrossRef]
25. Ha, B.Y.; Thirumalai, D. Electrostatic persistence length of a polyelectrolyte chain. *Macromolecules* **1995**, *28*, 577–581. [CrossRef]
26. Manghi, M.; Netz, R.R. Variational theory for a single polyelectrolyte chain revisited. *EPJE* **2004**, *14*, 67–77. [CrossRef] [PubMed]
27. Dobrynin, A.M. Electrostatic persistence length of semiflexible and flexible polyelec-trolytes. *Macromolecules* **2005**, *38*, 9304–9314. [CrossRef]
28. Everaers, R.; Milchev, A.; Yamakov, V. The electrostatic persistence length of polymers beyond the OSF limit. *Eur. Phys. J. E* **2002**, *8*, 3–14. [CrossRef] [PubMed]
29. Khokhlov, A.R.; Katchaturian, K.A. On the theory of weakly charged polyelectrolytes. *Polymer* **1982**, *23*, 1742–1750. [CrossRef]
30. Trizac, E.; Shen, T. Bending stiff charged polymers: The electrostatic persistence length. *EPL* **2016**, *116*, 18007.
31. de Gennes, P.G.; Pincus, P.; Velasco, R.M.; Brochard, F. Remarks on Polyelectrolyte Conformation. *J. Phys.* **1976**, *37*, 461–1473. [CrossRef]
32. Doi, M.; Edward, S.F. *The Theory of Polymer Dynamics*; Clarendon Press: Oxford, UK, 1986.
33. Essafi, W.; Lafuma, F.; Williams, C.E. Effect of Solvent Quality on the Behaviour of Highly Charged Polyelectrolytes. *J. Phys. II* **1995**, *5*, 1269–1275. [CrossRef]
34. Khokhlov, A. On the collapse of weakly charged polyelectrolytes. *J. Phys. A Math. Gen.* **1980**, *13*, 979. [CrossRef]
35. Borue, V.Y.; Erukhimovich, I.Y. A statistical theory of weakly charged polyelectrolytes: Fluctuations, equation of state and microphase separation. *Macromolecules* **1988**, *21*, 3240–3249. [CrossRef]
36. Leibler, L.; Joanny, J.F. Weakly charged polyelectrolytes in a poor solvent. *J. Phys.* **1990**, *51*, 545–557.
37. Rayleigh, L. On the equilibrium of liquid conducting masses charged with electricity. *Lond. Edinburgh Dublin Philos. Mag. J. Sci.* **1882**, *14*, 184–186. [CrossRef]
38. Dobrynin, A.V.; Rubinstein, M.; Obukhov, S. Cascade of transitions of polyelectrolytes in poor solvents. *Macromolecules* **1996**, *29*, 2974–2979. [CrossRef]
39. Kantor, Y.; Kardar, M. Excess charge in polyampholytes. *EPL* **1994**, *27*, 643. [CrossRef]
40. Barrat, J.L.; Joanny, J.F. Theory of polyelectrolyte solutions. *Adv. Chem. Phys.* **1996**, *94*, 1.
41. Dobrynin, A.; Rubinstein, M. Theory of Polyelectrolyte in Solutions and at Surfaces. *Prog. Polym. Sci.* **2005**, *30*, 1049–1118. [CrossRef]
42. Boroudjerdi, H.; Kim, Y.W.; Naji, A.; Netz, R.; Schlagberger, X.; Serr, A. Statics and Dynamics of Strongly Charged Soft Matter. *Phys. Rep.-Rev. Sect. Phys. Lett.* **2005**, *416*, 129–199. [CrossRef]
43. Kramarenko, E.Y.; Philippova, O.E.; Khokhlov, A.R. Polyelectrolyte networks as highly sensitive polymers. *Polym. Sci. Ser. C* **2006**, *48*, 1–20. [CrossRef]
44. Kabanov, A.V.; Bronich, T.K.; Kabanov, V.A.; Yu, K.; Eisenberg, A. Soluble Stoichiometric Complexes from Poly(N-ethyl-4-vinylpyridinium) Cations and Poly(ethylene oxide)-block-polymethacrylate Anions. *Macromolecules* **1996**, *29*, 6797–6802. [CrossRef]

45. Decher, G. Fuzzy Nanoassemblies: Toward Layered Polymeric Multicomposites. *Science* **1997**, *277*, 1232–1237. [CrossRef]
46. Wittmer, J.P.; Johner, A.J.; Joanny, J.F. Precipitation of polyelectrolytes in the presence of multivalent salts. *J. Phys. II Fr.* **1995**, *5*, 635–654. [CrossRef]
47. Lee, N.; Thirumalai, D. Dynamics of collapse of flexible polyelectrolytes in poor solvents. *Macromolecules* **2001**, *34*, 3446–3457. [CrossRef]
48. Koculi, E.; Lee, N.-K.; Thirumalai, D.; Woodson, S.A. Folding of the Tetrahymena ribozyme by polyamines: Importance of counterion valence and size. *J. Mol. Biol.* **2004**, *341*, 27–36. [CrossRef] [PubMed]
49. Netz, R.R.; Andelman, D. Neutral and charged polymers at interfaces. *Phys. Rep.-Rev. Sect. Phys. Lett.* **2003**, *380*, 1–95. [CrossRef]
50. Pincus, P. Colloid Stabilization with Grafted Polyelectrolytes. *Macromolecules* **1991**, *24*, 2912–2917. [CrossRef]
51. Borisov, O.V.; Zhulina, E.B. Structure of weakly charged polymer brushes: Monomer density profiles. *J. Phys. II Fr.* **1997**, *7*, 447–458.
52. Porcel, C.H.; Schlenoff, J.B. Compact Polyelectrolyte Complexes: Saloplastic Candidates for Biomaterials. *Biomacromolecules* **2009**, *10*, 2968–2975. [CrossRef]
53. Shamoun, R.F.; Reisch, A.; Schlenoff, J.B. Extruded Saloplastic Polyelectrolyte Complexes. *Adv. Funct. Mater.* **2012**, *22*, 923–1931. [CrossRef]
54. Reisch, A.; Roger, E.; Phoeung, T.; Antheaume, C.; Orthlieb, C.; Boulmedais, F.; Lavalle, P.; Schlenoff, J.B.; Frisch, B.; Schaaf, P. On the Benefits of Rubbing Salt in the Cut: Self-Healing of Saloplastic PAA/PAH Compact Polyelectrolyte Complexes. *Adv. Mater.* **2014**, *26*, 2547–2551. [CrossRef] [PubMed]
55. Diddens, D.; Bascnagel, J.; Johner, A. Microscopic Structure of Compacted Polyelectrolyte Complexes: Insights from Molecular Dynamics Simulations. *ACS Macro Lett.* **2019**, *8*, 123–127. [CrossRef] [PubMed]
56. Wittmer, J.; Johner, A.; Joanny, J.F. Random and alternating polyampholytes. *EPL* **1993**, *24*, 263. [CrossRef]
57. Moldakarimov, S.; Johner, A.; Joanny, J.-F. Charge relaxation in polyampholytes of various statistics. *Eur. Phys. J. E* **2003**, *10*, 303–318[CrossRef]
58. Rumyantsev, A.M.; Jackson, N.E.; Yu, B.; Ting, J.M.; Chen, W.; Tirrell, M.V.; de Pablo, J.J. Controlling Complex Coacervation via Random Polyelectrolyte Sequences. *ACS Macro Lett.* **2019**, *8*, 1296–1302. [CrossRef]
59. Rumyantsev, A.; Jackson, N.E.; de Pablo, J.J. Polyelectrolyte Complex Coacervates: Recent Developments and New Frontiers. *Annu. Rev. Condens. Matter Phys.* **2021**, *12*, 155–156. [CrossRef]
60. Sing, C.E.; Perry, S.L. Recent progress in the science of complex coacervation. *Soft Matter* **2020**, *16*, 2885–2914. [CrossRef]
61. Danielsen, S.P.; McCarty, J.; Shea, J.-E.; Delaney, K.T.; Fredrickson, G.H. Molecular design of self-coacervation phenomena in block polyampholytes. *Proc. Natl. Acad. Sci. USA* **2019**, *116*, 8224–8232. [CrossRef]
62. Rumyantsev, A.M.; Johner, A.; de Pablo, J.J. Sequence blockiness controls the structure of polyampholyte necklaces. *ACS Macro Lett.* **2021**, *10*, 1048–1054. [CrossRef]
63. Lee, N.K.; Jung, Y.; Johner, A.; Joanny, J.F. Globular Polyampholytes: Structure and Translocation. *Macromolecules* **2021**, *54*, 2394–2411. [CrossRef]
64. Chae, M.-K.; Lee, N.-K.; Jung, Y.; Johner, A.; Joanny, J.-F. Partially Globular Conformations from Random Charge Sequences. *ACS Macro Lett.* **2022**, *11*, 382–386. [CrossRef]
65. Chae, M.-K.; Lee, N.-K.; Jung, Y.; Johner, A.; Joanny, J.-F. Structure of a hydrophobic polyelectrolyte chain with a random sequence. *Macromolecules* **2022**, *55*, 6275–6285. [CrossRef]
66. Chae, M.-K.; Lee, N.-K.; Jung, Y.; Johner, A. Shape Fluctuations of Random Polyampholyte and Intrinsically Disordered Protein Sequences. *Macromolecules* **2023**, *56*, 785–793. [CrossRef]
67. Flory, P.J. *Statistical Mechanics of Chain Molecules*; Oxford University Press: New York, NY, USA, 1988.
68. Fleer, G.; Stuart, M.C.; Scheutjens, J.; Cosgrove, T.; Vincent, B. *Polymers at Interfaces*; Springer: London, UK, 1993.
69. Muthukumar, M. *Polymer Translocation*; CRC Press: Boca Raton, FL, USA, 2016.
70. Muthukumar, M. *Physics of Charged Macromolecules*; Cambridge University Press: Cambridge, UK, 2023.
71. Lee, N.-K.; Obukhov, S. Collapse dynamics of a polyelectrolyte. *EPL* **2004**, *66*, 350. [CrossRef]
72. Winkler, R. Collapse of polyelectrolyte macromolecules by counterion condensation and ion pair formation: A molecular dynamics simulation stud. *Phys. Rev. Lett.* **1998**, *80*, 3731. [CrossRef]
73. Netz, R.R. Nonequilibrium Unfolding of Polyelectrolyte Condensates in Electric Fields. *Phys. Rev. Lett.* **2003**, *90*, 128104. [CrossRef]
74. Ozmaian, M.; Makarov D.E. Long-lived metastable knots in polyampholyte chains. *PLoS ONE* **2023**, *18*, e0287200. [CrossRef]
75. Nylander, T.; Samoshina, Y.; Lindman, B. Formation of polyelectrolyte–surfactant complexes on surfaces. *Adv. Colloid Interface Sci.* **2006**, *123–126*, 105–123. [CrossRef]
76. Honig, B.; Nicholls, A. Classical Electrostatics in Biology and Chemistry. *Science* **1995**, *268*, 5214.
77. Hsieh, Y.C.; Delarue, M.; Orland, H.; Koehl, P. Analyzing the Geometry and Dynamics of Viral Structures: A Review of Computational Approaches Based on Alpha Shape Theory, Normal Mode Analysis, and Poisson-Boltzmann Theories. *Viruses* **2023**, *15*, 1366. [CrossRef] [PubMed]
78. Johner, A. Interaction between like-charged surfaces in water: Does mean-field (Poisson Boltzmann) always apply for monovalent counterions? *J. Club Condens. Matter Phys.* **2020** , *35*, 551–560.

79. Zwicker, D.; Decker, M.; Jaensch, S.; Jülicher, F. Centrosomes are autocatalytic droplets of pericentriolar material organized by centrioles. *Proc. Natl. Acad. Sci. USA* **2014**, *111*, E2636–E2645. [CrossRef]
80. Ismail, H.; Liu, X.; Yang, F.; Li, J.; Zahid, A.; Dou, Z.; Liu, X.; Yao, X. Mechanisms and regulation underlying membraneless organelle plasticity control. *J. Mol. Cell Biol.* **2021**, *13*, 239–258. [CrossRef]
81. Plimpton, S. Fast parallel algorithms for short-range molecular dynamics. *J. Comput. Phys.* **1995**, *117*, 1–19. [CrossRef]
82. Liao, Q.; Dobrynin, A.V.; Rubinstein, M. Counterion-correlation-induced attraction and necklace formation in polyelectrolyte solutions: Theory and simulations. *Macromolecules* **2006**, *39*, 1920–1938. [CrossRef]

MDPI AG
Grosspeteranlage 5
4052 Basel
Switzerland
Tel.: +41 61 683 77 34

Polymers Editorial Office
E-mail: polymers@mdpi.com
www.mdpi.com/journal/polymers

www.ingramcontent.com/pod-product-compliance
Lightning Source LLC
Chambersburg PA
CBHW061459210926
51850CB00008B/1761
9783725847877